APPLICATION OF NANOTECHNOLOGY IN MEMBRANES
FOR WATER TREATMENT

Sustainable Water Developments
Resources, Management, Treatment, Efficiency and Reuse

Series Editor

Jochen Bundschuh
University of Southern Queensland (USQ), Toowoomba, Australia
Royal Institute of Technology (KTH), Stockholm, Sweden

ISSN: 2373-7506

Volume 5

Application of Nanotechnology in Membranes for Water Treatment

Editors

Alberto Figoli

Institute on Membrane Technology, ITM-CNR, Rende (CS), Italy

Jan Hoinkis

Institute of Applied Research, Karlsruhe University of Applied Sciences, Karlsruhe, Germany

Sacide Alsoy Altinkaya

Department of Chemical Engineering, Izmir Institute of Technology, Urla, Izmir, Turkey

Jochen Bundschuh

Deputy Vice-Chancellor's Office (Research and Innovation) & Faculty of Health, Engineering and Sciences, University of Southern Queensland, Toowoomba, QLD, Australia & Royal Institute of Technology (KTH), Stockholm, Sweden

CRC Press
Taylor & Francis Group
Boca Raton London New York

CRC Press is an imprint of the
Taylor & Francis Group, an **informa** business

A BALKEMA BOOK

Published by:
CRC Press/Balkema
P.O. Box 447, 2300 AK Leiden, The Netherlands
e-mail: Pub.NL@taylorandfrancis.com
www.crcpress.com – www.taylorandfrancis.com

First issued in paperback 2020

© 2017 by Taylor & Francis Group, LLC
CRC Press is an imprint of Taylor & Francis Group, an Informa business

No claim to original U.S. Government works

ISBN 13: 978-0-367-57363-8 (pbk)
ISBN 13: 978-1-138-89658-1 (hbk)

Visit the Taylor & Francis Web site at
http://www.taylorandfrancis.com

and the CRC Press Web site at
http://www.crcpress.com

Typeset by MPS Limited, Chennai, India

Although all care is taken to ensure integrity and the quality of this publication and the information herein, no responsibility is assumed by the publishers nor the author for any damage to the property or persons as a result of operation or use of this publication and/or the information contained herein.

Library of Congress Cataloging-in-Publication Data

About the book series

Augmentation of freshwater supply and better sanitation are two of the world's most pressing challenges. However, such improvements must be done economically in an environmental and societally sustainable way.

Increasingly, groundwater – the source that is much larger than surface water and which provides a stable supply through all the seasons – is used for freshwater supply, which is exploited from ever-deeper groundwater resources. However, the availability of groundwater in sufficient quantity and good quality is severely impacted by the increased water demand for industrial production, cooling in energy production, public water supply and in particular agricultural use, which at present consumes on a global scale about 70% of the exploited freshwater resources. In addition, climate change may have a positive or negative impact on freshwater availability, but which one is presently unknown. These developments result in a continuously increasing water stress, as has already been observed in several world regions and which has adverse implications for the security of food, water and energy supplies, the severity of which will further increase in future. This demands case-specific mitigation and adaptation pathways, which require a better assessment and understanding of surface water and groundwater systems and how they interact with a view to improve their protection and their effective and sustainable management.

With the current and anticipated increased future freshwater demand, it is increasingly difficult to sustain freshwater supply security without producing freshwater from contaminated, brackish or saline water and reusing agricultural, industrial, and municipal wastewater after adequate treatment, which extends the life cycle of water and is beneficial not only to the environment but also leads to cost reduction. Water treatment, particularly desalination, requires large amounts of energy, making energy-efficient options and use of renewable energies important. The technologies, which can either be sophisticated or simple, use physical, chemical and biological processes for water and wastewater treatment, to produce freshwater of a desired quality. Both industrial-scale approaches and smaller-scale applications are important but need a different technological approach. In particular, low-tech, cost-effective, but at the same time sustainable water and wastewater treatment systems, such as artificial wetlands or wastewater gardens, are options suitable for many small-scale applications. Technological improvements and finding new approaches to conventional technologies (e.g. those of seawater desalination), and development of innovative processes, approaches, and methods to improve water and wastewater treatment and sanitation are needed. Improving economic, environmental and societal sustainability needs research and development to improve process design, operation, performance, automation and management of water and wastewater systems considering aims, and local conditions.

In all freshwater consuming sectors, the increasing water scarcity and correspondingly increasing costs of freshwater, calls for a shift towards more water efficiency and water savings. In the industrial and agricultural sector, it also includes the development of technologies that reduce contamination of freshwater resources, e.g. through development of a chemical-free agriculture. In the domestic sector, there are plenty of options for freshwater saving and improving efficiency such as water-efficient toilets, water-free toilets, or on-site recycling for uses such as toilet flushing, which alone could provide an estimated 30% reduction in water use for the average household. As already mentioned, in all water-consuming sectors, the recycling and reuse of the respective wastewater can provide an important freshwater source. However, the rate at which these water efficient technologies and water-saving applications are developed and adopted depends on the behavior of individual consumers and requires favorable political, policy and financial conditions.

Due to the interdependency of water and energy (water-energy nexus); i.e. water production needs energy (e.g. for groundwater pumping) and energy generation needs water (e.g. for cooling), the management of both commodities should be more coordinated. This requires integrated energy and water planning, i.e. management of both commodities in a well-coordinated form rather than managing water and energy separately as is routine at present. Only such integrated management allows reducing trade-offs between water and energy use.

However, water is not just linked to energy, but must be considered within the whole of the water-energy-food-ecosystem-climate nexus. This requires consideration of what a planned water development requires from the other sectors or how it affects – positively or negatively – the other sectors. Such integrated management of water and the other interlinked resources can implement synergies, reduce trade-offs, optimize resources use and management efficiency, all in all improving security of water, energy, and food security and contributing to protection of ecosystems and climate. Corresponding actions, policies and regulations that support such integral approaches, as well as corresponding research, training and teaching are necessary for their implementation.

The fact that in many developing and transition countries women are disproportionately disadvantaged by water and sanitation limitation requires special attention to this aspect in these countries. Women (including schoolgirls) often spend several hours a day fetching water. This time could be much better used for attending school or working to improve knowledge and skills as well as to generate income and so to reduce gender inequality and poverty. Absence of in-door sanitary facilities exposes women to potential harassment. Moreover, missing single-sex sanitation facilities in schools and absence of clean water contributes to diseases. This is why women and girls are a critical factor in solving water and sanitation problems in these countries and necessitates that men and women work alongside to address the water and wastewater related operations for improvement of economic, social and sustainable freshwater provision and sanitation.

Individual volumes published in the series are spanning the wide spectrum between research, development and practice in the topic of freshwater and related areas such as gender and social aspects as well as policy, regulatory, legal and economic aspects of water. It covers all fields and facets in optimal approaches to the:

- Assessment, protection, development and sustainable management of groundwater and surface water resources thereby optimizing their use.
- Improvement of human access to water resources in adequate quantity and good quality.
- Meeting of the increasing demand for drinking water, and irrigation water needed for food and energy security, protect ecosystems and climate and to contribute to a social and economically sound human development.
- Treatment of water and wastewater also including its reuse.
- Implementation of water efficient technologies and water saving measures.

A key goal of the series is to include all countries of the globe in jointly addressing the challenges of water security and sanitation. Therefore, we aim to a balanced choice of authors and editors originating from developing and developed countries as well as gender equality. This will help society to provide access to freshwater resources in adequate quantity and good quality, meeting the increasing demand for drinking water, domestic water and irrigation water needed for food security while contributing to social and economically sound development.

This book series aims to become a state-of-the-art resource for a broad group of readers including professionals, academics and students dealing with ground- and surface water resources, their assessment, exploitation and management as well as the water and wastewater industry. This comprises especially hydrogeologists, hydrologists, water resources engineers, wastewater engineers, chemical engineers and environmental engineers and scientists.

The book series provides a source of valuable information on surface water but especially on aquifers and groundwater resources in all their facets. Thereby, it covers not only the scientific and technical aspects but also environmental, legal, policy, economic, social, and gender aspects of groundwater resources management. Without departing from the larger framework of integrated groundwater resources management, the topics are centered on water, solute and heat transport in aquifers, hydrogeochemical processes in aquifers, contamination, protection, resources assessment and use.

The book series constitutes an information source and facilitator for the transfer of knowledge, both for small communities with decentralized water supply and sanitation as well as large industries that employ hundreds or thousands of professionals in countries worldwide, working in the different fields of freshwater production, wastewater treatment and water reuse as well as those concerned with water efficient technologies and water saving measures. In contrast to many other industries, suffering from the global economic downturn, water and wastewater industries are rapidly growing sectors providing significant opportunities for investments. This applies especially to those using sustainable water and wastewater technologies, which are increasingly favored. The series is also aimed at communities, manufacturers and consultants as well as a diversity of stakeholders and professionals from governmental and non-governmental organizations, international funding agencies, public health, policy, regulators and other relevant institutions, and the broader public. It is designed to increase awareness of water resources protection and understanding of sustainable water and wastewater solutions including the promotion of water and wastewater reuse and water savings.

By consolidating international research and technical results, the objective of this book series is to focus on practical solutions in better understanding ground- and surface water systems, the implementation of sustainable water and wastewater treatment and water reuse and the implementation of water efficient technologies and water saving measures. Failing to improve and move forward would have serious social, environmental and economic impacts on a global scale.

The book series includes books authored and edited by world-renowned scientists and engineers and by leading authorities in economics and politics. Women are particularly encouraged to contribute, either as author or editor.

Jochen Bundschuh
(Series Editor)

Editorial board

Table of contents

List of contributors

Sacide Alsoy Altinkaya — Department of Chemical Engineering, Izmir Institute of Technology, Urla, Izmir, Turkey

Jiannis Anastasopoulos — Foundation for Research and Technology-Hellas (FORTH) / Institute of Chemical Engineering Sciences (ICE-HT), Rio-Patras, Greece; Department of Chemical Engineering, University of Patras, Rio-Patras, Greece

Müşerref Arda — Department of Chemistry, Ege University, Izmir, Turkey

Arash Arianfar — Department of Chemical Engineering, Ege University, Izmir, Turkey

Lawrence Arockiasamy — King Abdullah Institute for Nano Technology, King Saud University, Riyadh, Kingdom of Saudi Arabia

Amaia Soto Beobide — Foundation for Research and Technology-Hellas (FORTH) / Institute of Chemical Engineering Sciences (ICE-HT), Rio-Patras, Greece

Federica Bisignano — Institute on Membrane Technology (ITM-CNR), National Research Council of Italy, Rende (CS), Italy

Marcel Boerrigter — LEITAT Technological Center, Terrassa, Spain

Marcella Bonchio — Institute on Membrane Technology, National Research Council, ITM-CNR, Padova Unit, Padova, Italy; Department of Chemical Sciences, University of Padova, Padova, Italy

Samuel Bunani — Department of Chemical Engineering, Ege University, Izmir, Turkey; Department of Chemistry, Ege University, Izmir, Turkey; Department of Chemistry, University of Burundi, Bujumbura, Burundi

Jochen Bundschuh — Deputy Vice-Chancellor's Office (Research and Innovation) & Faculty of Health, Engineering and Sciences, University of Southern Queensland, Toowoomba, Queensland, Australia; Royal Institute of Technology, Stockholm, Sweden

Ozgur Cakmakcı — Department of Environmental Engineering, Suleyman Demirel University, Isparta, Turkey; Department of Environmental Engineering, Mus Alpaslan University, Mus, Turkey

Mauro Carraro — Department of Chemical Sciences, University of Padova, Padova, Italy

Alfredo Cassano — Institute on Membrane Technology (ITM-CNR), National Research Council of Italy, Rende (CS), Italy

Evrim Celik — Department of Environmental Engineering, Suleyman Demirel University, Isparta, Turkey; National Research Center on Membrane Technologies, Istanbul Technical University, Istanbul, Turkey

Christiane Chaumette — Fraunhofer Institute for Interfacial Engineering and Biotechnology, Stuttgart, Germany

Heechul Choi	School of Environmental Science and Engineering, Gwangju Institute of Science and Technology, Gwangju, Korea
Carmela Conidi	Institute on Membrane Technology (ITM-CNR), National Research Council of Italy, Rende (CS), Italy
Giorgio De Luca	Institute on Membrane Technology (ITM-CNR), National Research Council of Italy, Rende (CS), Italy
Minxia Ding	Université de Rennes 1, Institut des Sciences Chimiques de Rennes (UMR CNRS 6226), Rennes, France
Enrico Drioli	Institute on Membrane Technology, National Research Council, ITM-CNR, Rende (CS), Italy
Özdemir Egemen	Faculty of Fisheries, Ege University, Izmir, Turkey
Elmuntaser Eltayeb	Department of Chemical Engineering, Ege University, Izmir, Turkey
Mirko Faccini	LEITAT Technological Center, Terrassa, Spain
Alberto Figoli	Institute on Membrane Technology (ITM-CNR), Institute on Membrane Technology, National Research Council of Italy, Rende (CS), Italy
Giulia Fiorani	Institute on Membrane Technology, National Research Council, ITM-CNR, Padova Unit, Padova, Italy
Bartolo Gabriele	Department of Chemistry and Chemical Technologies, University of Calabria, Rende (CS) Italy
Francesco Galiano	Institute on Membrane Technology, National Research Council, ITM-CNR, Rende (CS), Italy
Abaynesh Yihdego Gebreyohannes	Institute on Membrane Technology, National Research Council of Italy, Rende (CS), Italy
Aziz Ghoufi	Université de Rennes 1, Institut de Physique de Rennes (UMR CNRS 6251), Rennes, France
Lidietta Giorno	Institute on Membrane Technology, National Research Council of Italy, Rende (CS), Italy
Nidal Hilal	Centre for Water Advanced Technologies and Environmental Research (CWATER); Swansea University, Swansea, United Kingdom
Jan Hoinkis	Institute of Applied Research (IAF), Karlsruhe University of Applied Sciences, Karlsruhe, Germany
Ilkay Isguder	Department of Environmental Engineering, Suleyman Demirel University, Isparta, Turkey
Daniel Johnson	Qatar Environment and Energy Research Institute (QEERI), Doha, Qatar
Nalan Kabay	Department of Chemical Engineering, Ege University, Izmir, Turkey
Theodoros Karachalios	NanoThinx S.A., Rio-Patras, Greece
Mehmet Kitis	Department of Environmental Engineering, Suleyman Demirel University, Isparta, Turkey; National Research Center on Membrane Technologies, Istanbul Technical University, Istanbul, Turkey
Derya Y. Koseoglu-Imer	Faculty of Civil Engineering, Department of Environmental Engineering, Istanbul Technical University, Maslak, Istanbul, Turkey; National Research Center on Membrane Technologies, Istanbul Technical University, Istanbul, Turkey

Katerina Kouravelou   NanoThinx S.A., Rio-Patras, Greece

İsmail Koyuncu   National Research Center on Membrane Technologies, Istanbul Technical University, Maslak, Istanbul, Turkey; Faculty of Civil Engineering, Department of Environmental Engineering, Maslak, Istanbul, Istanbul Technical University, Istanbul, Turkey

Tiziana Marino   Institute on Membrane Technology, National Research Council, ITM-CNR, Rende (CS), Italy

Gloria Modugno   Department of Chemical Sciences, University of Padova, Padova, Italy

Taylan Ö. Pek   ITOB-OSB Tekeli, Menderes, Izmir, Turkey

Richard Renou   Université de Rennes 1, Institut des Sciences Chimiques de Rennes (UMR CNRS 6226), Rennes, France

Rene Ruby-Figuero   Institute on Membrane Technology (ITM-CNR), National Research Council of Italy, Rende (CS), Italy

Gökhan Sert   Department of Chemical Engineering, Ege University, Izmir, Turkey; Faculty of Fisheries, Ege University, Izmir, Turkey

Silvia Simone   Institute on Membrane Technology, National Research Council, ITM-CNR, Rende (CS), Italy

Anthony Szymczyk   Université de Rennes 1, Institut des Sciences Chimiques de Rennes (UMR CNRS 6226), Rennes, France

George A. Voyiatzis   Foundation for Research and Technology-Hellas (FORTH) / Institute of Chemical Engineering Sciences (ICE-HT), Rio-Patras, Greece

Nevzat O. Yigit   Department of Environmental Engineering, Suleyman Demirel University, Isparta, Turkey; National Research Center on Membrane Technologies, Istanbul Technical University, Istanbul, Turkey

Mithat Yüksel   Department of Chemical Engineering, Ege University, Izmir, Turkey

Haochen Zhu   State Key Laboratory of Pollution Control and Resources Reuse, College of Environmental Science and Engineering, Tongji University, Shanghai, P.R. China

Foreword by Ahmad Fauzi Ismail

Population growth and industrialization have caused an increase in the global freshwater demand. These activities lead to water pollution, which later treated its main sources. Therefore, an advance technology is urgently needed to resolve this issue. In this context, a hybrid technology of membrane and nanotechnology has been explored to create sustainability of human needed and environmental conservation.

The book "Application of Nanotechnology in Membranes for Water Treatment" edited by Alberto Figoli, Jan Hoinkis, Sadice Alsoy Altinkaya and Jochen Bundschuh not only provides a series of innovative solutions for water reclamation via advanced membrane technology but also serves as a medium to promote international networking for the development of advanced membrane technology for Universal well-being.

In the current era of global collaboration, it is of primary importance to gather the efforts of individual research that holds high potential to be translated into real working system for community benefits. In this way it can promote the networking among different research institutions in order to achieve the common goal of finding the solutions to address the grand challenge in supplying freshwater and better sanitation system that is economically, environmentally and societally sustainable. Therefore, it is very fortunate to have this book published timely. The editors have successfully gathered some translational-able research studies from all over the world to present the innovative solutions towards the mentioned challenges.

There are various books published in the similar topic that emphasized on the application of membrane based technology for water and wastewater treatment. However, this book is unique for the reason that the chapters were contributed by established researchers all around the globe based on their recent research findings. In addition, this book provides a holistic coverage of membrane based development for water treatment, from the membrane preparation and characterizations to the performance for a specific process and application. Since that water scarcity has become a global risk and one of the most serious concerns for the scientific community in this century, the publication of this book is therefore significant as one of the medium for a good reference of an alternative solution in water reclamation. Personally, I strongly encourage engineers, scientists, professors and graduates students to read and keep this book as a guideline for the research, development and commercialization of membrane-based technology for water treatment.

Ahmad Fauzi Ismail, PhD., FASc., CEng., FIChemE.
Deputy Vice Chancellor (Research and Innovation)
Universiti Teknologi Malaysia
Johor Bahru, Johor, MALAYSIA
March 2017

Editors' foreword

There is an alarming shortage of water with climate change in many places across the world and this will probably get worse in the upcoming years more and more. Therefore, the decreasing total amount of available global freshwater supply has to be well preserved by efficient, sustainable and cost-effective technologies. In this context, water contaminated from industrial, agricultural and municipal usage has to be efficiently treated since, according to the World Health Organization (WHO), the most dangerous threat for health of mankind emerging within the next years is polluted water.

How could this be achieved? Membrane Bioreactors (MBR) represent a new technology in water treatment and in combination with subsequent Nanofiltration (NF) or Reserve Osmosis (RO) membranes are well established preferably in industrial waste water treatment. However, a major concern is the fouling of the membrane, the reversible and irreversible blocking of pores by colloidal/organic foulants. European Commission proposes the development of novel membranes or combination of membranes for water technologies in both FP7 and H2020 Research and Innovation programmes. The project BioNexGen was co-financed by the European Commission within the scope of the 7th Framework Programme. The main focus of BioNexGen was to develop a new class of functional low fouling membranes for MBR technology with high water flux and high rejection of low-molecular weight organics like pesticides, pharmaceuticals, polycyclic aromatic hydrocarbons and inorganic salts.

In the BioNexGen project, different strategies, implementing nanotechnological tools, novel engineering and process intensification concepts have been developed in order to improve anti-fouling properties and the overall performance of the new NF MBR hybrid approach. In particular, the research has been devoted mainly to the: *a*) Formation of nanostructured membranes through aligned carbon nanotubes (CNTs) with tailored physical and chemical properties for high-flux performances; *b*) Preparation of hydrophilic nanostructured membranes through polymerisable bicontinuous microemulsions (PBM) with antimicrobial or catalytic activity, *c*) Preparation of hydrophilic nanostructure layers by means of Layer-by-Layer technique; *d*) Functionalising of nanostructured membranes with antimicrobial species like silver compounds and/or photocatalysts.

The proposed strategies allowed controlling the functionalisation of the membrane materials with respect to pore and surface characteristics which are drivers for the performance of the membrane in the bioreactor. The membrane manufacturing methods have been optimised with respect increasing membrane stability, production efficiency and reducing solvent and additive usage and hence costs of the materials. An international conference (Nanomemwater Application of Nanotechnology in Membranes for Water Treatment) was organized by Izmir Institute of Technology on October 11–15, 2013, Turkey and held as the summit for the project BioNexGen. This book is the followout of this Nanomemwater conference.

This book is divided in three parts: Part I contains the studies on "Membrane developing and characterization" and it reports the formation of nanostructured polymeric membranes by polymerized bicontinuous membranes (PBM), Chapter 1 and by Layer-by-Layer technique for water treatment (Chapter 2), specifically developed in BioNexGen project. In particular, in chapter 1, the different PBM structures reported in literature have been described and their potentialities and applications in different fields, such as gas separation, direct methanol fuel cell and wastewater treatment, have been deeply investigated, too. In Chapter 2, first the use of LbL deposition for preparing nanofiltration, reverse osmosis, forward osmosis and pervaporation membranes is discussed through extensive literature search. In the second part, part of the results obtained throughout the BioNexGen project are shown to illustrate the potential of LbL technique in mitigating bacterial fouling.

Then, mixed matrix polymeric membranes and their application complete the first part of the book. Chapter 3 gives a general overview on the fabrication and application areas of mixed matrix flat-sheet membranes. Chapter 4 focused on a different approach on how the different signal responsive interactions within the membranes or its surrounding can be tuned and monitored in controlled environments in order to modulate reversible enzyme immobilization. In particular, nano-designed magnetic responsive MBRs have been demonstrated to hold a bigger future prospect to design the next generation responsive membranes towards advanced functions, within the logic of designing more complex membrane systems that are capable of mimicking nature. Chapter 5 describes the detailed study developed with BioNexGen on the inclusion of Carbon Nanotubes (CNTs) into porous polymeric membranes, having the objective to overcome the immanent limitation of counterbalance between flux and selectivity. The exceptionally high aspect ratios of CNTs at such quite small dimensions and the molecularly slick, chemically inert hydrophobic graphitic walls constitute collaterals for transport applications. Combined to their nano-scale internal diameters they compose a unique phenomenon of speedy transfer and high selectivity. In fact, in the literature, liquid flow through membranes composed of an array of aligned CNTs is reported to be four to five orders of magnitude faster than is predicted from conventional fluid flow theory. This is attributed to the almost frictionless interface at the CNT wall. In the present chapter, the effect of inclusion of CNTs on the morphology and permeation properties of membranes is investigated with the help of scanning electron microscopy (SEM), atomic force microscopy (AFM), water flux, permeation tests and contact angle measurements. Chapter 6 also deals with mixed matrix CNTs blended polyethersulfone (PES) membranes by using different ratios of N-methyl-2-pyrrolidone (NMP) to dimethylformamide (DMF). The CNTs membranes have been used for water treatment and anti-fouling protein tests have been reported. Chapter 7 reports first an overview on the photocatalytic process, TiO_2 oxidative action, the environmental/human health risks associated with the use of this semiconductor, as well as an investigation of the TiO_2 nanoparticles synthesis methods. Then, some examples evidencing the possibility to efficiently apply TiO_2-based polymeric membranes for disinfection applications are described. Finally, the difficulties in the optimization of the TiO_2-polymeric casting solution preparation, which still limits the use of TiO_2-based membranes on an industrial scale, are presented. To complete the discussion on TiO_2-based membranes, Chapter 8, refers to the most recent advances regarding the preparation and application of TiO_2 mixed matrix membranes with particular emphasis on water and wastewater treatment applications.

Chapter 9 reports the use of innovative nanosized metal oxides and polyoxometalates in membranes and in water treatment processes thanks to their antimicrobial and catalyst properties. The first part ends with Chapter 10, which describes the Atomic-Force Microscopy (AFM) tool applied to the development of polymer and mixed matrix filtration membranes, surface analysis, and to obtain useful information about the mechanisms of biofouling of membrane surfaces, and aid in the development of novel biofouling-resistant surfaces.

Modeling and simulation approach in membranes, applied to water treatment, are reported in the second part of the book. In chapter 11, a short illustration of the potentialities of molecular dynamics (MD) simulations in the investigation of water and ion transport through model nanoporous membranes is discussed. In chapter 12, *ab initio* modeling is illustrated in order to investigate particular properties of some promising nanostructures, such as CNTs studied in the context of the BioNexGen project with the aim of proposing novel mixed matrix polymer membranes. Specifically, the rejection by multi-walled CNTs (MW-NTs) of low molecular weight organic solutes coming from industrial wastewater, such as cosmetics and textiles, and olive oils has been investigated in depth, together with their water permeability.

The last part of the book is about membrane processes and chapter 13 deals with the recovery bioactive compounds in citrus wastewater by membrane operations. In the first section of the chapter an overview of traditional processes in the citrus industry is provided. Then, selected applications related to the use of pressure-driven membrane operations and the immobilization of β-cyclodextrins in polymeric membranes are analyzed and discussed as innovative approach for specifically removing molecules of interest. Finally, Membrane Bio Reactor (MBR) process in water and wastewater treatment and reuse is discussed and integrated membrane processes such as MBR and NF/RO is reported too.

We believe that this book will provide the readers with a thorough understanding of the different available approaches for manufacturing membranes both with innovative polymeric systems and inorganic nano-materials which could give enhanced functionalities, catalytic and antimicrobial activities in order to improve the performance of the existing membranes. In addition, the molecular dynamic simulation approaches in the book will be helpful for the readers who are interested in investigating the potentials of new nanomaterials and enhancing the optimization of membrane material and structural features membranes.

We hope that this book will help all readers, professionals, academics and non-specialists, as well as key institutions that are working on membrane technology and water treatment projects. It will be useful for leading decision and policy makers, water sector representatives and administrators, policy makers from the governments, business leaders, business houses in water treatment, and engineers/scientists from both industrialized and developing countries as well. It is expected that this book will become a standard, used by educational institutions, and Research and Development establishments involved in the respective issues.

Alberto Figoli
Jan Hoinkis
Sadice Alsoy Altinkaya
Jochen Bundschuh
(editors)
February 2017

About the editors

Alberto Figoli (1970, Italy) got his PhD degree at Membrane Technology Group, Twente University (NL) in 2001. He is Senior Researcher at Institute on Membrane Technology (ITM-CNR) in Rende (CS), Italy, since 2001.

Alberto Figoli is expert in the field of membrane technology, particularly in *membrane preparation and characterisation* and *application in water treatment.*

In 1996, he obtained his Master Degree in Food Science and Technology, at the Agriculture University of Milan. Then, he worked for about 1 year at Quest International Nederland B.V. (ICI), Naarden (The Netherlands) at the Process Research Group, on a pilot plant for aromatic compounds extraction using the pervaporation membrane technology. He has been granted twice by National Research Council of Italy (CNR) to the "Short Term Mobility Programme", as visiting researcher of the "Environmental Protection Agency of United States (USEPA)" at the "Sustainable Technology Division" in Cincinnati (USA). The research was devoted to pervaporation studies in the environmental field.

In the last years, he has been involved in several European and National projects (as scientific responsible or principal investigator for ITM-CNR) in the field of Membrane Technology. In particular, he has been the scientific coordinator of several National projects with Industries and a Marie Curie EU Project. He is author of more than 100 scientific peer-review papers and chapters published in international journals and books. He is the editor of three books and author of two patents on membrane technology.

In 2015, he has been elected in the board, as Council Member, of the European Membrane Society, EMS, for the period 2015–2019. He is responsible for the Awards and Summer Schools.

Jan Hoinkis (1957, Germany) conducted a doctorate in thermodynamics at the University of Karlsruhe, Germany (now Karlsruhe Institute of Technology). After completion of his thesis he moved to the Swiss company Ciba-Geigy, where he was working as head of a R&D group on process development in the field of fine chemicals production with focus on environmentally friendly technologies. He has been working since 1996 as a professor at the Karlsruhe University of Applied Sciences giving lectures in chemistry, thermodynamics as well as environmental process engineering. His R&D work is focused on water treatment and water reuse with special attention on sensor-based membrane technologies. He was involved in several national and international R&D projects. In 2008 he was appointed Scientific Director of the Institute of Applied Research, which is the central research facility at the Karlsruhe University of Applied Sciences.

Sacide Alsoy Altinkaya (1969, Turkey) is a professor in Chemical Engineering at Izmir Institute of Technology. She has received PhD degree in Chemical Engineering from The Pennsylvania State University, USA.

Her research experience in the field of membrane science and technology include modeling of membrane formation by phase inversion technique and membrane separation processes, preparation, characterization and application of polymeric membranes in the field of drug delivery, food packaging, biomedical and environmental applications. She has collaborated with different universities, industrial partners in Europe and in USA as well and has been the scientific coordinator of national and international projects supported by European Union, National Science Foundation of USA (NSF) and French Foreign Ministry. She is the member of European Membrane Society. She attended many international conferences with oral/poster presentations, gave lectures as invited speaker and taught many courses not only on membranes but also on transport phenomena, heat and mass transfer and mathematical modeling in engineering.

She has been the reviewer for many international journals, industrial projects, the author of several scientific publications and holds a US patent. She received "L'Oréal Turkey For Woman In Science Young Woman Scientist Award" for her achievements in membrane development.

Jochen Bundschuh (1960, Germany), finished his PhD on numerical modeling of heat transportin aquifers in Tübingen in 1990. He is working in geothermics, subsurface and surface hydrology and integrated water resources management, and connected disciplines. From 1993 to1999 he served as an expert for the German Agency of Technical Cooperation (GTZ, now GIZ) and as a long-term professor for the DAAD (German Academic Exchange Service) in Argentine. From 2001 to 2008 he worked within the framework of the German governmental cooperation (Integrated Expert Program of CIM; GTZ/BA) as adviser in mission to Costa Rica at the Instituto Costarricense de Electricidad (ICE). Here, he assisted the country in evaluation and development of its huge low-enthalpy geothermal resources for power generation. Since 2005, he is an affiliate professor of the Royal Institute of Technology, Stockholm, Sweden. In 2006, he was elected Vice-President of the International Society of Groundwater for Sustainable Development (ISGSD). From 2009–2011 he was visiting professor at the Department of Earth Sciences at the National Cheng Kung University, Tainan, Taiwan.

Since 2012, Dr. Bundschuh is a professor in hydrogeology at the University of Southern Queensland, Toowoomba, Australia where he leads the Platform for Water in the Nexus of Sustainable Development working in the wide field of water resources and low/middle enthalpy geothermal resources, water and wastewater treatment and sustainable and renewable energy resources. In November 2012, Prof. Bundschuh was appointed as president of the newly established Australian Chapter of the International Medical Geology Association (IMGA).

Dr. Bundschuh is author of the books "Low-Enthalpy Geothermal Resources for Power Generation" (2008) (CRC Press/Balkema, Taylor & Francis Group) and "Introduction to the Numerical Modeling of Groundwater and Geothermal Systems: Fundamentals of Mass, Energy and Solute Transport in Poroelastic Rocks". He is editor of 18 books and editor of the book series "Multiphysics Modeling", "Arsenic in the Environment", "Sustainable Energy Developments" and "Sustainable Water Developments" (all CRC Press/Balkema, Taylor & Francis Group). Since 2015, he is an editor in chief of the Elsevier journal "Groundwater for Sustainable Development".

Acknowledgements

The editors thank the authors of the chapters for their participation, discussion and contribution. The editors and authors thank also the technical people of Taylor & Francis Group, for their cooperation and the excellent typesetting of the manuscript.

CHAPTER 1

Polymerizable microemulsion membranes: from basics to applications

Francesco Galiano, Bartolo Gabriele, Jan Hoinkis & Alberto Figoli

1.1 INTRODUCTION

Microemulsions are thermodynamically stable and optical isotropic dispersions made up of two immiscible liquids (oil and water phases) stabilized by the presence of a surfactant. Microemulsions are completely transparent and macroscopically homogeneous. However, on a microscopic level, due to the interface created from the surfactant located between the water and oil domains, they are heterogeneous systems (Solans and Garcìa-Selma, 1997).

In contrast to emulsions, microemulsions form spontaneously when the components are mixed. The transparency of the system is due to the considerably smaller microstructures in comparison to the wavelength of visible light making them not observable through an optical microscope (Haegel et al., 2001). In Table 1.1 the main differences between microemulsions and emulsions are reported. Depending on the oil (o)/water (w) ratio, microemulsions can be differentiated into: o/w microemulsions (where the oil droplets are dispersed in a water phase), w/o microemulsions (where the water droplets are dispersed in an oil continuous phase), and bicontinuous microemulsions (where oil and water domains are randomly dispersed in order to form an interconnected network).

Due to the presence of water and oil microdomains within the same system, microemulsions are suitable to solubilize hydrophilic and lipophilic materials that exhibit relatively low solubility both in water and organic solvents.

Nowadays microemulsions find application in different fields, such as liquid-liquid extractions, catalysis, coatings and textile finishing, cosmetics and pharmaceuticals (as drug delivery systems) (Agrawal and Agrawal, 2012). In Figure 1.1, the current different fields of application of microemulsions, taken from registered patents, are shown.

According to patent data (Fig. 1.1), the pharmaceutical field is the predominant area (almost 35%) in microemulsion applications. In particular, microemulsions are used as drug delivery systems (for transdermal, oral, transmucosal and extra-vascular administration), due to their ability to efficiently solubilize lipophilic drugs as well as their potential effect on topic and systemic drug bioavailability (Levy and Sintov, 2008). The cosmetic field represents 20% of applications. Microemulsions are mainly used for preshampoos formulations (Araujo and Gesztesi, 2010) and general cosmetic skin care products (Diec et al., 1998; Hwang, 2013). The biomedical field follows with 16% of applications. In this area, microemulsions formulations suitable for artificial cornea, contact lens applications (Yong Choi and Ying, 2012), blood gas controls and calibrator production (Feil, 1992) have been developed. Microemulsions are also used in other fields such as: thinner paint production (Hawes et al., 2012), porous composites, catalytic materials (Texter and Yan, 2010), fluoropolymers production (Hegenbarth et al., 1996) and in some diesel fuel compositions in order to reduce harmful diesel emissions (Bock et al., 1995).

The aim of the present chapter is to illustrate the polymerization of polymerizable microemulsion systems for the production of materials, microlatexes, and membranes. A brief history of microemulsions to-date will be described as well as the breakthroughs reached up to now in the

1

Table 1.1. Comparison between emulsions and microemulsions.

Parameter	Emulsions	Microemulsions
Appearance	Cloudy/milky	Transparent
Formation	Form only if mechanical energy is applied	Form spontaneously
Stability	Not thermodynamically stable	Thermodynamically stable
Droplet size	0.1–10 μm	10–100 nm
Structure	Spherical droplets of o/w or w/o	Depends on composition and on the properties of the interfacial surfactant film (spherical droplets of o/w or w/o, lamellar, bicontinuous)
Surfactant	Low concentration of surfactant is required	High concentration of surfactant is required (up to 20/30 wt%)

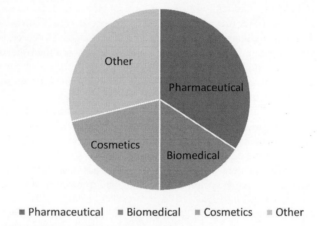

■ Pharmaceutical ■ Biomedical ■ Cosmetics ■ Other

Figure 1.1. Diagram of microemulsions main fields of application (Patents data). References: Pharmaceutical field: Acosta-Zara and Yuan (2009), Bidgood *et al.* (1992), Constantinides (1994), D'Cruz *et al.* (2000), Dennis *et al.* (2003a, 2003b), Diec *et al.* (1998), Dietz *et al.* (1997, 1999), Kim (2009), Levy and Sintov (2008), Pather *et al.* (2010), Von (1997); Biomedical field: Dietz *et al.* (1997, 1999), Drewes *et al.* (2008), Feil (1992), Inlow and Maiorella (1998), Yong Choi and Ying (2012); Cosmetic field: Araujo and Gesztesi (2010), Barrow and Slavtcheff (1995), Hamachi *et al.* (2004), Halloran (1992), Hwang (2013), Kim (2009), Linn (1987), Rataj (2014); Other fields: Bock *et al.* (1995), Chittofrati *et al.* (1995), Figoli *et al.* (2014), Hawes *et al.* (2012), Hegenbarth *et al.* (1996), Hazbun and Schon (1988), Liu and Stammer (2011), Mondin *et al.* (1998), Nawrath *et al.* (2005), Seiders (1987), Texter and Yan (2010), Van *et al.* (2002), Von (1997).

field of microemulsion polymerization and the main fields of application. The following sections will report on the structure, polymerization and formulation of microemulsions, while the concluding section will illustrate paradigmatic case studies.

1.2 A BRIEF HISTORY OF MICROEMULSIONS TO-DATE

The term microemulsion was used for the first time by Schulman *et al.* (1959) to describe a multiphase system formed by a monomer, water, a surfactant and often a co-surfactant.

Even before Schulman, several attempts were made in order to try to predict the different microemulsion types (Mehta and Kaur, 2011). The first concept was developed by Bancroft (1913), and it is known as Bancroft's rule. It states that oil-soluble emulsifiers lead to the formation of w/o emulsions, while water-soluble emulsifiers lead to the formation of o/w emulsions.

The first important application of microemulsions started in the early 1980s in order to improve oil recovery from natural oil reservoirs (Shah, 1981). As a consequence of the oil crisis, microemulsions systems had started to be extensively studied and investigated. In particular, the process of microemulsion polymerization started to attract the first interests for the production of stable latexes (with a nanosized structure) which could be applied in microencapsulation (as drug delivery systems), in biomedical applications and in the production of different polymers (Candau, 1992).

Later on, extensive studies were thus conducted on microemulsion polymerization in order to obtain latexes (or membranes) with different structures. In 1997 Li *et al.* (1997) investigated the polymerization of polymerizable bicontinuous microemulsions (PBM) as coating materials for hollow fiber membranes. In particular, the microemulsion was formulated by using methyl methacrylate (MMA) as monomer, 2-hydroxyethyl methacrylate (HEMA) as co-surfactant, acryloyloxyundecyldimethylammonium acetate (AUDMAA) as surfactant, and water. The developed coated hollow-fibers showed different PEG separations based on the bicontinuous microemulsion composition together with higher permeation rates in comparison to membranes prepared with pure bicontinuous microemulsions. The interest in microemulsion polymerization continued during the 1990's resulting in several in-depth studies by *Gan et al.* Microporous polymeric materials were for instance produced by Gan *et al.* (1995) through a polymerization of PBMs. Transparent solid materials were obtained by using the zwitterionic surfactant AUDMAA, preserving the bicontinuous original structure even during the polymerization. By increasing the concentration of water (20–45 wt%), the porosity of the microemulsion was also increased. In a second paper, Gan *et al.* (1997) studied the polymerization of MMA by using the non-ionic polymerizable surfactant ω-methoxy poly(ethylene oxide)$_{40}$ undecyl-α-methacrylate (PEO-R-MA-40) for the production of microporous membranes. Even in this case, the different content of water within the microemulsion was found to be related to different pore dimensions. The effects of anionic and cationic surfactants were evaluated by Chieng *et al.* (1996) using sodium dodecyl sulfate (SDS) and dodecyltrimethylammonium bromide (DTAB) for the preparation of polymeric membranes. The cationic surfactant DTAB allowed transparent membranes with smaller pore sizes (less than 100 nm) to be obtained, in comparison to membranes prepared with SDS (up to 3 μm). A co-surfactant, usually represented by a short-chain alcohol, was usually applied in order to prepare microemulsions. Burban (1995), however, by means of double-chained surfactants, showed that it was possible to avoid the use of the co-surfactant for the polymerization of MMA. Liu *et al.* (2000) prepared NF membranes through a polymerization of acrylonitrile as monomer and PEO as the polymerizable surfactant. The effect of water concentration on pore radius was evaluated; it ranged from 0.38 to 2.4 nm. Membranes prepared from w/o microemulsions were tested in pervaporation (PV) for ethanol/water mixtures separation, presenting a higher affinity for water with a separation factor of about 20.

The mechanism of microstructures formation during bicontinuous microemulsion polymerization was studied by Peinado *et al.* (2006) by scanning calorimetry and fluorescence. In particular, the photopolymerization of MMA and methylaminoethyl methacrylate, stabilized by the surfactants cetyltrimethylammonium bromide (CTAB) and sodium dodecyl sulfate (SDS), was investigated. Even if microemulsions were designed for preparing nanoporous materials, mesoporous structures were actually obtained. This behavior was explained by the phase separation phenomenon which occurred during the polymerization. The type of surfactant also influenced the final morphology, with the production of a material with a more open structure when CTAB was used. In Table 1.2, a summary of literature data on polymerizable microemulsions is reported.

Table 1.2. Summary of some polymerizable microemulsion data from literature.

Monomer [wt%]	Water [wt%]	Surfactant [wt%]	Co-surfactant [wt%]	Initiator	Application	Reference
MMA (6–57 wt%)	4–40 wt%	SDS (1–14 wt%)	HEMA (20–54 wt%)	DBK (UV)	–	Chieng et al. (1995)
MMA (18 wt%)	30–36 wt%	SDS (4–10 wt%)	HEMA (42%)	APS-TMEDA, (redox)	–	Chieng et al. (1996)
MMA (18 wt%)	28–33.2 wt%	DTAB (6.8–12 wt%)	HEMA (42%)	APS-TMEDA, (redox)	–	Chieng et al. (1996)
MMA (30–48 wt%)	20–50 wt%	AUDMAA (17.5–28 wt%)	HEMA (30–48 wt%)	APS-TMEDA, (redox)	–	Li et al. (1996)
MMA (30–48 wt%)	20–50 wt%	AUTMAB (17.5–28 wt%)	HEMA (30–48 wt%)	APS-TMEDA, (redox)	–	Li et al. (1996)
MMA (20–60 wt%)	20–40 wt%	AUDMAA (20–40 wt%)	–	DMPA (UV)	–	Gan et al. (1995)
AN (32.5–45 wt%)	10–35 wt%	PEO-R-MA-40 (32.5–45 wt%)	–	APS-TMEDA, (redox)	Pervaporation EtOH/H$_2$O	Liu et al. (2000)
MMA (7.5–20 wt%)/ NiPAAm (0–25 wt%)	23–33 wt%	C1-PEO-C11-MA-40 (35 wt%)	HEMA (5–20 wt%)	DMPA (UV)	Potential construction of dressings and cell-delivery systems for wound healing	Wang et al. (2006)
MMA (28–34 wt%)	15–30 wt%	Na11-EAAU (21–25.5 wt%)	HEMA (21–25.5 wt%)	DBK (UV)	–	Gan et al. (1994)
MMA (8 wt%)	48–54 wt%	C12TAB (6–12 wt%)	HEMA (32 wt%)	DBK (UV)	–	Gan and Chew (1997)
MMA/HEMA (30–48 wt%)	20–50 wt%	AUDMAA (17.5–28 wt%)	MMA/HEMA (7:3) (30–48 wt%)	APS-TMEDA, (redox)	–	Gan and Chew (1997)
MMA/HEMA (7:3) (30–48 wt%)	20–50 wt%	AUTMAB (17.5–28 wt%)	MMA/HEMA (7:3) (30–48 wt%)	APS-TMEDA, (redox)	–	Gan and Chew (1997)
MMA (13–38.6 wt%)	55.9–81.8 wt%	AOA/SDS (5.2–10.2 wt%)	–	γ-ray	–	Chen and Zhang (2007)
Isopropyl myristate	10–50 wt%	Caprylocaproyl macrogolglycerides	Polyglycerol oleate	Interfacial polymerization	Possibility to serve as templates for nanoparticles formation	Graf et al. (2008)
AN (22–34 wt%)	25 wt%	AIPS (30 wt%)	SPM (10–22 wt%)	DMPA (UV)	Direct methanol fuel cell application	Lim et al. (2007)
MMA (21–32 wt%)	29–41 wt%	AUTMAB (25–37 wt%)	HEMA (0–10 wt%)	APS-TMEDA, (redox); DMPA (UV)	Gas separation	Figoli (2001)
MMA (21 wt%)	41 wt%	AUTEAB (25 wt%)	HEMA (10 wt%)	APS-TMEDA, (redox)	Coating for water treatment	Galiano et al. (2015)

AIPS: 3-((11-acryloyloxyundecyl)imidazoyl)propyl sulfonate; AN: acrylonitrile; AOA: 12-acryloxy-9-octadecenoic acid; APS: ammonium persulfate; AUTEAB: acryloyloxyun-decyltriethylammonium bromide; AUTMAB: Acryloyloxylundecyltrimethylammonium bromide; DBK: dibenzyl ketone; DMPA: 2,2-dimethoxy-2-phenylacetophenone; Na11-EAAU: 11-(N-ethylacrylamido)undecanoate; NiPAAm: N-isopropylacrylamide; PEG: polyethylene glycol; SPM: 3-sulfopropylmethacrylate; TMEDA: N,N,N,N-tetramethylethylenediamine.

1.3 STRUCTURES OF POLYMERIZABLE MICROEMULSIONS

As stated in the introduction, depending on their composition, polymerizable microemulsions can be divided into oil-in-water (o/w), water-in-oil (w/o) and bicontinuous microemulsions. The type of microemulsion structure depends on the combination of surfactant and/or co-surfactant, oil and water.

1.3.1 *Oil-in-water (o/w) polymerizable microemulsions*

In o/w microemulsions, the polymerization of styrene and MMA was mainly studied. These systems were prepared using non-ionic, cationic (e.g. CTAB) and anionic (e.g. SDS) surfactants. One of the limitations of these systems was the very low solubility of the monomer, normally not exceeding a concentration of 8–10 wt%. For this reason, high concentrations of surfactants were required, making the process very expensive. Furthermore, the latexes often showed instability and turbidity, as a result of the incompatibility between the polymer formed and the co-surfactant. Often, in fact, the alcohols used as co-surfactants were solvent for the monomer but non-solvents for the final polymer. The latexes obtained were usually in the range of 20 to 60 nm in diameter (Chow and Gan, 2005).

1.3.2 *Water-in-oil (w/o) polymerizable microemulsions*

Acrylamide (AM), HEMA and acrylic acid (AA) were the most common water-soluble monomers for w/o (or inverse) microemulsion preparation. The latexes obtained were formed by swollen water polymer particles dispersed in the continuous organic phase (Candau, 1995). The water-soluble monomers usually played also the role of co-surfactant by enhancing the monomer solubilization through an increase in the flexibility and the fluidity of the interfaces (Chow and Gan, 2005). For instance, hydrocarbon solutions of dialkyl sulfosuccinates surfactants, where the oil phase was toluene, benzene, decane or heptane, did not require any addition of co-surfactants (Candau, 1995).

1.3.3 *Polymerizable bicontinuous microemulsions (PBMs)*

In bicontinuous microemulsions, the amount of monomer that can be incorporated in the microemulsion can reach 25 wt%, producing stable and clear microlatexes with a very small particle size. The organic and aqueous phases coexist in an interconnected network with surfactant molecules localized at the interface of water-oil domains. MMA and styrene are the most used monomers for bicontinuous microemulsion preparation. A co-surfactant, such as AA and HEMA, is commonly used and non-ionic, cationic, zwitterionic and anionic surfactants are employed. A cross-linking agent, such as ethylene glycol dimethacrylate (EGDMA), can be included into the bicontinuous microemulsion formulation to consolidate the final structure. Since all the components of the microemulsion are polymerizable, they can be copolymerized forming a strong and resistant network (O'Donnell and Kaler, 2007).

1.4 MICROEMULSION PREPARATION AND ITS TERNARY PHASE DIAGRAM

As described, in microemulsion formulation two immiscible liquids (water and oil) are brought together by a surfactant in order to form a single-phase system (Agrawal and Agrawal, 2012). Microemulsions are usually prepared by adding short-chain alcohols (butanol, pentanol or hexanol) acting as co-surfactants. The formation of microemulsions requires specific and proper amounts of the corresponding components and they are generally prepared by a phase titration method, which can be represented with the help of a ternary phase diagram. The construction of

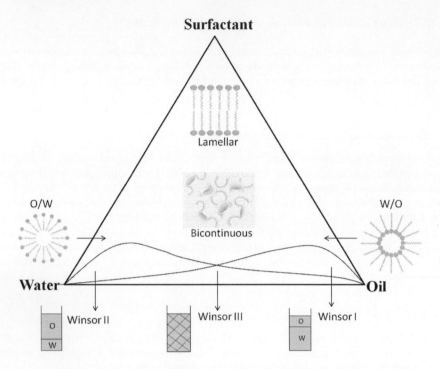

Figure 1.2.   Ternary phase diagram representing some phase equilibria in multicomponent systems.

a phase diagram allows the determination of the water dilutability of the system and the range of compositions constituting a monophasic region (Mehta and Kaur, 2011).

As shown in Figure 1.2, microemulsions can be in the form of oil-swollen micelles dispersed in a continuous water phase (o/w), or water-swollen droplets dispersed in an oil phase (w/o). In the intermediate region, the structure is not globular anymore, but it possesses a sponge-like structure (bicontinuous microemulsion) where the oil and water domains coexist in interconnected domains. At low surfactant concentrations, different Winsor phase equilibria can be identified: Winsor I (o/w), Winsor II (w/o) and Winsor III (bicontinuous microemulsion).

As shown in Figure 1.2, at high oil concentration, reverse micelles capable of dissolving water molecules are formed. w/o microemulsions are formed as a result of the further addition of water to the system. Here, water exists in the oil phase as droplets surrounded by the surfactant molecules. Upon further dilution with water, a crystalline liquid region may be observed with the formation of double layers of surfactant (lamellar structure). Finally, as the amount of water increases, water forms a continuous phase containing droplets of oil stabilized by surfactant molecules (o/w microemulsions). The bicontinuous microemulsion forms in the middle phase, which is in equilibrium with almost pure water and oil phases, respectively.

1.5   MICROEMULSION POLYMERIZATION

Free radical polymerization of polymerizable microemulsions has been widely investigated over the past few decades for the production of membranes. Most of the studies focused on the o/w microemulsion polymerization. The basic sequence of events describing the general mechanism of o/w microemulsion polymerization is shown in Figure 1.3. Two stages (defined as Intervals) can be identified in microemulsion polymerization: Interval I, in which particle nucleation occurs, and Interval II, in which particle growth occurs.

Figure a Formation of initiating radicals in the aqueous medium

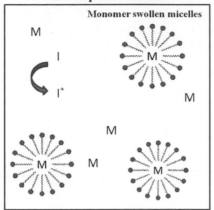

I	Initiator molecule
I*	Initiator radical
M	Monomer molecule
MPM*	Oligomeric radicals
P	Polymer chain
～～●	Surfactant molecule

Figure b Nucleation of monomer-swollen micelles

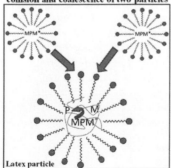

Figure c Growth of latex particles by diffusion of monomer molecules

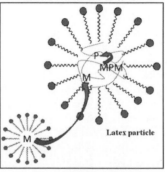

Figure d Growth of latex particles by collision and coalescence of two particles

Figure e Polymer particles formation

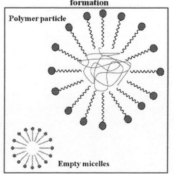

Figure 1.3.   Schematic representation of o/w microemulsion polymerization (adapted from O'Donnell and Kaler, 2007).

Interval I: Once initiators are added to the microemulsion, they produce, by their decomposition in the aqueous phase, radical species able to promote a reaction (Fig. 1.3a). Thus, radical initiators start reacting with water-soluble monomer molecules dispersed in the aqueous phase to form oligomeric radicals. At this point, particle nucleation occurs (Fig. 1.3b).

Interval II: In the second stage, the oligomeric radicals, being more hydrophobic than the monomer from which they originated, enter more favorably into the monomer-swollen micelles present in the aqueous phase. Polymerization propagates within the monomer-swollen micelles starting the process of latex particles formation. The growth of latex particles is due to the diffusion of monomer molecules from monomer-swollen micelles (Fig. 1.3c) and to the collision and subsequent coalescence between two particles (Fig. 1.3d).

If the transfer of a radical from the propagating polymer to a monomer molecule occurs, the monomeric-formed radical can continue to propagate within the latex particle or it can exit the particle and enter a micelle starting a new propagation event and a new latex particle formation. Polymerization continues until the monomer is totally consumed. The final latex consists of surfactant-stabilized polymer particles and empty micelles formed by the excess surfactant (O'Donnell and Kaler, 2007) (Fig. 1.3e).

The porosity of polymeric materials obtained by microemulsion polymerization is strictly related to the microemulsion composition. The water concentration, in fact, can have a great effect on the resulting morphology of the final membrane. Water concentrations between 20 and 50 wt% usually lead to open cell structures, while water concentrations below 20 wt% lead to the formation of closed-cell structures.

1.6   SURFACTANTS

Surfactants are amphiphilic molecules consisting of a hydrophilic water-soluble head and a hydrophobic oil-soluble tail. They are able to lower the surface tension of a liquid, facilitating the wetting of surfaces or the miscibility among different liquids. The stabilizing behavior of surfactants depends on their molecular structure (geometry), the interactions they undergo with water and/or oil and the elastic properties of the interfacial film they form. Since they play a crucial role in microemulsion formulation, the different types of surfactants are presented here.

Surfactants can be divided into ionic and non-ionic surfactants depending on the presence of a charge. Generally, ionic surfactants consist of an anionic headgroup, such as SO_3^-, or a cationic headgroup like $-NMe_3^+$. Surfactants that bear both a positively and a negatively charged group are zwitterionic and thus neutral overall. Non-ionic surfactants consist generally of an AB block copolymer type structure. Those most commonly used are built up of ethylene blocks (hydrophobic moiety) and ethyleneoxide blocks (hydrophilic moiety) (Fig. 1.4).

In microemulsion preparation, both types of polymerizable and non-polymerizable surfactants are used for the production of nanostructured materials. However, the structures of the materials obtained with non-polymerizable surfactants can be easily destroyed by chemical or physical forces. For this reason, the use of polymerizable surfactants is generally preferred. In fact, polymerizable surfactants can be polymerized and cross-linked with other microemulsion components, due to the presence of their polymerizable end. In this way, a strong and resistant network can be obtained, which is able to retain its original structure (Miller *et al.*, 1999).

1.7   CASE STUDIES

Microemulsions are very well-known systems and have been successfully applied in different fields (cosmetic and pharmaceutical areas in particular). However, there are still not many works reported in the literature on the practical applications of the membranes obtained by microemulsion polymerization.

In this section, three of the works carried out on the application of membranes obtained by microemulsions polymerization (in particular, in bicontinuous structure) are discussed. In the following case studies, the membranes obtained by polymerization of PBMs have been tested in gas separation as a supported liquid membrane (SLM), in membrane bioreactor (MBR) technology for wastewater treatment and in direct methanol fuel cell (DMFC) applications.

An anionic surfactant

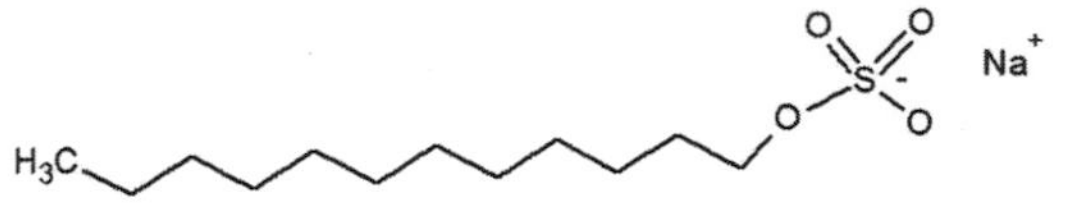

Sodium dodecyl sulfate
(SDS)

A cationic surfactant

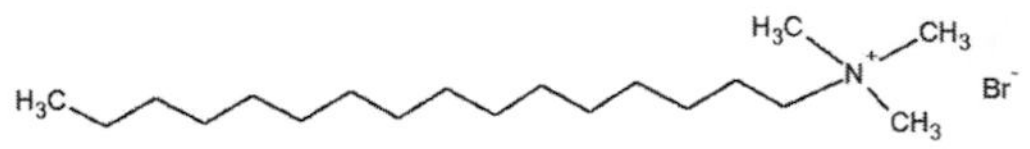

Cetyltrimethylammonium
bromide (CTAB)

A zwitterionic surfactant

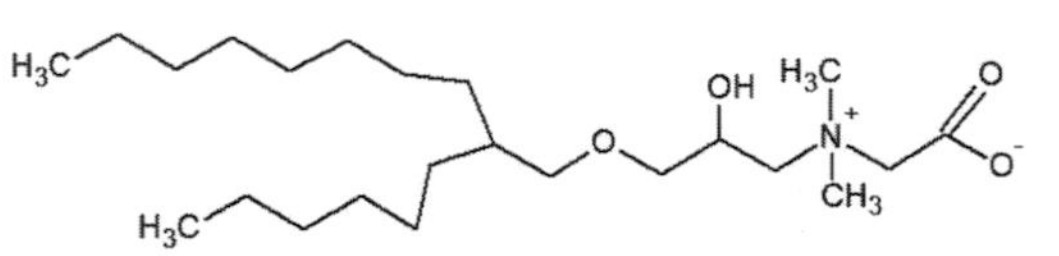

Guerbet alcohol hexadecyl glycidyl
ether glycine betaine

A non-ionic surfactant

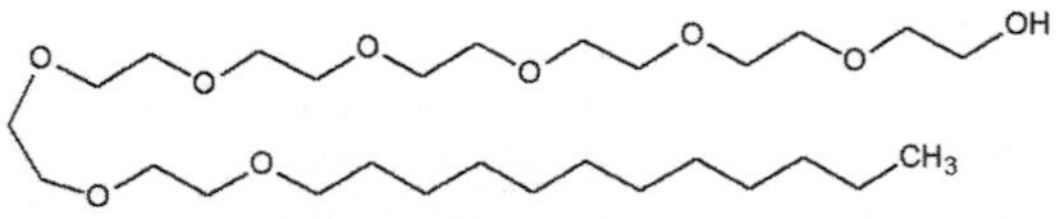

Octaethylene glycol
monododecyl ether

A polymerizable (cationic) surfactant

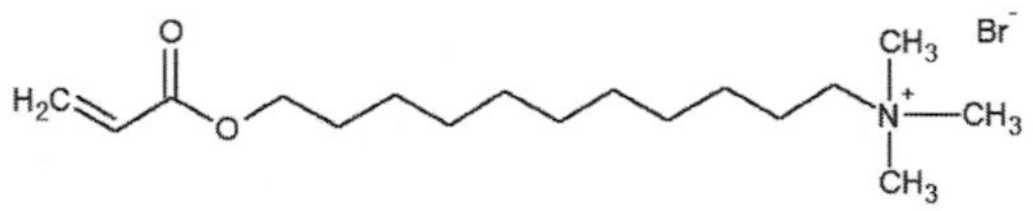

Acryloxylundecyltrimethylammonium
bromide (AUTMAB)

Figure 1.4.   Different structures of surfactants.

1.7.1   *PBM in gas separation*

In the work of Figoli (2001), PBM membranes were prepared, characterized and applied as "nano liquid membranes" in N_2/O_2 separation. For the first time, PBM membrane performances were investigated, for the facilitated transport of oxygen, with and without oxygen carriers. Figure 1.5 shows a scheme of the liquid membrane loaded with a mobile carrier for the required separation.

Bicontinuous microemulsions were prepared by synthesizing two polymerizable surfactants: the cationic surfactant AUTMAB and the zwitterionic surfactant AUDMAA. The use of a poly-merizable surfactant enables the covalent binding of the surfactant into the polymeric matrix, thus preventing its leaching and the restructuring of the interfacial film during the polymerization. In this way, the original structure of the microemulsion can be retained even after the polymerization. MMA and HEMA were used as monomer and co-surfactant, respectively. The microemulsions were polymerized both by redox and UV initiators. The nanostructured microemulsions thus

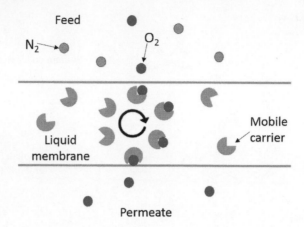

Figure 1.5.   Scheme of transport of gas molecules through a liquid membrane loaded with a mobile carrier.

produced were then characterized by different techniques. From SEM analyses, the bicontinuous microemulsion structure was observed. The average pore size was of about 60–70 nm for membranes containing HEMA, while, a pore size of about 4 nm was found (by thermoporometry analyses) for PBM membranes without co-surfactant. The addition of the co-surfactant is able to tailor the size dimension of the bicontinuous structure; in fact, increasing the concentration of HEMA, the channel width size decreases from 60–70 nm (without HEMA) to 4 nm (maximum HEMA concentration). Electrical resistance measurements demonstrated the interconnectivity of the pores within the PBM membranes.

The oxygen facilitated transport through PBM membranes was performed by impregnating the membranes in an aqueous solution containing a new type of water-soluble carrier based on a Co-porphyrin system with oxygen affinity. The selectivity (O_2/N_2) obtained was about 2.8, and the resulting PBM membrane was stable for more than 3 weeks with a humidified gas feed stream.

The main advantages of the PBM membranes, used as SLM, lie in their nanometer pore size which can guarantee a good stability against significant transmembrane pressure gradients and loss of the liquid membrane phase (Figoli, 2010).

1.7.2   *PBM in wastewater treatment*

The possibility of using PBM membranes in the field of wastewater treatment was recently studied by Galiano *et al.* (2015). PBM membranes, applied as coating material on commercial polyethersulfone (PES) membranes, presented some peculiar aspects, which made them ideal candidates for water treatment processes. The preparation of the bicontinuous microemulsion was based on the synthesis of the cationic surfactant acryloyloxyundecyltriethylammonium bromide (AUTEAB) (Figoli, 2014). Due to its cheaper synthesis cost, AUTEAB was used instead of the most common surfactant AUTMAB, usually reported in the literature for making PBM membranes (Gan and Chew, 1997; Li *et al.*, 1996). MMA, HEMA and EGDMA were used as oil phase, co-surfactant and cross-linker respectively. Redox initiators were used for the polymerization reaction. The prepared bicontinuous microemulsion was then cast on PES ultrafiltration membranes and left to polymerize on their surface. The resulting composite PBM membrane was fully characterized, and the results compared with the PES uncoated membrane.

The bicontinuous structure of the polymerized microemulsion was observed from the SEM image (Fig. 1.6). The surface of the PBM membrane was characterized by the random distribution of polymerized polymer channels and non-polymerized water channels forming the pores of the membrane. The dimension of the water channels was, thus, directly related to the pore size of the membrane falling in the range of 30–50 nm in width.

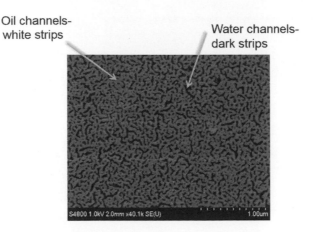

Figure 1.6.   SEM picture of PBM membrane surface (Galiano *et al.*, 2015).

In particular, it was found that PBM membranes presented a higher hydrophilic character in comparison to uncoated PES membranes. Accordingly, the contact angle decreased of about 30% (from 68° of PES membrane to about 47° of PBM membrane). Furthermore, PBM membranes presented, as evidenced by AFM analyses, a 17 times smoother surface. A higher hydrophilicity and a smoother surface are two important aspects that contribute to the anti-fouling properties exhibited by PBM membranes. Organic compounds, in fact, are more inclined to establish hydrophobic interactions with the membrane surface, causing their accumulation and the formation of a cake layer. For this reason, membranes with a more hydrophilic character can reduce the deposition of organic materials, thus limiting fouling. Furthermore, smoother surfaces contribute to the prevention of fouling, by reducing the accumulation of foulants into the "valleys" typical of rough materials. The particular structure of the PBM membrane, characterized by interconnected water and oil polymerized channels, also played a significant role in determining its anti-fouling properties. The possibility of having "slotted" pores (represented by the water non-polymerized channels) instead of circular pores, made the PBM membranes less inclined to fouling. As already studied by Bromley *et al.* (2002) the pore geometry can have a significant impact on the formation of fouling. In particular, it was found that slotted pores can limit the particle bridging over the pores responsible for cake deposition and fouling formation. The anti-fouling properties of the PBM membranes were finally successfully proved by filtration tests carried out with the model foulant humic acid (HA) and compared with uncoated PES membranes.

Although PBM membranes showed a lower permeability (when only pure water was applied for filtration tests) in comparison to PES membranes due to the extra mass transport resistance of the PBM layer, they presented a lower reduction in permeability when the solution with HA was applied. The lower tendency of being affected by fouling guaranteed a more constant permeability in time and therefore higher performances in comparison to PES membranes, as can be seen from Figure 1.7.

Another important facet of PBM membranes, was related to their antimicrobial activity, due to the presence of the copolymerized surfactant AUTEAB. As other quaternary ammonium salts, AUTEAB showed an important antimicrobial activity, which was maintained even after the polymerization of the microemulsion on the PES membrane surface. Clearly, antimicrobial membranes, such as PBM, are significantly less prone to the biofouling phenomenon.

All these properties made PBM membranes ideal candidates for applications in MBR technology or in other wastewater treatment processes where membranes with anti-fouling and anti-biofouling properties are highly desired. The scale-up of PBM membranes was then carried out and the prepared membranes were assembled in an MBR module (Fig. 1.8).

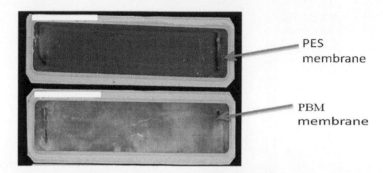

Figure 1.7.   Fouling effect on PES and PBM-coated membrane after treatment with HA (Galiano *et al.*, 2015).

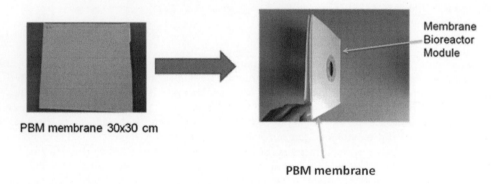

Figure 1.8.   From flat-sheet membrane to MBR module.

When PBM membranes were tested in MBR technology for the treatment of model textile wastewater, they showed their great potential to be successfully applied in this particular membrane process (Deowan, 2014; Galiano, 2013). A model textile dye solution containing two dyes (Acid Red 4 and Remazol Brilliant Blue) was used as model wastewater. The MBR system with commercial PES membranes was run for more than 140 days (Deowan *et al.*, 2016). During the period of the experiment, the membrane module was first cleaned and then replaced due to a drastic decrease in membrane performances caused by the excessive fouling. The same procedure was carried out with coated PBM membranes. In this case, the MBR system was run for more than 100 days, during which no cleaning or membrane replacement was needed. The results obtained with MBR confirmed the benefits of using the novel PBM-coated membranes in comparison to uncoated commercial PES membranes (Deowan *et al.*, 2016). The intrinsic antimicrobial activity (due to the presence of the cationic surfactant) made the PBM membranes highly resistant to biofouling and therefore perfectly usable in a membrane bioreactor (MBR) process, where biological sludge is applicable to The constant water permeability was an indirect proof of membrane resistance to biofouling. Furthermore, no evidence of formation of microorganism colonies was observed on the PBM membrane surface after their removal from the MBR. MBR tests demonstrated how the lifecycle of PBM membranes was significantly enhanced in comparison to uncoated membranes. Finally, based on these results, less cleaning efforts (due to the anti-fouling and anti-biofouling properties) will be required, reducing, consequently, the related costs. A constant water permeability over time and a good rejection of organic compounds were important benefits achieved by PBM membranes. An increase in the efficiency and durability of the membrane was also obtained. The good quality of the filtrate was proved by COD rejection of about 95%, which was comparable to the commercial PES membranes. Moreover, PBM-coated membranes showed 10% higher blue dye removal efficiency compared to the commercial ones (Deowan *et al.*, 2016). Additionally, due to their anti-fouling properties, PBM membranes may require a lower aeration rate, thus allowing a further reduction in operational costs.

1.7.3   *PBM in DMFC*

In the study of Lim *et al.* (2007) a nanostructured proton exchange membrane (PEM) was developed by the polymerization of a bicontinuous microemulsion. The low methanol permeability exhibited by the prepared PEM membranes made them suitable to be applied in processes such as DMFC. The zwitterionic surfactant AIPS was synthesized and applied for the dispersion of acrylonitrile and water, and SPM was added as co-surfactant contributing to the stabilization of the microemulsion. Microemulsion samples, prepared with the appropriate concentration of components, were thus photopolymerized to obtain transparent and resistant polymeric films. Polymerized membranes presented interconnected hydrophilic channels (1.5–2 nm) and a continuous amphiphilic matrix. The proton conductivity of prepared membranes was found to be dependent on the quantity of hydrophilic channels present in the membrane. For this reason, an increase in proton conductivity ($0.127\,\mathrm{S\,cm^{-1}}$) was observed for the membranes prepared with the higher content of SPM (22 wt%), which showed higher water uptake levels. Furthermore, PEM presented a low methanol permeability (about 1.5×10^{-6}) with a reduction of 60% in comparison to Nafion 112 membrane permeability. This behavior was explained by the very small interconnected channels of the PEM. The DMFC single cell loaded with PEM showed a power output of 15–$20\,\mathrm{mW\,cm^{-2}}$ with single cell performances increasing with the increase of SPM content in the microemulsion formulation. Furthermore, stability tests carried out demonstrated the stability of PEM membranes in the DMFC operation environment.

1.8   CONCLUSIONS

The preparation of polymeric materials by microemulsion polymerization can be considered as an important technique for the production of materials with controlled and constrained structures. The possibility to retain, after polymerization, the original microemulsion structure allows the production of latexes with fixed and desired geometry. Moreover, the possibility of using polymerizable surfactants further increases the stability of the produced membranes making them highly resistant and avoiding the restructuration of the polymer chains. These materials can thus find successful application in ultrafiltration and nanofiltration processes, surface coatings, adhesives, paints, food and pharmaceutical processing, as some of the most important examples of application. The case studies presented here provide clear evidence that the polymerization of microemulsions can still be recognized as an attractive technology and can offer new prospectives for the preparation of novel and promising materials with innovative characteristics. However, these PBM-coated membranes are still only available on a lab scale and therefore, future efforts should be directed to the scaling-up of the coating. A possible strategy could be the use of existing coating technology, such as the UV-based polymerization process, to make the system commercially viable.

REFERENCES

Acosta-Zara E.J. & Yuan S. (2009) Linker-based lecithin microemulsion delivery vehicles. EP 2120871 A1.
Agrawal, O.P. & Agrawal, S. (2012) An overview of new drug delivery system: microemulsion. *Asian Journal of Pharmaceutical Sciences*, 2(1), 5–12.
Araujo, M.K. & Gesztesi, J.L. (2010) Cosmetic microemulsion. EP 1670421 B1.
Bancroft, W.D. (1913) The theory of emulsification V. *Journal of Physical Chemistry*, 17, 501–519.
Barrow, S.R. & Slavtcheff, C.S. (1995) Hydroalcoholic cosmetic microemulsions, WO 1995003772 A1.

Bidgood, A.M., Musson, D.P.E.D.G. & Olejnik, O. (1992) Aqueous ophthalmic microemulsions of tepoxalin. EP 0480690 A1.

Bock, J., Grimes, P.G., Pace, S.J., Robbins, M.L., Sexton, M.D. & Smith, A.K. (1995) Microemulsion diesel fuel compositions and method of use. EP 0475620 B1.

Bromley, A.J., Holdich, R.G. & Cumming, I.W. (2002) Particulate fouling of surface microfilters with slotted and circular pore geometry. *Journal of Membrane Science*, 196, 27–37.

Burban, J.H., He, M. & Cussle, E.L. (1995) Organic microporous materials made by bicontinuous microemulsion polymerization. *AlChE Journal*, 41, 907–914.

Candau, F. (1992) Polymerization in microemulsions. In: Paleos, C.M. (ed) *Polymerisation in organised Media*. Gordon and Breach Science Publishers, Philadelphia, PA. pp. 215–282.

Chen, J. & Zhang, Z. (2007) Radiation-induced polymerization of methyl methacrylate in microemulsion with high monomer content. *European Polymer Journal*, 43, 1188–1194.

Chieng, T.H., Gan, L.M. & Chew, C.H. (1995) Morphology of microporous polymeric materials by polymerization of methyl methacrylate and 2-hydroxyethyl methacrylate in microemulsions. *Polymer*, 36, 1941–1946.

Chieng, T.H., Gan, L.M. & Te, W.K. (1996) Porous polymeric membranes by bicontinuous microemulsion polymerization: effect of anionic and cationic surfactants. *Polymer*, 37, 5917–5925.

Chittofrati, A., Tentorio, A. & Visca, M. (1995) Use of electrically conductive oil-in-water microemulsions based on perfluorinated compounds as catholytes in electrochemical processes. EP 0346932 B1, 1995.

Chow, P.Y. & Gan, L.M. (2005) Microemulsions polymerizations and reactions. *Advances in Polymer Science*, 175, 257–298.

Constantinides, P.P. (1994) Therapeutic microemulsions. WO 1994008605 A1.

D'Cruz, O., Li, M., Uckun, F.M. & Yiv, S. (2000) Gel-microemulsion formulations. WO 2000056366 A1.

Dennis, D.M., Gravenstein, N., Modell, J.H., Morey, T.E. & Shah, D. (2003a) Microemulsion and micelle systems for solubilizing drugs. EP 1305005 A2.

Dennis, D.M., Gravenstein, N., Model, J.H., Morey, E.T. & Shah, D.O. (2003b) Novel microemulsion and micelle systems for solubilizing drugs. WO 2003015823 A1.

Deowan, S.A. (2014) *Development of Membrane Bioreactor (MBR) Process applying novel low fouling Membranes*, PhD Thesis, University of Calabria, Rende, Italy.

Deowan, S.A., Galiano, F., Hoinkis, J., Gabriele, B., Johnson, D., Altinkaya, S.A., Bartolo, G., Hilal, N., Drioli, E. & Figoli, A. (2016) Novel low fouling membrane bioreactor (MBR) for industrial wastewater treatment. *Journal of Membrane Science*, 510, 524–532.

Diec, K.H., Meier, W. & Schreiber, J. (1998) Cosmetic or dermatological microemulsion based gels. WO 1998015254 A1.

Dietz, M.T., Lu, Y.Y., Uy, R. & Young, C.I. (1997) Polymerized microemulsion pressure sensitive adhesive compositions and methods of preparing and using same. WO 1997005171 A1.

Dietz, M.T., Lu, Y.Y., Uy, R. & Young, C.I. (1999) Use of bicontinuous microemulsions as pressure sensitive adhesives. EP0741765B1.

Drewes, S., Mentrup, E., Pooth, R. & Wimmer, T. (2008) Use of an aqueous micro-emulsion for the preparation of a formulation for the treatment of adipose diseases. EP 1970051 A1.

Feil, M.C. (1992) Thermodynamically-stable aqueous perfluorocarbon microemulsion useful as blood gas control or calibrator. EP 0272040 B1.

Figoli, A. (2001) *Synthesis Nf nanostructured mixed Matrix Membrane for facilitated Gas Separation*. PhD Thesis, University of Twente, Twente, The Netherlands.

Figoli, A. (2010) Liquid membrane in gas separations. In: Kislik, V.D. (ed) *Liquid Membranes: Principle and Applications in Chemical Separations and Wastewater Treatment*. Elsevier, Amsterdam. pp. 327–356.

Figoli, A., Hoinkis, J., Gabriele, B., De Luca, G., Galiano, F. & Deowan, S.A. (2014) Bicontinuous microemulsion polymerized coating for water treatment. Patent application PCT/EP2014/070603= WO2014/EP070603.

Galiano, F. (2013) *Preparation and Characterisation of Polymerizable nicontinuous Microemulsion Membranes for Water Treatment Application*. PhD Thesis, University of Calabria, Rende, Italy.

Galiano, F., Figoli, A., Deowan, S.A., Johnson, D., Altinkay, S.A., Veltri, L., De Luca, G., Mancuso, R., Hilal, N., Gabriele, B. & Hoinkis, J. (2015) A step forward to a more efficient wastewater treatment by membrane surface modification via polymerizable bicontinuous microemulsion. *Journal of Membrane Science*, 482, 103–114.

Gan, L.M. & Chew, C.H. (1997) Microporous polymer composites from microemulsion polymerization. *Colloids and Surfaces* A: *Physicochemical and Engineering Aspects*, 123–124, 681–693.

Gan, L.M., Chieng, T.H., Chew, C.H. & Ng, S.C. (1994) Microporus polymeric materials from microemulsion polymerization. *Langmuir*, 10, 4022–4026.

Gan, L.M., Li, T.D., Chew, C.H. & Teo, W.K. (1995) Microporous polymeric materials from polymerization of zwitterionic microemulsions. *Langmuir*, 11, 3316–3320.

Gan, L.M., Liu, J., Poon, L.P. & Chew, C.H. (1997) Microporous polymeric composites from bicontinuous microemulsion polymerization using a polymerizable nonionic surfactant. *Polymer*, 38, 5339–5345.

Graf, A., Jack, K.S., Whittaker, A.K., Hoo, S.M. & Rades, T. (2008) Protein delivery using nanoparticles based on microemulsions with different structure-types. *European Journal of Pharmaceutical Sciences*, 33, 434–444.

Haegel, F.-H., Schlüpen, J., Schultze, J.W., Winkels, S. & Stromberg, C. (2001) Anodic polymerization of thiophene derivatives from microemulsions and liquid crystals. *Electrochimica Acta*, 46, 3973–3984.

Halloran, D.J. (1992) Optically clear hair care compositions containing silicone based microemulsions. EP 0514934 A1.

Hamach, T., Ozaki, M. & Tanaka, H. (2004) Polyorganosiloxane micro-emulsion composition and raw material for cosmetics. EP 1406575 A2.

Hawes C.L., Timoth, E.S. & Teague, G. (2012) Microemulsion paint thinner. US 8257484 B1.

Hazbun, E.A. & Schon, S.G. (1988) Fire resistant microemulsions containing phenyl alcohols as cosurfactants. US 4770670 A.

Hegenbarth, J., Jian-Guo, C., Wu, H.S. & Xin-Kang, C. (1996) Microemulsion polymerization systems for fluoromonomers. WO 1996022315 A1.

Inlow, D. & Maiorella, B. (1989) Lipid microemulsions for culture media. WO 1990003429 A1.

Kim, S. (2009) Sugar-based surfactant microemulsions containing essential oils for cosmetic and pharmaceutical use. WO 2009029046 A1.

Levy, H. & Sintov, A. (2010) Pharmaceutical compositions based on a microemulsion. WO 2008096351 A1.

Li, T.D., Gan, L.M., Chew, C.H., Teo, W.K. & Gan, L.H. (1996) Preparation of ultrafiltration membranes by direct microemulsion polymerization using polymerizable surfactants. *Langmuir*, 12, 5863–5868.

Li, T.D., Gan, L.M., Chew, C.H., Teo W.K. & Gan, L.H. (1997) Hollow-fiber membranes coated with polymerizable bicontinuous microemulsions. *Journal of Membrane Science*, 133, 177–187.

Lim, T.H., Tham, M.P., Zhaolin, L., Hong, L. & Guo, B. (2007) Nano-structured proton exchange membranes molded by polymerizing bi-continuous microemulsion. *Journal of Membrane Science*, 290, 146–152.

Linn, E.E.-C., West, M.P. & York, T.O. (1987) Skin moisturizing microemulsions. EP 0226337 A1.

Liu, J., Teo, W.K., Chew, C.H. & Gan, L.M. (2000) Nanofiltration membranes prepared by direct microemulsion copolymerization using poly(ethylene oxide) macromonomer as a polymerizable surfactant. *Journal of Applied Polymer Science*, 77, 2785–2794.

Liu, Y. & Stammer, A. (2011) Preparation of silicone microemulsions. EP 2268255 A1.

Mehta, S.K. & Kaur, G. (2011) Microemulsions: thermodynamic and dynamic properties. In: Tadashi, M. (ed) *Thermodynamics*. InTech. Available from: http://www.intechopen.com/books/thermodynamics/microemulsions-thermodynamic-and-dynamic-properties [accessed December 2016].

Miller, S.A., Ding, J.H. & Gin, D.L. (1999) Nanostructured materials based on polymerizable amphiphiles. *Current Opinion in Colloid & Interface Science* 4, 338–347.

Mondin, M., Andries, N. & Massaux, J. (1998) Microemulsion/all purpose liquid cleaning composition based on EO-PO nonionic surfactant. US 5854193 A.

Nawrath, A., Sottmann, T. & Strey, R. (2005) Microemulsions and use thereof as a fuel. WO 2005012466 A1.

O'Donnell, J. & Kaler, E.W. (2007) Microstructure, kinetics, and transport in oil-in-water microemulsion polymerizations. *Macromolecular Rapid Communications*, 28, 1445–1454.

Pather, S.I., Gupte, S.V., Khankari, K., Hontz, J. & Kumbale, R. (2010) Microemulsions as solid dosage forms for oral administration. EP 1104290 B1.

Peinado, C., Bosch, P., Martìn, V. & Corrales, T. (2006) Photoinitiated polymerization in bicontinuous microemulsions: fluorescence monitoring. *Journal of Polymer Science* A: *Polymer Chemistry*, 44, 5291–5303.

Rataj, N.V. & Aubry, J.M. (2014) Volatile aqueous microemulsions of perfumes and essential oils based on the use of solvosurfactants. WO 2014080150 A1.

Schulman, J.H., Staeckenius, W. & Prince, L.M. (1959) Mechanism of formation and structure of micro emulsions by electron microscopy. *The Journal of Physical Chemistry*, 63, 1677–1680.

Seiders, R.P. (1987) Microemulsions containing sulfolanes. US H366 H.

Shah, D.O. (ed) (1981) *Surface Phenomena in enhanced Recovery*. Plenum Press, New York.

Solans, C. & Garcìa-Selma, M.J. (1997) Surfactants for microemulsions. *Current Opinion in Colloid & Interface Science*, 2, 464–471.

Texter, J. & Yan, F. (2010) Nanoporous and microporous solvogels and nanolatexes by microemulsion polymerization. US 7759401 B2.

Wang, L.S., Chow, P.Y., Phan, T.T., Lim, J. & Yang, Y.Y. (2006) Fabrication and characterization of nanostructured and thermosensitive polymer membranes for wound healing and cell grafting. *Advanced Functional Materials*, 16, 1171–1178.

Van, R.C.J.M., Vogelaar, L., Nijdam, W., Barsema, J.N. & Wessling, M. (2002) Method of making a product with a micro or nano sized structure and product. WO 2002043937 A2.

Von, C.C. (1997) Microemulsions for use as vehicles for administration of active compounds. WO 1997009964 A1.

Yong Choi, E.P. & Ying, J.Y. (2012) Forming copolymer from bicontinuous microemulsion comprising monomers of different hydrophilicity. EP 2408828 A1.

CHAPTER 2

The use of layer-by-layer technique in the fabrication of thin film composite membranes for water treatment

Sacide Alsoy Altinkaya

2.1 INTRODUCTION

Membranes provide advantages over other water treatment technologies, such as disinfection, distillation, or media filtration, due to their ease of operation, which does not require chemical additives, thermal inputs, or the regeneration of spent media. Ideal separation membranes should overcome the trade-off between selectivity and permeability while providing long-term stability. To achieve this task, it is necessary to form an ultra-thin defect free active separation layer on top of a highly permeable, mechanically strong support membrane. The layer-by-layer (LbL) assembly technique has great potential for the fabrication of ultra-thin defect free membranes with improved permeation and rejection properties, since the thickness and morphology of the membrane can be easily controlled.

The layer-by-layer technique consists of the alternate deposition of oppositely charged poly-electrolytes on a support followed by rinsing after each deposition step to remove excess and weakly adsorbed polyelectrolytes (Fig. 2.1).

Although the LbL assembly is usually applied on charged supports through electrostatic interactions, non-electrostatic interactions such as hydrogen bonding (Erel-Unal and Sukhishvili, 2008;

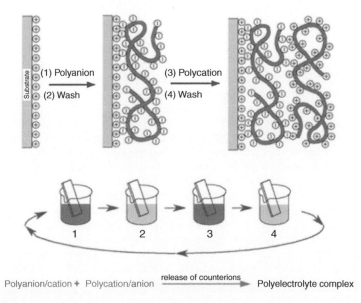

Figure 2.1. Schematic representation of layer-by-layer self-assembly of polyelectrolytes. Adapted from Ronge *et al.* (2014) with the permission of The Royal Society of Chemistry.

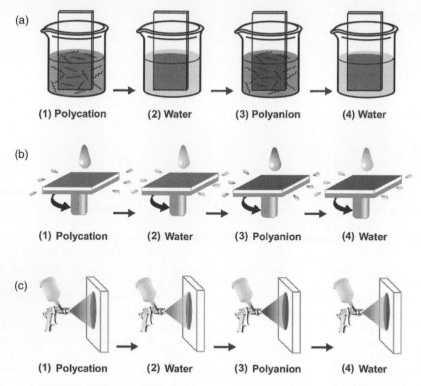

Figure 2.2.   Schematic representation of different processes used for LbL coating of polyelectrolytes: (a) dip coating, (b) spin coating and (c) spray coating. Adapted from Li *et al.* (2012) with the permission of The Royal Society of Chemistry.

Quinn *et al.*, 2007; Stockton and Rubner, 1997; Sukhishvili and Granick, 2000), hydrophobic interactions (Kotov, 1999; Quinn *et al.*, 2007; Schonhoff, 2003) and chemical cross-linking (Tieke, 2011) can also be used to drive multilayer assembly. In the case of neutral or hydrophobic substrates, either polyelectrolytes are modified with suitable groups or support is modified to enhance a good anchoring of the LbL assembly.

The LbL coating of polyelectrolytes on various substrates can be applied with dip coating, spray coating and spin coating (Fig. 2.2).

Among these techniques, dip coating is the most commonly used, however, the overall process is long since the method is mainly diffusion-controlled and rinsing steps between the depositions are required to prevent the formation of polyelectrolyte complexes, hence, flocculation on the surface. To decrease tedious labor work, dynamic coating machines were developed in which polyelectrolyte layers are deposited *in-situ* by alternatively circulating oppositely charged polyelectrolyte solutions. In spray coating, the polyelectrolyte solutions are sprayed over the substrate and the excess solution is drained by gravity. Compared to dip coating, the deposition time during spray coating is reduced by almost two orders of magnitude; hence, its scale-up is possible through automated spray coaters. During spin coating, polyelectrolyte adsorption is controlled by centrifugal forces, viscous forces and electrostatic interactions. Highly ordered internal structures in the polyelectrolyte layers are obtained with spin coating as a result of fast rearrangement of the polymer chains on the substrate and desorption of weakly adsorbed chains (Chiarelli *et al.*, 2001; Cho *et al.*, 2001; Fadhillah *et al.*, 2012). Although the deposition time is short in spin coating, relatively small surface areas are coated with this technique.

The LbL approach allows a precise control over the layer thickness by varying the polyelectrolyte types, the number of layers deposited and the adsorption conditions. The amount of polyelectrolyte deposited, and therefore the thickness of the layers, increases either linearly or exponentially depending on the deposition conditions. For instance, a linear growth was observed for the poly(styrene sulfonate) (PSS)/poly(diallyldimethylammonium) (PDADMAC) system at 25°C while the increase was exponential at 55°C (Salomäki *et al.*, 2005). Strong attraction between the polyelectrolytes and charge compensation results in the linear growth of the layers (Ladam *et al.*, 2000; Lavalle *et al.*, 2002; Porcel *et al.*, 2006, 2007). It was shown that in linearly growing films the layer thickness is typically several nanometers per bilayer due to charge compensation, which causes electrostatic repulsion and limits polyelectrolyte adsorption (Caruso *et al.*, 1999; Ladam *et al.*, 2000). In the case of exponential growth mode, in- and out-diffusion of the polyelectrolytes causes the deposition of layers with micrometer thickness.

The separation characteristics and functional properties of the LbL modified membranes depend on the structure and morphology of the assembly. Many works in the literature were aimed at investigating the influences of various parameters on the structure of the deposited layers and they will be discussed in the following section.

2.2   THE INFLUENCES OF PREPARATION CONDITIONS ON THE STRUCTURE AND MORPHOLOGY OF THE LBL MODIFIED MEMBRANES

The preparation conditions, such as pH, ionic strength, temperature, polyelectrolyte concentration, adsorption time, polyelectrolyte pairs and support on which layers are deposited, influence the structure of the layers (Table 2.1). Among these parameters, ionic strength plays a key role in controlling the thickness, stability and structure of the layers. Increasing the ionic strength of the polyelectrolyte solution leads to thicker layers (Arys *et al.*, 1998; Belyaev *et al.*, 1998; Bijlsma *et al.*, 1997; Clark *et al.*, 1997; Dubas and Schlenoff, 1999; Hoogeveen *et al.*, 1996a; Kolarik *et al.*, 1999; Krasemann and Tieke, 2000; Laurent and Schlenoff, 1997; Lehr *et al.*, 1995; Lvov *et al.*, 1993, 1997, 1998; 1999; Schlenoff *et al.*, 1998) and higher roughness of the membranes (Kolarik *et al.*, 1999). At high ionic strengths, polyelectrolyte charges are screened, which suppresses the electrostatic attraction between the polyelectrolytes causing the formation of more coiled and loopy structures. The pH of the polyelectrolyte solution determines both the charge density of the weak adsorbing polyelectrolytes and the surface charge density, thus, pH is used to control the morphology of the LbL assembly constructed from weak polyelectrolytes (Yoo *et al.*, 1998). Below a minimum charge density, the charge reversal is not sufficient for the formation of multilayers (Glinel *et al.*, 2002; Schoeler *et al.*, 2002; Steitz *et al.*, 2001; Voigt *et al.*, 2003). The adsorption of highly charged polyelectrolytes results in thin layers with flat chain conformations and thicker, loopier type structures are observed when they have a lower charge (Yoo *et al.*, 1998). With increasing surface charge density and polyelectrolyte concentration in dipping solution, more polyelectrolyte adsorption occurs to overcompensate for the charges on the surface. At low polyelectrolyte concentration, the conformation of the adsorbed layer is flat and at high concentrations a thicker coiled conformation is observed (Voigt *et al.*, 2003). The temperature of the polyelectrolyte solution is another important parameter that influences the LbL process. It was shown that the layers deposited at high temperatures are significantly thicker than the assembly fabricated at room temperature (Büscher *et al.*, 2002; Tan *et al.*, 2003). This can be attributed to the higher chain mobility at elevated temperatures leading to an increase in the number of loops and tails attached to the support. The choice of polyelectrolyte pairs affects the structural features of the multilayer assembly, such as thickness, porosity, roughness, hydrophilicity, swelling degree and mechanical properties. The chain stiffness and the balance between hydrophobicity and hydrophilicity of the polyelectrolytes determine the thickness of the assembly (Klitzing, 2006). As the hydrophobicity of the polyelectrolyte pairs increases, more stable and thinner polyelectrolyte multilayers are obtained, which are less prone to swelling by salt solutions (Dubas and Schlenoff, 2001; Harris and Bruening, 2000; Miller and Bruening, 2005). The choice of support is also important for constructing LbL assembly, since the adsorption of polyelectrolytes is strongly influenced by the support. The influence of chemical properties of substrate is limited to the first few bilayer depositions. However, two physical properties, which are charge density and the relative dielectric permittivity of the substrate, can influence the multilayer morphology up to a thickness in the micrometer range (Guillaume-Gentil *et al.*, 2011). Guillaume-Gentil *et al.* (2011) found that the surface morphology of the multilayer assembly is not influenced by the physical properties of the underlying substrate for a strongly interacting polyelectrolyte couple, poly(allylamine hydrochloride)/poly(styrene sulfonate) (PAH/PSS). In contrast, they observed continuous, flat layers of the weakly interacting pair poly(L-lysine)/hyaluronic acid (PLL/HA) on substrates of low dielectric permittivity and inhomogeneous droplet-films on substrates of high dielectric permittivity. A good support for LbL assembly should have a low surface roughness and a high surface charge density (Guillaume-Gentil *et al.*, 2011). In addition, the supports should have high chemical, thermal and mechanical stability and should provide sufficient permeability and hydrophilicity (Tieke *et al.*, 2001).

2.3   THE STABILITY OF LBL DEPOSITED LAYERS

The enthalpy change during polyelectrolyte complexation is small and the multilayer formation is entropically driven, hence thermodynamically highly favorable (Schlenoff *et al.*, 2008). The increase in entropy occurs due to the displacement of counterions as soon as polyelectrolytes adsorb on an oppositely charged polyelectrolyte. The Gibbs free energy change for the LbL deposition process is the sum of three terms. The first term represents the attraction energy between the surface charge groups and the opposite charges on the polyelectrolyte and this is known as intrinsic compensation. The second term is a conformational term, which is a penalty for the adsorption due to loss of entropy, and finally the third term corresponds to segment-segment repulsive energy and is another penalty for adsorption. It is obvious that when the sum of the repulsive and conformation energies exceeds the attraction energy, then desorption of deposited layers is favored. The stability is predominantly controlled by the strength of interactions between the polyelectrolytes, which is determined by the charge density of the polyelectrolytes and the ionic strength of the solution. When a salt is present in the solution contacting multilayers, salt counterions participate in charge neutralization called extrinsic compensation, which is shown with the following reversible ion exchange reaction (Dubas and Schlenoff, 2001):

$$Pol^+Pol_m^- + Na_{aq}^+ + Cl_{aq}^- \Leftrightarrow Pol^+Cl_m^- + Pol^-Na_m^+$$

$$K = \frac{y^2}{(1-y)[NaCl]} = \left(\frac{y^2}{[NaCl]_{aq}^2}\right)_{y\to 0} \tag{2.1}$$

where m refers to LbL deposited layers, K is the equilibrium constant for this reaction and y refers to polyelectrolyte multilayers in extrinsically compensated form. It is clear that at high salt concentrations in solution, the extrinsic charge compensation (between $Pol^+Cl_m^-$ and $Pol^-Na_m^+$) becomes dominant over the intrinsic one ($Pol^+Pol_m^-$) between polyelectrolytes due to increased charge screening along the polyelectrolyte chain. This causes conformational change resulting in the formation of more coiled and loopy structures. In conclusion, high ionic strength of the solution in contact with the layers and the low charge density of polyelectrolytes leads to enhanced chain mobility, hence the detachment of layers by weakening electrostatic attraction between polyelectrolytes and inducing conformational change of the layers (Hoogeveen *et al.*, 1996b; Linford *et al.*, 1998). pH is also an important factor on the stability of layers assembled from weak polyelectrolytes since they have a pH-dependent charge. At certain pH values the charge density of polyelectrolytes decreases substantially and layers detach from each other (Brynda and Houska, 1996, 1998; Harris *et al.*, 1999).

Table 2.1.   The influences of preparation conditions on the morphology of the LbL modified membranes.

Preparation condition	Influence on the morphology of modified membrane	Reference
Ionic strength of polyelectrolyte solution	Increasing ionic strength leads to thicker layers and higher roughness of the membranes if the layers are stable; in the case of instability of layers with high ionic strength, thinner layers are obtained due to decomposition of loosely attached layers	Arys *et al.* (1998), Belyaev *et al.* (1998), Bijlsma *et al.* (1997), Clark *et al.* (1997), Dubas and Schlenoff (1999), Hoogeveen *et al.* (1996a), Kolarik *et al.* (1999), Krasemann and Tieke (2000), Laurent and Schlenoff (1997), Lehr *et al.* (1995), Lvov *et al.* (1993, 1997, 1998, 1999), Schlenoff *et al.* (1998)
pH of polyelectrolyte solution	pH influences the charge density of weak polyelectrolytes Thin layers with flat chain conformations are observed when polyelectrolytes are highly charged At the deposition pH corresponding to the average of the pKa values of the polycation and polyanion, maximum density of ionic cross-links in the assembly is achieved	Glinel *et al.* (2002), Schoeler *et al.* (2002), Steitz *et al.* (2001), Voigt *et al.* (2003)
Temperature of polyelectrolyte solution	Deposition at temperatures greater than room temperature results in thicker layers compared to assembly formed at room temperature	Büscher *et al.* (2002), Tan *et al.* (2003)
Polyelectrolyte concentration	Increasing polyelectrolyte concentration leads to thicker layers At low concentrations, the conformation of adsorbed layers is rather flat and high concentrations result in short trains and long tails at the surface of the assembly	Voigt *et al.* (2003)
Polyelectrolyte pair	As the hydrophobicity of polyelectrolyte pairs increases, more stable and thinner layers are obtained Polyelectrolytes with a high charge density form a high degree of ionic cross-linking, hence showing low swelling	Dubas and Schlenoff (2001), Harris and Bruening (2000), Klitzing *et al.* (2006), Miller and Bruening 2005)
Type of support	The chemical properties of support influences the deposition of the first few layers For weakly interacting layers, continuous, flat layers are observed on substrates of low dielectric permittivity and inhomogeneous droplet-films on substrates of high dielectric permittivity Higher charge density of support does not influence the morphology but allows more adsorption and increases stability of layers	Guillaume-Gentil *et al.* (2011), Tieke *et al.* (2001)

LbL-assembled membranes used for water treatment are exposed to harsh conditions; hence, the stability and long-term performance of these membranes should be determined. Recently, de Grooth *et al.* (2015) modified two types of support membranes, non-ionic polyether sulfone (PES) (*MWCO* = 100 kDa) and ionic sulfonated polyether sulfone (SPES) (*MWCO* = 10 kDa), with the primary polycation poly (allylamine)hydrochloride (PAH) and the quaternary poly-cation poly(diallyldimethylammonium) (PDADMAC) and polyanion polystyrene sulfonic acid (PSS, *Mw* = 100 kDa). The physical stability of the membranes was determined by measuring the

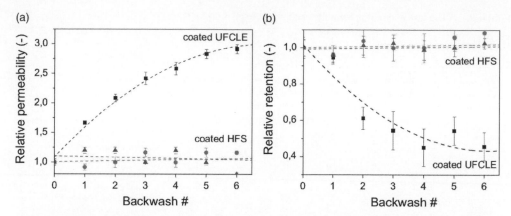

Figure 2.3.   Change in (a) permeability, (b) MgSO$_4$ rejection of (PDADMAC/PSS)$_n$ coated PES membranes during backwash cycles. Both permeability and rejection values were normalized with respect to initial values. UFCLE: Poly(ether sulfone) (PES) membrane (MWCO:100 kDa), HFS: PES membrane coated with a separation layer of sulfonated poly(ether sulfone) (SPES) (MWCO:10 kDa). Reprinted from de Grooth *et al.* (2015) with permission from Elsevier.

permeability and MgSO$_4$ retention of the membranes after each backwash cycle. Figure 2.3 shows that the permeability of the PES coated membranes (Code of the membrane: UFLCE) increased almost threefold after six backwashes, while the rejection of MgSO$_4$ decreased from 84% to 38%.

The changes in the permeability and retention were attributed to the detachment of the polyelectrolytes. The multilayer assembly formed on the negatively charged SPES support (Code of the membrane: HFS) membrane did not show any change in either the permeability or the retention of ions at the end of several backwash cycles. This result clearly showed the key role of surface charge density of support membranes for an increased physical stability of the LbL coated membranes. For testing chemical stability, SPES modified membranes were exposed to sodium hypochlorite solution, which is commonly used for the chemical cleaning of membranes. Figure 2.4 shows that the hypochlorite stability of the membranes is significantly influenced by the polycation type. The membranes coated with PDADMAC/PSS showed 4–5 times higher stability of the permeability and retention compared to PAH/PSS coated membranes.

Saeki *et al.* (2013) investigated the influences of two cross-linking methods, amine and silane coupling reactions, on the stability of LbL-assembled nanofiltration (NF) membranes in high ionic strength conditions and chlorine treatment. Both amine and silane coupling reactions improved the stability of the membrane performance under high ionic strength conditions, however, only the silane coupling reaction prevented the loss of deposited layers under harsh chlorine exposure.

2.4   THE LBL COATED MEMBRANES AND THEIR APPLICATIONS

LbL deposition of polyelectrolytes is used as an effective modification technique for improving the anti-fouling properties of ultrafiltration (UF) and microfiltration (MF) membranes. Most commercial UF and MF membranes are manufactured from hydrophobic polymers, which are prone to organic fouling as a result of hydrophobic interactions. Polyelectrolytes are usually hydrophilic, hence, coating/impregnating membranes with polyelectrolytes changes their surface properties from hydrophobic to hydrophilic. LbL assembly on the membranes is also constructed to adjust the permeability of molecules or ions that can permeate across the membranes. By changing the pH and ionic strength of the polyelectrolyte solution used for building layers, it is possible to balance intrinsic and extrinsic charge compensation, and therefore to control the pore size of the

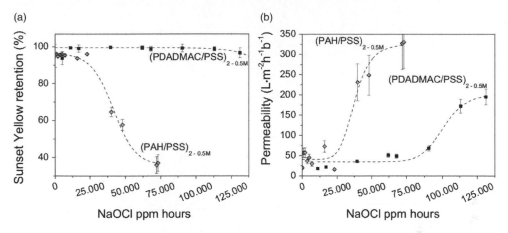

Figure 2.4. The change in (a) retention (b) permeability of (PAH/PSS)₂ and (PDADMAC/PSS)₂ coated membranes stored in sodium hypochlorite solution at pH 8. Reprinted from de Grooth *et al.* (2015) with permission from Elsevier.

resulting membrane. LbL assembly is considered as a promising alternative in manufacturing membranes with sufficiently high flux and high selectivity for various applications. The use of LbL deposition for preparing water treatment membranes is discussed in the following section.

2.4.1 *Nanofiltration*

The LbL deposition of polyelectrolytes on suitable supports can be used in the preparation of nanofiltration membranes for water softening, partial desalination, selective removal and separation of metal ions and charged/non-charged species (Table 2.2). The LbL method offers several advantages for NF membranes. Firstly, polyelectrolyte layers have high water adsorption capacities allowing them to achieve a high selectivity in the passage of water relative to the passage of salt. Secondly, the presence of ionic groups facilitates the permeation of water while rejecting ions with the same charges due to Donnan exclusion effects. In addition, the thickness, charge density and composition of the selective layer can be precisely adjusted by carefully selecting the deposition conditions and polyelectrolyte pairs. As a result, the selectivity, flux and fouling rates can be easily controlled. Nanofiltration membranes reject molecules with a molecular weight larger than 1000 g/mole; hence, the selective layer should have pore sizes smaller than 2 nm to allow the permeation of only small molecules. The thickness of the LbL formed selective layer is on the order of nanometer range, which allows a reasonably high permeation rate in addition to high selectivity. Polyelectrolytes with high charge densities, including PSS and SPEEK as polyanions and PAH, PDADMAC and PEI as polycations, are commonly used to prepare dense layers required for nanofiltration membranes. In the study by Krasemann and Tieke (2000) the number of layers was increased up to 90 in order to obtain desirable rejection properties. The separation factors for Na^+/Mg^{2+} and Cl^-/SO_4^{2-} were found to be 112.5 and 45, respectively, when 60 layers of PAH/polystyrenesulfonate were deposited on PAN/PET supporting membranes treated with oxygen plasma. The same group assembled 60 layers of the dicopper complex of 18-azacrown-N6 (hexacyclen, 1,4,7,10,13,16-hexaazaoctadecane, aza6) and polyvinylsulfate (PVS) on the PAN/PET support membrane for the desalination of aqueous electrolyte solutions under nanofiltration conditions (El-Hashani *et al.*, 2008). The rejection of chloride and sulfate ions were 45.7 and 91.9% at the operating pressure of 2.5 MPa; on the other hand, the flux of the resulting membranes was very low due to significant mass resistance created by the layers. Lajimi *et al.* (2004) coated cellulose acetate membranes with alginate and chitosan layers and determined

Table 2.2. Recent studies published on the application of the LbL coated membranes in nanofiltration.

Application	PE pair/# of bilayers	Support	Permeability	Selectivity	Reference
Separation of Cl⁻ from SO_4^{2-} and Separation of Na⁺ from Mg^{2+}	PAH/PSS $n=60$	PAN/PET	NaCl: 22.5×10^{-6} cm s⁻¹ (with salt) NaCl: 44.4×10^{-6} cm s⁻¹ (without salt) Na_2SO_4: 0.5×10^{-6} cm s⁻¹ (with salt) Na_2SO_4: 4.5×10^{-6} cm s⁻¹ (without salt) $MgCl_2$: 0.2×10^{-6} cm s⁻¹ (with salt) $MgCl_2$: 2.9×10^{-6} cm s⁻¹ (without salt)	Na^+/Mg^{2+}: 112.5 (with salt) Na^+/Mg^{2+}: 15.1 (without salt) Cl^-/SO_4^{2-}: 45 (with salt) Cl^-/SO_4^{2-}: 9.9 (without salt)	Krasemann and Tieke (2000)
Separation of Cl⁻ from SO_4^{2-}	dicopper complex of 18-azacrown-$N6$/PVS $n=60$	PAN/PET	8.5 ± 1 L m⁻² h⁻¹ MPa⁻¹	6.7	El-Hashani et al. (2008)
Separation of $MgSO_4$ from NaCl	CHI/ALG $n=25$	CA: 20 wt % Acetone: Formamide: 2:1 Annealing temperature: 70°C	NaCl: 4 L m⁻² h⁻¹ (0.34 M) $MgSO_4$: 3.5 L m⁻² h⁻¹ (0.017 M)	$MgSO_4$/NaCl: 1.63	Lajimi et al. (2004)
Separation of sugars and salts	PSS/PDADMAC $n=5.5$	Alumina	Pure water flux: 1.6 ± 0.2 m³ m⁻² day⁻¹	Glycerol/glucose: 2.0 ± 0.3 Glucose/sucrose: 9 ± 3 Glucose/raffinose: 50 ± 20	Miller and Bruening (2004)
	PSS/CHI $n=4.5$		Pure water flux: 1.5 ± 0.3 m³ m⁻² day⁻¹	Glycerol/glucose: 1.7 ± 0.3 Glucose/sucrose: 6 ± 1 Glucose/raffinose: 19 ± 5	
	PSS/PDADMAC (0.5 M NaCl in polyelectrolyte solution) $n=5.5$		1.6 ± 0.2 m³ m⁻² day⁻¹	NaCl/sucrose: 7 ± 2	
	PSS/PDADMAC (0.1 M NaCl in polyelectrolyte solution)		2.6 ± 0.5 m³ m⁻² day⁻¹	NaCl/sucrose: 15 ± 5	

Separation	Multilayer	Substrate	Flux	Selectivity	Reference
Separation of Cl^- from SO_4^{2-}	[PSS/PDADMAC]$_n$/ PSS	PES (MWCO: 50 kDa)	1.7 ± 0.1 ($n=2$) 1.6 ± 0.1 ($n=4$)	2.7 ± 0.3 ($n=2$) 32 ± 6 ($n=4$)	Malaisamy and Bruening 2005
Separation of Cl^- from SO_4^{2-}	(PSS/PDADMAC)$_n$ PSS	Au-coated Si	4.2 ± 0.2 ($n=3$) m^3 m^{-2} day^{-1} (0.5 M NaCl in the last layer) 2.7 ± 0.2 ($n=3$) m^3 m^{-2} day^{-1} (1 M NaCl in the last layer)	2.7 ± 0.5 ($n=3$) m^3 m^{-2} day^{-1} (0.5 M NaCl in the last layer) 26.4 ± 0.9 ($n=3$) m^3 m^{-2} day^{-1} (1 M NaCl in the last layer)	Hong et al. (2006a)
Separation of Cl^- from F^- and Br^-	(PSS/PDADMAC)$_4$ PSS 0.5 M NaCl in the last layer	Alumina	3.5 ± 0.2 m^3 m^{-2} day^{-1} 3.7 ± 0.3 m^3 m^{-2} day^{-1}	Cl^-/F^-: 3.4 ± 0.2 Cl^-/Br^-: 3.8 ± 0.1	Hong et al. (2007)
Separation of chloride from phosphate	[PSS/PDADMAC]$_n$ PSS	Alumina	2.9 ± 0.1 m^3 m^{-2} day^{-1} ($n=3.5$) 2.4 ± 0.1 m^3 m^{-2} day^{-1} ($n=4.5$)	4.3 ± 0.1 ($n=3.5$) 48.3 ± 5.8 ($n=4.5$)	Hong et al. (2009)
Separation of anions	[PDADMAC/ SPEEK]$_n$ $n=10$	Hydrolyzed polyacrylonitrile (PAN-H)	0.09 m^3 m^{-2} h^{-1} MPa^{-1} nm^{-1} (0 M NaCl) 0.12 m^3 m^{-2} h^{-1} MPa^{-1} nm^{-1} (0.1 M NaCl) 0.06 m^3 m^{-2} h^{-1} MPa^{-1} nm^{-1} (0.5 M NaCl)	Cl^-/HPO_4^{2-}: 10 (0 M NaCl) 12 (0.1 M NaCl) 20 (0.5 M NaCl) Cl^-/SO_4^{2-}: 10 (0 M NaCl) 18 (0.1 M NaCl) 13 (0.5 M NaCl) SO_4^{2-}/HPO_4^{2-}: 1 (0 M NaCl) 1 (0.1 M NaCl) 1 (0.5 M NaCl)	Ahmadiannamini et al. (2010)
Separation of Na^+ from Mg^{2+}	[PSS/PDADMAC]$_4$ PSS Salt concentrations in the deposition solution of each layer: 0.5 M NaCl [PSS/PDADMAC]$_4$ PSS/PAH	Alumina	1.74 ± 0.03 m^3 m^{-2} day^{-1} 1.04 ± 0.15 m^3 m^{-2} day^{-1}	1.6 ± 0.1 16.2 ± 1.7	Ouyang et al. (2008)
Separation of Fe^{3+} from Fe^{2+}	[PAA/CTAC]$_n$	P(AN-co-EDAMA)		1.07 ± 0.03 ($n=1$) 2.10 ± 0.04 ($n=3$)	Fu and Kobayashi (2010)

n: number of bilayers.

the optimal number of bilayers as around 15/20 based on permeation rate/salt retention and the separation factor of monovalent/divalent salts. Major improvement in the LbL coated membranes was achieved by the group of M.L. Bruening when the number of layers deposited on porous supports were reduced from several tens to just a few ones (Hong and Bruening, 2006; Hong et al., 2006a, 2006b, 2007, 2009; Malaisamy and Bruening, 2005; Miller and Bruening, 2004; Ouyang et al., 2008; Shan et al., 2010). It was shown that 2.5 bilayers of poly(styrenesulfonate) (PSS)/protonated poly(allylamine) (PAH) or 3.5 bilayers of PSS/poly(diallyldimethylammonium chloride) (PDADMAC) were enough to reduce the molecular weight cut-off of polyethersulfone ultrafiltration supports from 50 kDa to <500 Da (Malaisamy and Bruening, 2005). 50 kDa support membranes modified with 3.5 bilayers of PSS/PDADMAC exhibited 95% rejection of SO_4^{2-} and a chloride/sulfate selectivity of 27, while 4.5 bilayer PSS/PAH coatings showed a glucose/raffinose selectivity of 100. The authors clearly demonstrated that the molecular weight cut-off value of a support membrane should be smaller than 100 kDa to convert UF membranes to NF membranes with a few layers of polyelectrolyte deposition. Hong et al. (2006a) reported a 96% rejection of SO_4^{2-} and Cl^-/SO_4^{2-} selectivity of 26 with the $(PSS/PDADMAC)_3PSS$ layers deposited on porous alumina, which also allowed a 2.7-fold higher flux along with a greater passage of Cl^- compared to the commercial NF270 membranes. The $(PSS/PDADMAC)_{4.5}PSS$ membrane on the same support exhibited Cl^-/F^- and Br^-/F^- selectivities >3 (Hong et al., 2007) and a chloride/phosphate selectivity of 48 (Hong et al., 2009) along with solution fluxes that are higher than those of the commercial NF 270 and NF 90 membranes. Ahmadiannamini et al. (2010) deposited poly(diallyldimethylammonium) chloride and sulfonated poly(ether ketone) (SPEEK) on hydrolyzed polyacrylonitrile (PAN-H) supports and tested the performances of the resulting membranes for anion separation application. The highest selectivities for Cl^-/HPO_4^{2-} and Cl^-/SO_4^{2-} separation were around 20 while for the SO_4^{2-}/HPO_4^{2-} separation the selectivity was approximately 1 when the number of bilayers were 10. Ouyang et al. (2008) reported that LbL modified NF membranes composed of PSS/PAH or hybrid PSS/PDADMAC+ PSS/PAH layers are effective in selectively removing Mg^{2+} from solutions containing NaCl and $MgCl_2$. On the other hand, pure PSS/PDADMAC coating and commercial NF270 membranes show only relatively low rejections of Mg^{2+}. Na^+/Mg^{2+} selectivity increased with the magnitude of surface charge when the outermost layer of the assembly was PAH. Fu and Kobayashi (2010) showed the multilayer membranes prepared from cetyl trimethyl ammonium chloride (CTAC) and poly(acrylic acid) (PAA) alternately deposited onto negatively and positively charged polyacrylonitrile copolymer membranes made of poly(acrylonitrile-co-acrylic acid) [P(AN-co-AA)] and poly(acrylonitrile-co-N,N-dimethylethyl ammonium bromide) [P(AN-co-EDAMA)] can effectively remove Fe^{3+} from Fe^{2+}. M.L. Bruening's group also demonstrated the effectiveness of LbL modified NF membranes in separating neutral species. For example, they reported that PSS/PAH layers on a porous alumina substrate can effectively separate neutral amino acids according to their sizes and the flux of these membranes are comparable to that of commercial NF membranes (Hong and Bruening, 2006). In another work, they showed that multilayer membranes assembled from a 4.5-bilayer poly(styrenesulfonate) (PSS)/poly(allylamine hydrochloride) (PAH) coating on porous alumina exhibits NaCl/sucrose selectivities of 130 and NaCl/dye selectivities >2200 (Hong et al., 2006b). Wang et al. (2009) deposited 3.5 bilayers of SPEEK and PEI on PAN support membranes. It was shown that NaCl rejection of the membranes can be significantly increased by assembling the layers under pressure (Fig. 2.5).

They investigated the influence of the molecular weight of PEI on the permeability and salt rejection of the membranes. The flux of polyelectrolyte multilayer membranes from both low and high molecular weight PEI decreased with the increased layers, however, no salt was rejected when low molecular weight PEI was used (Fig. 2.6) due to the looser structure of the selective layer. This result suggested that when low molecular weight PEI is used, a higher number of layers are required to achieve salt rejection.

These membranes exhibited not only better rejection but also higher anti-fouling properties in comparison to commercial membranes NTR 7450 (Fig. 2.7).

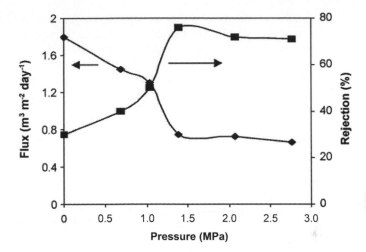

Figure 2.5.   The influence of pressure applied during deposition of SPPEK/PEI layers on the flux and salt rejection of modified PAN membranes. Reprinted from Wang *et al*. (2009) with permission from Elsevier.

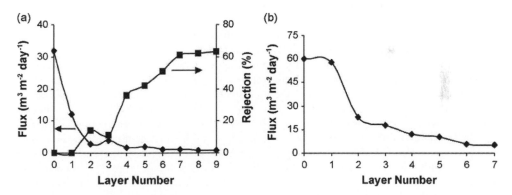

Figure 2.6.   The influence of molecular weight of cationic polyelectrolyte, PEI, on the flux and salt rejection of modified PAN membranes. (a) Branched PEI (*Mw* = 25000), (b) Low *Mw* PEI (*Mw* = 800). Reprinted from Wang *et al*. (2009) with permission from Elsevier.

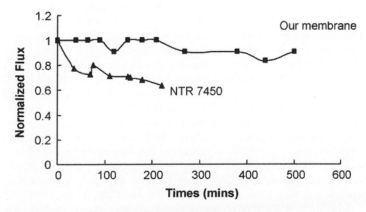

Figure 2.7   Flux decline during dead-end filtration of humic acid ($1\,g\,L^{-1}$ in $1\,mM$ CaCl$_2$, pressure = 1.379 MPa). Reprinted from Wang *et al*. (2009) with permission from Elsevier.

Shan *et al.* (2010) reported an improvement in the separation properties of LbL modified membranes with the deposition of colloids on the surface of the membranes and demonstrated the feasibility of regenerating the multilayer coating. Ba *et al.* (2010) designed a positively charged nanofiltration membrane by chemical modification of P84 copolyimide using branched polyethylenimine (PEI) and coated the surface with a neutral polymer, polyvinyl alcohol (PVA) and negatively charged polymers, polyacrylic acid (PAA) and polyvinyl sulfate (PVS). In the case of PVS-coated membranes, flux recovery after acid cleaning was almost 100% due to a decreased charge interaction between foulants and the membrane surface. The flux restoration in the PAA-coated membrane was not as good as that in the PVS-coated membrane since acid treatment made the PAA-coated membrane more positively charged. LbL coated polymeric membranes were also successfully applied for solvent-resistant nanofiltration. Li *et al.* (2008, 2010) coated the PAN support membrane with LbL deposition of (poly(diallyldimethylammonium chloride) and sulfonated poly(ether ether ketone) and the modified membranes were successfully applied in the filtrations of organic solvents, including aprotic solvents like THF and DMF. Coated membranes showed comparable selectivity but better solvent stability and higher fluxes in comparison to commercial membranes.

2.4.2   *Reverse osmosis*

Reverse osmosis membranes are commonly used for removing dissolved salts and other minerals from sea water and brackish water. The most commonly used membranes for desalination are thin film composite membranes consisting of a polyamide based active layer providing ion selectivity deposited on a porous polysulfone support layer. Polyamide is not resistant to chlorine treatment, thus, most of the research focused on developing RO membranes for desalination that cannot lose their rejection properties when exposed to chlorine. The LbL deposition of polyelectrolytes on porous support membranes is one of the approaches used in the literature to obtain dense reverse osmosis (RO) membranes (Table 2.3). Jin *et al.* (2003) modified PAN/PET porous support membranes with alternating electrostatic layer-by-layer adsorption of polyvinylamine (PVA) and polyvinyl sulfate (PVS). The rejections of the resulting 60 bilayer membranes for NaCl and Na_2SO_4 were 93.5 and 98.5% at 4.0 MPa, respectively. The performances of the membranes were also tested with sea water and the rejections for Mg^{2+}, Ca^{2+} and Na^+ were determined as 98, 96.4 and 74.5%, respectively (Toutianoush *et al.*, 2005). Park *et al.* (2010) coated commercial polysulfone support membranes with poly(allylamine hydrochloride) (PAH) and poly(acrylic acid) (PAA) layers and tested the performance of the resulting membranes for desalination of sea water. The layers were deposited at the pH conditions allowing a low charge density of each polyelectrolyte. Thermal cross-linking and switching phenomena of the charge density were used to improve ion rejection rates. An ion rejection rate >99% was achieved with 10 to 20 bilayers of PAH pH 7.5/PAA pH 3.5 at the pH of sea water, 8.1 after the recycling process (Fig. 2.8). Enhanced ion rejection rates with recycling of the permeate was due to converting the high ion concentration of the feed solution into low concentrated solutions.

Excellent ion rejection rates were obtained with a high degree of internal charge pairing within the multilayers, where the polyelectrolytes are at a highly charged state at pH 8.1. Although the rejection rate was high and comparable to that of the commercial membrane, the fluxes through these membranes were low. Fadhillah *et al.* (2013) fabricated thin film composite membranes by depositing alternate layers of poly(allyl amine hydrochloride) (PAH) and poly(acrylic acid) (PAA) on a polysulfone support membrane. Successful coating of the support membrane was confirmed by AFM images (Fig. 2.8), which indicated that the polyelectrolyte deposition reduced the roughness of the PSF membranes. They measured significantly higher surface roughnesses for two commercial RO membranes, Hydronautics SWC and ESPA (136.56 ± 15.95 nm and 103.67 ± 9.98 nm, respectively), compared to PAH/PAA modified membranes (11.26 ± 1.80 nm and 10.33 ± 1.58 nm for 60 and 120 [PAH/PAA] bilayers, respectively) (Fig. 2.9). The smoothness created by LbL deposition of polyelectrolytes is an important advantage, since rough surfaces are more susceptible to fouling.

Table 2.3. Recent studies published on the application of the LbL coated membranes in reverse osmosis.

Application	PE pair/# of bilayers	Support	Permeability/flux	Rejection [%]	Reference
Desalination	PVA/PSS $n = 60$	PAN/PET	$1137\,\mathrm{mL\,m^{-2}\,h^{-1}\,MPa^{-1}}$	NaCl: 93.5%; Na$_2$SO$_4$: 98.5% $p = 4.0\,\mathrm{MPa}$	Jin et al. (2003)
Desalination	PVA/PSS $n = 60$	PAN/PET	$40 \pm 2\,\mathrm{L\,m^{-2}\,h^{-1}\,MPa^{-1}}$	Mg^{2+}: 98.0 ± 1% Ca^{2+}: 96.4 ± 1% Na$^+$: 74.5 ± 0.8% (using non-diluted sea water) $p = 4.0\,\mathrm{MPa}$	Toutianoush et al. (2005)
Desalination	[PAH pH 7.5/PAA pH 3.5]$_n$ $n = 10$	PSf	$40\,\mathrm{L\,m^{-2}\,h^{-1}}$ before thermal annealing $8\,\mathrm{L\,m^{-2}\,h^{-1}}$ after thermal annealing: $T = 180°\mathrm{C}$ $t = 1\,\mathrm{h}$	NaCl: 50% before thermal annealing NaCl: 80% after thermal annealing NaCl: >99% (after recycling process)	Park et al. (2010)
Desalination	[PAH/PAA]$_n$	PSf	$30\,\mathrm{L\,m^{-2}\,h^{-1}}$ ($n = 60$) $15\,\mathrm{L\,m^{-2}h^{-1}}$ ($n = 120$)	NaCl: 58% ($n = 60$) NaCl: 65.5% $n = 120$	Fadhillah et al. (2013)
Desalination	[PAH/MTM]$_n$		$25.5\,\mathrm{L\,m^{-2}\,h^{-1}}$ ($n = 9$) $8.3\,\mathrm{L\,m^{-2}\,h^{-1}}$ ($n = 18$)	NaCl: 30% ($n = 9$) NaCl: 81% ($n = 18$)	Choi et al. (2014)
Desalination	[PSS/PAH]$_n$ [sPSF/PAH]$_n$ $n = 10$		$8\,\mathrm{L\,m^{-2}\,h^{-1}}$ $6\,\mathrm{L\,m^{-2}\,h^{-1}}$	NaCl: 90% NaCl: 97%	Cho et al. (2015)
Desalination	[PSS/PAH]$_n$	Commercial RO membrane (ES-20)	$4.54 \times 10^{-5}\,\mathrm{L\,m^{-2}\,h^{-1}\,Pa^{-1}}$ $2.17 \times 10^{-5}\,\mathrm{L\,m^{-2}\,h^{-1}\,Pa^{-1}}$	$p = 2400\,\mathrm{kPa}$ $C_{\mathrm{NaCl}} = 2000\,\mathrm{mg\,L^{-1}}$ NaCl: 98.2% NaCl: 99.4%	Ishigami et al. (2012)

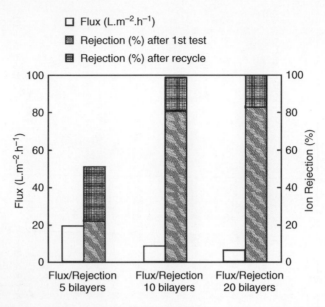

Figure 2.8.   Ion rejection and flux of PAH pH 7.5/PAA pH 3.5 modified polysulfone membranes before and after recycle operation. Adapted from Park *et al.* (2010) with the permission of The Royal Society of Chemistry.

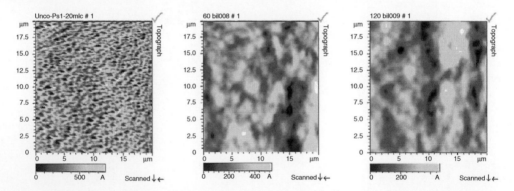

Figure 2.9.   AFM images of (a) bare PSF, (b) [PAH/PAA]$_{60}$, (c) [PAH/PAA]$_{120}$ deposited PSF membranes. Reprinted from Fadhillah *et al.* (2013) with permission from Elsevier.

The salt rejections of 120 bilayers and 60 bilayers PAH/PAA deposited membranes were determined as 65.5 and 58%, respectively, while the water fluxes were 15 and 30 L m^{-2} h^{-1}. The membranes remained stable within 40 h of exposure to 2000 mg L^{-1} sodium chloride solution, which corresponds to typical RO operational conditions. Choi *et al.* (2014) used a hybrid combination of [poly(allylamine hydrochloride) (PAH)/montmorillonite (MTM)]$_n$ layers to prepare RO desalination membranes. By inserting PAH/PAA layers between two adjacent PAH/MTM layers, the flux was maintained at ~7.5 ± 0.5 L m^{-1} h^{-1}, comparable to commercial membranes, while the salt rejection was around ~75 ± 2.5%. In addition, the stability of the membrane against the chlorine attack was enhanced, as confirmed by the similar ion rejection value (~74 ± 5.0%) after the NaOCl treatment. Cho *et al.* (2015) reported the deposition of PSS/PAH and sPSf/PAH on polysulfone support membranes. To facilitate a reduction in membrane pore size and swelling, and

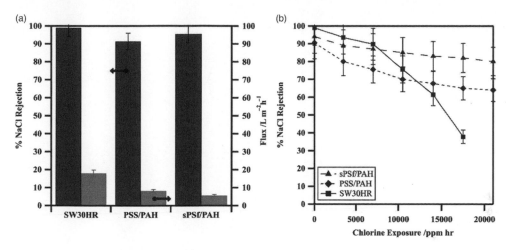

Figure 2.10. (a) NaCl rejection (black columns) and flux (gray columns) of crosslinked membranes, (b) NaCl rejection of crosslinked membranes as a function of chlorine exposure time. The rejection values for commercial membranes, SW30HR, are also shown. ($p = 2400\,kPa$; NaCl concentration $= 2000\,mg\,L^{-1}$). Adapted from Cho *et al.* (2015) with permission of John Wiley and Sons.

hence to increase the rejection of monovalent salts, they crosslinked PAH layers with gluteralde-hyde. The resultant membranes exhibited NaCl rejection of more than 95% and greater resistance to extended chlorine exposure than commercial aromatic polyamide membranes (Fig. 2.10).

The LbL deposition of polyelectrolytes was also applied on commercial thin film composite RO membranes to improve their anti-fouling and rejection properties. Polyamide, which forms the active layer of most commercial RO membranes, is negatively charged under typical operating conditions (pH > 4) due to carboxyl groups on the membrane surface, thus, polyamide based RO membranes are especially prone to fouling by cationic contaminants. Zhou *et al.* (2009) modified commercial polyamide thin film composite membranes by the electrostatic self-assembly of polyethyleneimine and the modified membrane showed significantly improved resistance to cationic foulants due to enhanced electrostatic repulsion and the increased surface hydrophilicity. Xu *et al.* (2011) used chitosan for modification and the resulting polyamide/chitosan membrane showed higher flux and salt rejections than the original polyamide membrane. Ishigami *et al.* (2012) deposited PSS and PAH on the commercial polyamide membrane and investigated the effect of the layer number on the water permeability, sodium chloride (NaCl) rejection and the anti-fouling capability of the membranes. Their results demonstrated increased salt rejection and anti-fouling capability with the increased layer number due to enhanced hydrophilicity and smoothed surface morphology.

2.4.3 *Forward osmosis*

Recently, forward osmosis has received significant attention for water treatment due to its low energy requirement. LbL modified membranes have high water permeability and rejection against divalent ions, high solvent resistance and thermal stability, which make these membranes suitable for forward osmosis (Table 2.4). In addition, their long-term stability can be provided through chemical cross-linking methods. Saren *et al.* (2011) deposited 1, 3, and 6 bilayers of PAH and PSS polyelectrolyte pairs on PAN support and tested the resulting membranes for forward osmosis. For an ideal FO membrane, the substrate should have a low mass transfer resistance. To achieve high porosity and a finger-like structure in the support membrane with a high hydrophilicity, preparation conditions were adjusted. The salt rejection of the membranes increased while both solvent and solute fluxes decreased by increasing the number of bilayers from 1 to 3. The performance of

Table 2.4. Recent studies published on the application of the LbL coated membranes in forward osmosis.

Application	PE pair/# of bilayers	Support	Solvent flux	J_w/J_s	Reference
Removal of MgCl$_2$	[PAH/PSS]$_n$	PAN	27.7 L m^{-2} h^{-1} ($n=1$) AL-FS 28.7 ($n=3$) AL-FS 22 ($n=6$) AL-FS	23.8 mM ($n=1$) 6.3 ($n=3$) 3.2 ($n=6$)	Saren et al. (2011)
Removal of MgCl$_2$	[PAH/PSS]$_n$ (not crosslinked)	PAN AL-DS	30 L m^{-2} h^{-1} ($n=1$) AL-DS 80 L m^{-2} h^{-1} ($n=3$) AL-DS Feed: DI DS: 0.5 M MgCl$_2$	45 mM ($n=1$) 2.7 mM ($n=3$)	Qiu et al. (2011)
Removal of MgCl$_2$	[PAH/PSS]$_n$ (crosslinked)	PAN AL-DS	27 L m^{-2} h^{-1} ($n=1$) AL-DS 60 L m^{-2} h^{-1} ($n=3$) AL-DS Feed: DI DS: 0.5 M MgCl$_2$	60 mM ($n=1$) 1 mM ($n=3$)	
Removal of MgCl$_2$	[PAH/PSS]$_n$ $n=3$ (not crosslinked)	PAN	23.33 ± 1.3 L m^{-2} h^{-1} ($n=1$) AL-DS AL-DS Feed: DI DS: 0.5 M MgCl$_2$	0.355 ± 0.050 g L^{-1}	Duong et al. (2013)
	[PAH/PSS]$_3$ (crosslinked)		15.80 ± 0.072 L m^{-2} h^{-1} AL-DS Feed: DI DS: 0.5 M MgCl$_2$	0.117 ± 0.021 g L^{-1}	
Removal of MgCl$_2$	[PAA-PSS/PAH]$_n$	PAI-PEI	28 L m^{-2} h^{-1} ($n=3$) 18 L m^{-2} h^{-1} ($n=6$) 13 L m^{-2} h^{-1} ($n=10$) AL-DS Feed: DI DS: 0.5 MgCl$_2$	0.07 0.068 0.073	Cui et al. (2013)
Removal of MgCl$_2$	[PSS/PAH]$_2$	PES	73 L m^{-2} h^{-1} (AL-DS) 21.5 L m^{-2} h^{-1} (AL-DI) Feed: DI DS: 0.5 M MgCl$_2$	0.06 g L^{-1} (AL-DS) 0.03 g L^{-1} (AL-DI)	Liu et al. (2015)
Removal of from NaCl	[MPD-TMC]$_{10}$	PAN-PEI-PAA	24.6 L m^{-2} h^{-1} AL-DS Feed: DI DS: 0.5 NaCl	0.1 g L^{-1}	Kwon et al. (2015)

AL-DS: active layer of the membrane is in contact with draw solution; AL-FS: active layer of the membrane is in contact with feed solution; DI: deionized water the 3 bilayer membrane was found to be higher than those of the other FO membranes reported in the literature when the active layer of the membrane faced the feed solution. In the following publications by the same group, the effect of the cross-linking of the polyelectrolyte layers on the substrate morphology and structure, the separation layer, water permeability, salt rejection, the FO water flux and solute flux performance was investigated (Qiu *et al.*, 2011; Qi *et al.*, 2012). Both the uncrosslinked and crosslinked membranes showed high fluxes, but on the other hand, the cross-linking improved the salt rejection of the membranes. It was shown that the negative impact of internal concentration polarization can be minimized by controlling the number of polyelectrolyte bilayers. Duong *et al.* (2013) investigated the influences of salt concentration and deposition time, in addition to cross-linking conditions on the FO performances of similar PAH/PSS coated PAN membranes. The membranes were crosslinked by a combination of chemical and photo cross-linking techniques and their results suggested that with this technique it is possible to achieve reasonable water flux and reverse salt fluxes even with one bilayer. Setiawan *et al.* (2013) fabricated dual layer hollow fibers by a dry jet-wet spinning technique using poly(amide-imide) (PAI) and polyethersulfone as the selective layer on the outer side and the porous support layer on the inner side, respectively. The outer layer of the substrate was modified by polyethyleneimine (PEI) cross-linking followed by multilayer polyelectrolyte depositions. Their results suggested that a combination of polyelectrolyte cross-linking and LbL deposition can be effectively used to prepare NF-like FO membranes. Cui *et al.* (2013) demonstrated that the pH value and salt concentration in polyelectrolyte solutions and the heterogeneity of polyelectrolyte solutions play an important role on LbL FO performance. LbL membranes prepared from the most heterogeneous polyelectrolyte solutions showed the highest reverse salt flux. When the charge densities of both polycation and polyanion were high, then LbL layers had a tight structure resulting in both low water flux and reverse salt flux. Liu *et al.* (2015) deposited the polyanion PSS and polycation PAH, PDADMAC and PEI alternately on either the inner surface (id-LbL) or outer surface (od-LbL) of the polyethersulfone (PES) hollow fiber substrates to prepare NF and FO membranes. Two layer id-LbL membranes demonstrated an increased water flux and reduced salt leakage compared to od-LbL membranes, hence, they concluded that the inner surface of the hollow fiber support membrane is more suitable for LbL deposition than the outer surface. Recently, Kwon *et al.* (2015) developed a molecular layer-by-layer (mLbL) technique based on the alternative cross-linking of monomers to fabricate thin film composite forward osmosis membranes. In their approach, firstly a hydrolyzed negatively charged PAN support membrane was alternately coated with cationic PEI and then anionic PAA. Next, the PEI/PAA interlayer-coated HPAN support was dipped into MPD and then into TMC solutions to form a selective polyamide layer and these steps were repeated until the desired number of layers was built. Figure 2.11 shows the surface and cross section images of support and modified membranes.

The HPAN support membrane has uniform pores of 15 nm in diameter and consists of a very thin dense skin layer on top of a finger-like pore structure (Figs. 2.11a and 2.11d). The roughness of the IP-assembled PA membrane was much higher (33.8 ± 2.1 nm) than that of the mLbL-assembled PA membrane (mLbL-10) (6.9 ± 1.6 nm). High water fluxes and low specific salt fluxes are required to maximize the performance of FO membranes. In this respect, compared to commercial membranes, HTI-CTA, and IP-assembled PA membranes, the mLbL-10 membrane showed the best FO performance (Fig. 2.12).

AL-DS: active layer of the membrane is in contact with draw solution; AL-FS: active layer of the membrane is in contact with feed solution; DI: deionized water

However, poor stability of the membranes at high ionic strengths was reported as the primary drawback for the mLbL-assembled membranes.

2.4.4 *Pervaporation*

Pervaporation can be used as an alternative to distillation for separating organic compounds from water. LbL modified membranes usually have high hydrophilicity and a strong affinity towards water since they carry a high surface charge density. In addition, they are ionically crosslinked,

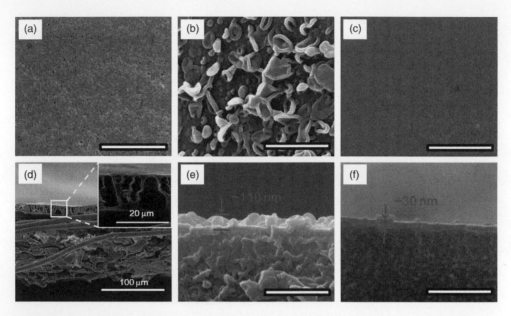

Figure 2.11.   Surface and cross sectional images of (a) and (d) the HPAN support, (b) and (e) the IP-assembled TFC membrane and (c) and (d) the mLbL-assembled TFC membrane (mLbL-10). Scale bar = 500 nm for all images except (d). Reprinted from Kwon *et al.* (2015) with permission from Elsevier.

limiting a high degree of swelling of the membrane and the selectivity of the thin active layer can be adjusted with the deposition conditions. These properties make these membranes attractive for pervaporation (Table 2.5). In the earliest studies carried out by Kraseman and his colleagues (Krasemann and Tieke, 1998; Krasemann *et al.*, 2001), more than 60 bilayers were deposited on porous PAN/PET support membranes. In the case of PAH/PSS pairs, selectivity of the modified membrane in the separation of ethanol from water significantly increased when the number of layers increased and the membrane was annealed at temperatures above 60°C (Krasemann and Tieke, 1998). In addition to PAH/PSS pairs, PDADMAC/PSS, P4VP/PSS, CHI/PSS, PAH/DEX, PEI/PSS, PAH/PVS, PEI/PVS and PVA/PVS pairs were also used in the modification. Optimum water enrichment was obtained, with the PVA/PVS and PEI/PVS modified membranes carrying the highest charge (Krasemann *et al.*, 2001). As shown in Figure 2.13, at small charge densities (i.e. small ρ_c values) leading to low cross-linking densities in the deposited layers, the flux is high and water concentration in the permeate is small. With the increased charge density of polyelectrolytes, the structure of the multilayer membrane becomes much tighter; as a result, flux decreases, but transport of water is favored.

Zhu *et al.* (2006, 2007) reported good performance in the separation of water from isopropanol using less than 10 layers of polyethylenimine and poly(acrylic acid) deposited on the microporous polyacrylonitrile substrate membrane. The reduction in the number of layers without compromising the selectivity was achieved using a relatively dilute concentration of the polyelectrolytes in the first few cycles of deposition, followed by depositions with more concentrated polyelectrolyte solutions. To ensure the well extended conformation of polyelectrolytes, polyelectrolyte concentrations were selected far below the critical overlapping concentration (Zhu *et al.*, 2006). It was also found that polyelectrolytes should have a high charge density and should be larger in size than the pore size of the support membrane in order to form a defect free membrane (Zhu *et al.*, 2007). Zhang *et al.* (2007) modified hydrolyzed PAN support membranes with a few polyethyleneimine (PEI) and polyacrylic acid (PAA) layers deposited under pressure for the pervaporation separation

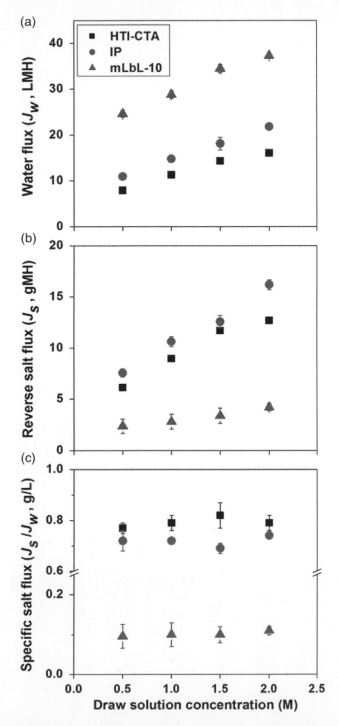

Figure 2.12.   (a) FO water flux (J_w), (b) reverse salt flux (J_s) and (c) specific salt flux (J_s/J_w) of the commercial HTI-CTA, IP-assembled TFC (IP) and mLbL-assembled TFC (mLbL-10) membranes as a function of draw solution NaCl concentrations. $T = 25 \pm 0.5°C$. Feed solution: DI water. Reprinted from Kwon *et al.* (2015) with permission from Elsevier.

Table 2.5. Recent studies published on the application of the LbL coated membranes in pervaporation.

Application	PE pair/# of bilayers	Support	Flux	Water content in permeate/ separation factor (α)	Reference
Separation of ethanol from water	[PAH/PSS]$_{60}$	PAN/PET	230 g m^{-2} h^{-1} $T = 58.5°C$	$\alpha = 70$	Krasemann and Tieke (1998)
Separation of ethanol from water	[PVA/PVS]$_{60}$	PAN/PET	water content in feed: 6.2% (w/w) 431 g m^{-2} h^{-1} (without NaCl) 98.5 g m^{-2} h^{-1} (with NaCl) 1270 g m^{-2} h^{-1} (without NaCl) 790 g m^{-2} h^{-1} (with NaCl) $T = 58.5°C$ water content in feed: 6.2% (w/w)	70% (w/w) (without NaCl) 94.0% (w/w) (with NaCl) 28.5% (w/w) (without NaCl) 35% (w/w) (with NaCl)	Krasemann et al. (2001)
	[PAH/PSS]$_{60}$				
Separation of isopropanol from water	[PEI/PAA]$_{10}$	PAN	989 g m^{-2} h^{-1} $T = 70°C$ water content in feed: 8.6% (w/w)	96.2% (w/w)	Zhu et al. (2006)
Separation of isopropanol from water		PAN	300 g m^{-2} h^{-1} post-treatment temperature: 50°C 600 g m^{-2} h^{-1} post-treatment temperature: 23°C post-treatment time*: 12 hour $T = 50°C$	98% (w/w) 90% (w/w)	Zhu et al. (2007)
Separation of ethanol from water	[PEI/PAA]$_2$PEI	PAN Hydrolysis of PAN with 2N NaOH, 60 min at $T = 65°C$	water content in feed: 9% (w/w) 226 g m^{-2} h^{-1} $T = 40°C$ water content in feed: 5% (w/w)	95.51% (w/w)	Zhang et al. (2007)
Dehydration of ethylene glycol	PEI/PAA	Commercial TFC polyamide	400 g m^{-2} h^{-1} $T = 40°C$ water content in feed: 9.2% (w/w)	98% (w/w)	Xu et al. (2010)
Separation of ethanol from water	[PEI/PAA]$_{4.5}$TiO$_2$ $C_{TALH} = 0.125\%$ (w/w) TALH was hydrolyzed with PEI to generate TiO$_2$ nanoparticles in-situ on the multilayer surface	PAN Hydrolysis of PAN with 2M NaOH, 30 min at $T = 60°C$	400 g m^{-2} h^{-1} $T = 60°C$ water content in feed: 5% (w/w)	98% (w/w)	Gong et al. (2014)

Separation	Membrane	Substrate	Flux / Conditions	Selectivity	Reference
Separation of alcohols from water	[BPDA-ODA-DABA/PDADMAC]$_{10}$ BPDA-ODA-DABA The film was not heated	Alumina	Ethanol: 35 ± 2 kg m^{-2} h^{-1} Isopropanol: 8.2 ± 3 kg m^{-2} h^{-1} $T = 50°C$ water content in feed: 10% (w/w)	Ethanol: 43.1 ± 2.1% (w/w) Isopropanol: 64.1 ± 7.6% (w/w)	Sullivan and Bruening (2005)
	[BPDA-ODA-DABA/PDADMAC]$_{10}$ BPDA-ODA-DABA The film was heated at 250°C for 2 hours.	Alumina	Ethanol: 4.3 ± 1 kg m^{-2} h^{-1} Isopropanol: 2.1 ± 0.5 kg m^{-2} h^{-1} $T = 50°C$ water content in feed: 10% (w/w)	Ethanol: 68 ± 3.1% (w/w) Isopropanol: 98 ± 1.1% (w/w)	
Separation of isopropanol from water	Two negatively charged polyelectrolyte complex colloidal nanoparticles (PEC$^-$) and one positively charged nanoparticle (PEC$^+$) (PDMC/CMCNa PEC$^-$), (PDADMAC/CMCNa PEC$^-$) and (PDADMAC/PSS PEC$^+$)	Polyamide reverse osmosis membrane	Flux: 2.75 kg m^{-2} h^{-1} (PEC$^+$/PDADMAC$^-$ CMCNa PEC$^-$)$_5$ Flux: 1.4 kg m^{-2} h^{-1} (PEC$^+$/PDADMAC$^-$ CMCNa PEC$^-$)$_{20}$ Flux: 2.4 kg m^{-2} h^{-1} (PEC$^+$/PDMC$^-$ CMCNa PEC$^-$)$_4$ Flux: 1.24 kg m^{-2} h^{-1} (PEC$^+$/PDMC$^-$ CMCNa PEC$^-$)$_{20}$ $T = 50°C$ water content in feed: 10% (w/w)	85% (w/w) 97% (w/w)	Zhao et al. (2010)

*Post-treatment time was 12 hour for both measurements.

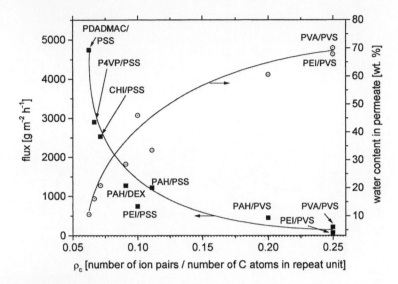

Figure 2.13.   Flux and water content in permeate as a function of the charge density (ρ_c) of the polyelectrolyte pair. $T = 58.5°C$; Feed: 6.2% (w/w) water; number of bilayers: 60. Reprinted from Krasemann *et al.* (2001) with permission from Elsevier.

of water/alcohol mixtures. To avoid the swelling of the layers, PEI was dissolved in ethanol solution instead of in pure water. It was shown that the hydrolysis conditions applied for modifying the support membrane strongly affects the pervaporation performance; in particular, the integrity of the first polyion layer was influenced by the charge density of the support. Xu *et al.* (2010) reduced the number of bilayers needed to form a permselective pervaporation membrane by depositing polyethyleneimine and poly(acrylic acid) on an interfacially polymerized polyamide membrane supported on a microporous polysulfone substrate instead of directly depositing on polysulfone support. It was shown that the separation factor of the membranes dramatically increased from 15–20 to 240–300 by increasing the number of bilayers from 1 to 2 at a feed water concentration of 3.5 and 9.2 wt%.

The rate of increase in the selectivity became slower beyond two bilayers (Fig. 2.14).

Wang *et al.* (2010) evaluated the salt, pH- and oxidant-responsive pervaporation behaviors of LbL modified membranes by post-treating with sodium chloride, hydrochloric acid, sodium hydroxide and sodium hypochlorite aqueous solutions. It was shown that high salt concentrations, extreme pH values and strong oxidants caused irreversible change in the structure, and therefore deteriorated the membrane performance. Gong *et al.* (2014) focused on obtaining superhydrophilic pervoparation membranes by growing TiO_2 nanoparticles on the surface of a layer-by-layer-assembled poly(ethyleneimine)/poly(acrylic acid) multilayer. Controversially, Sullivan and Bruening (2005) used mildly hydrophobic polyimides to limit swelling and, with the combination of cross-linking and annealing, pervaporation performance was significantly improved reaching a very high selectivity of 6100 for the separation of water from 90% isopropanol. Zhao *et al.* (2010) prepared negatively and positively charged colloidal nanoparticles and used them as LbL building blocks on the porous support membranes. The membranes exhibited a good pervaporation performance for the dehydration of isopropanol, and also durability and high permeate fluxes, although the film thicknesses were in the range of 500–3000 nm. Alternatively, polyelectrolyte complexes preformed in solution were also used as building blocks in constructing a multilayer assembly (Jin *et al.*, 2010; Zhao *et al.*, 2008). These membranes exhibited high pervaporation performance and stability, which were attributed to the cross-linking structure, highly hydrophilic character and homogeneity of the membranes. Yin *et al.* (2010) deposited polyelectrolyte layers under vibrating conditions and Zhang *et al.* (2008) applied an

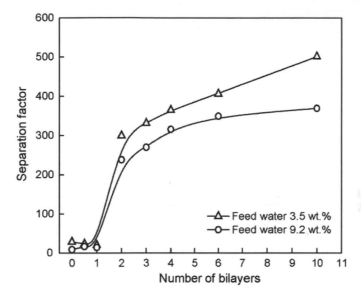

Figure 2.14. The change of separation factor as a function of the number of polyelectrolyte bilayers in the membrane. Temperature 40°C. Reprinted from Xu *et al.* (2010) with permission from Elsevier.

electric field during the deposition. Both approaches improved the pervaporation performance due to depositing the chains in a more ordered way and resulting in the formation of dense and uniform structures.

2.5   THE USE OF LBL TECHNIQUE TO MITIGATE BACTERIAL FOULING

Among various types of fouling, biological fouling is the most difficult one to remove from the surface. To prevent bacterial growth and biofilm formation, antibacterial nanomaterials are either blended into membrane casting solution, covalently attached to the surface or mixed into the selective layer of thin film composite membranes. Silver nanoparticles are commonly used in mitigating bacterial fouling. The challenge in preparing silver-based membranes is to locate silver nanoparticles on the surface of the membrane at required amounts and to control the release of silver from the membrane. Recently, we have shown that these challenges could be overcome by the use of the LbL technique through incorporating $AgCl/TiO_2$ xerogels into the polyelectrolyte layer (Kaner *et al.*, 2015). In this study, the LbL method was employed to modify commercial PES membranes through successive adsorption of chitosan and alginate as cationic and anionic polyelectrolytes. To mitigate bacterial fouling, pure, PEG mixed and PEGylated $AgCl/TiO_2$ xerogels were incorporated solely in the top layer of the LbL assembly. Table 2.6 lists the membranes prepared for biofouling studies. Selected SEM images for the membranes are shown in Figure 2.15.

While the overall morphology of bare and modified PES membranes is similar, AFM images indicate that the roughness of the modified membranes was found to be higher due to the presence of large agglomerates only seen in low magnification SEM images.

Figure 2.16 shows that the initial permeability of all $AgCl/TiO_2$ incorporated single layer membranes (encoded 0.1-$(CHI)_1$-0.5; 0.1-$(CHI)_1$-1; 0.1-$(CHI)_1$-5) was improved by 29% compared to the base membrane, regardless of the amount of $AgCl/TiO_2$, due to the increase in surface roughness and pore diameter of these membranes.

The water permeabilities of 1.5 bilayer pure $AgCl/TiO_2$ or PEG mixed $AgCl/TiO_2$ incorporated membranes were found to be the same as that of bare PES membranes. On the other hand, adding PEGylated $AgCl/TiO_2$ xerogels into the last layer of the assembly resulted in a 25% improvement

Table 2.6.   Single layer and 1.5 bilayer LbL modified membranes prepared by incorporating $AgCl/TiO_2$ xerogels in the last layer of the assembly. Reprinted from Kaner *et al.* (2015) with permission from Elsevier.

LBL layers	Polyelectrolyte concentration	Modification agent	Abbreviation
CHI	$0.1\,g\,L^{-1}$	–	$0.1\text{-}(CHI)_1$
CHI	$0.1\,g\,L^{-1}$	$0.5\,g\,L^{-1}\,AgCl/TiO_2$	$0.1\text{-}(CHI)_1\text{-}0.5Ag$
CHI	$0.1\,g\,L^{-1}$	$1\,g\,L^{-1}\,AgCl/TiO_2$	$0.1\text{-}(CHI)_1\text{-}1Ag$
CHI	$0.1\,g\,L^{-1}$	$5\,g\,L^{-1}\,AgCl/TiO_2$	$0.1\text{-}(CHI)_1\text{-}5Ag$
CHI/ALG/CHI	$0.1\,g\,L^{-1}$	–	$0.1\text{-}(CHI/ALG)_1CHI$
CHI/ALG/CHI	$0.1\,g\,L^{-1}$	$0.5\,g\,L^{-1}\,AgCl/TiO_2$	$0.1\text{-}(CHI/ALG)_1CHI\text{-}0.5Ag$
CHI/ALG/CHI	$0.1\,g\,L^{-1}$	$1\,g\,L^{-1}\,AgCl/TiO_2$	$0.1\text{-}(CHI/ALG)_1CHI\text{-}1Ag$
CHI/ALG/CHI	$0.1\,g\,L^{-1}$	$2.5\,g\,L^{-1}\,AgCl/TiO_2$	$0.1\text{-}(CHI/ALG)_1CHI\text{-}2.5Ag$
CHI/ALG/CHI	$0.1\,g\,L^{-1}$	$0.5\,g\,L^{-1}\,AgCl/TiO_2 +$ $0.05\,g\,L^{-1}\,PEG$	$0.1\text{-}(CHI/ALG)_1CHI\text{-}0.5Ag\text{-}0.05PEG$
CHI/ALG/CHI	$0.1\,g\,L^{-1}$	$0.5\,g\,L^{-1}\,PEGylated$ $AgCl/TiO_2$	$0.1\text{-}(CHI/ALG)_1CHI\text{-}PEGylated\ 0.5Ag$

in the initial water permeability and much better biological fouling resistance compared to the base PES membrane. Higher flux through 1.5 bilayer membranes incorporated with PEG modified $AgCl/TiO_2$ xerogels was associated with the higher surface roughness of this membrane (12 nm), where the PEG-mixed $AgCl/TiO_2$-incorporated 1.5 bilayer membrane and the base PES both have a surface roughness of 2.5 nm. Better biofouling resistance of the modified membranes incorporated with PEGylated $AgCl/TiO_2$ xerogels was attributed to uniform distribution of the Ag, Cl, Ti and O elements and a higher percentage of Ag and Ti on the surface as shown by EDX analysis (Kaner *et al.*, 2015).

2.6   CHALLENGES AND FUTURE PERSPECTIVES

Thin film composite membranes are attractive for water treatment applications since each layer can be prepared from different materials and optimized independently to achieve the desired separation tasks. The layer-by-layer deposition of polyelectrolytes on porous supports offers several advantages to form thin film composite membranes. First of all, the method allows the preparation of an ultra-thin, defect free selective layer on the nm scale, thus, compromise between selectivity and permeability can be overcome and membranes with both high flux and high selectivity can be designed. A wide variety of polyelectrolytes and a broad processing window helped to precisely tune the morphology of the membranes to achieve desirable structural properties. Furthermore, the production of these membranes in an aqueous environment is an important advantage compared to other methods, due to environmental concerns and sustainability. Many studies have shown that LbL modified membranes can be potentially used for important water treatment processes, such as nanofiltration, reverse osmosis, forward osmosis and pervaporation.

Research efforts were aimed at reducing the number of layers required to reach desired performances and increasing the stability of layers. Significant progress has been recorded to resolve these issues; however, further work is needed, especially to test the long-term stability of these membranes under real operating conditions. The application of the LbL technique in the lab is easy; however, detailed investigation should be carried out to determine the relationship among the deposition conditions, morphology and transport properties of the membranes. The most important challenge for the LbL modified membranes is their economical large scale production. With recent advances in the deposition method and deposition time and successful results obtained from pilot studies, most probably the process will be commercialized in the near future. Not only the function of the selective layer but also the role of the support layer is important for the overall

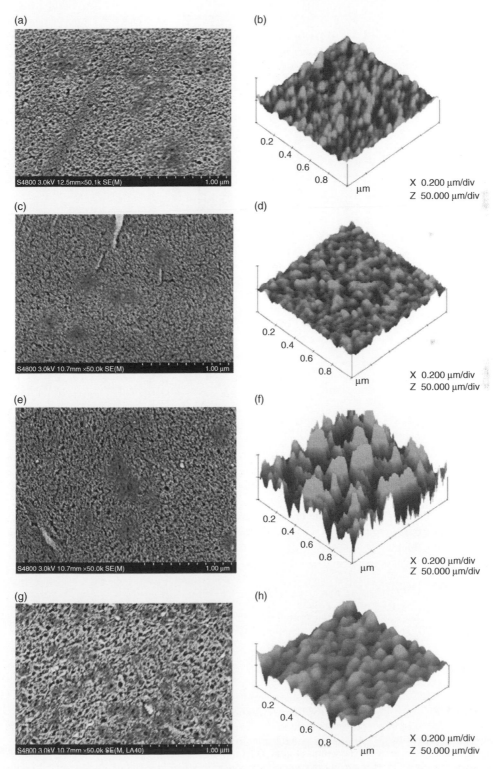

Figure 2.15.   SEM images at 50,000 × magnification and AFM images of the base and LbL modified membranes. (a) and (b) Base PES; (c) and (d) 0.1-(CHI)$_1$; (e) and (f) 0.1-(CHI)$_1$-0.5Ag; (g) and (h) 0.1-(CHI/ALG)$_1$CHI-PEGylated 0.5 Ag. Reprinted from Kaner *et al*. (2015) with permission from Elsevier.

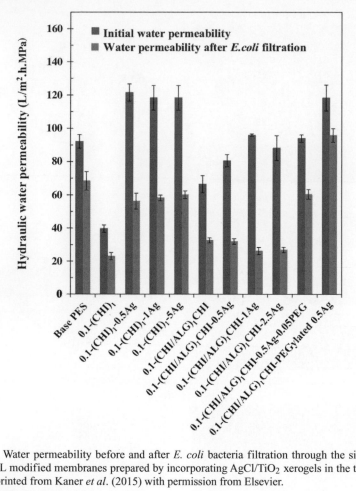

Figure 2.16.   Water permeability before and after *E. coli* bacteria filtration through the single layer and 1.5 bilayer LbL modified membranes prepared by incorporating AgCl/TiO$_2$ xerogels in the top layer of the assembly. Reprinted from Kaner *et al.* (2015) with permission from Elsevier.

performance of thin film composite membranes. The development of highly permeable support membranes from cheap polymers with uniform pores is required to achieve desirable separation with a minimum number of layers.

ACKNOWLEDGMENT

This research has received funding from the European Union's Seventh Framework Programme (FP7/2007–2013) under grant agreement no 246039.

NOMENCLATURE

PSS	poly(styrene sulfonate)
PDADMAC	poly(diallyldimethylammonium)
PES	polyether sulfone
SPES	sulfonated polyether sulfone
PAH	poly (allylamine)hydrochloride
SPEEK	sulfonated poly(etherketone)
PEI	poly(ethyleneimine)

PVS	polyvinylsulfate
PAN	polyacrylonitrile
PET	polyethylene terephthalate
PVA	polyvinyl alcohol
PAA	polyacrylic acid
MTM	montmorillonite
sPSf	sulfonated polysulfone
PLL	poly(L-lysine)
HA	hyaluronic acid
CTAC	cetyl trimethyl ammonium chloride
P(AN-co-AA)	poly(acrylonitrile-co-acrylic acid)
P(AN-co-EDAMA)	poly(acrylonitrile-co-N,N-dimethylethyl ammonium bromide)
PAI	poly(amide-imide)
DEX	dextran
CHI	chitosan
P4VP	poly(4-vinylpyridine)
ALG	alginate
PSf	polysulfone
MPD	m-phenylenediamine
TMC	trimesoyl chloride

REFERENCES

Ahmadiannamini, P., Li, X., Goyens, W., Meesschaert, B. & Vankelecom, I.F.J. (2010) Multilayered PEC nanofiltration membranes based on SPEEK/PDDA for anion separation. *Journal of Membrane Science*, 360, 250–258.

Arys, X., Jonas, A.M., Laguitton, B., Laschewsky, A., Legras, R. & Wischerhoff, E. (1998) Ultrathin multilayers made by alternate deposition of ionenes and polyvinylsulfate: from unstable to stable growth. *Thin Solid Films*, 327–329, 734–738.

Ba, C., Ladner, D.A. & Economy, J. (2010) Using polyelectrolyte coatings to improve fouling resistance of a positively charged nanofiltration membrane. *Journal of Membrane Science*, 347, 250–259.

Belyaev, V.V., Tolstikhina, A.L., Stepina, N.D. & Kayushina, R.L. (1998) Structure of ordered polyelectrolyte films from atomic-force microscopy and X-ray reflectivity data. *Crystal Reports*, 43, 124–128.

Bijlsma, R., van Well, A.A. & Cohen Stuart, M.A. (1997) Characterization of self-assembled multilayers of polyelectrolytes. *Physica* B, 234, 254–255.

Brynda, E. & Houska, M. (1996) Multiple alternating molecular layers of albumin and heparin on solid surfaces. *Journal of Colloid and Interface Science*, 183, 18–25.

Brynda, E. & Houska, M. (1998) Preparation of organized protein multilayers. *Macromolecular Rapid Communications*, 19, 173–176.

Büscher, K., Graf, K., Ahrens, H. & Helm, C.A. (2002) Influence of adsorption conditions on the structure of polyelectrolyte multilayers. *Langmuir*, 18, 3585–3591.

Caruso, F., Lichtenfeld, H., Donath, E. & Möhwald, H. (1999) Investigation of electrostatic interactions in polyelectrolyte multilayer films: binding of anionic fluorescent probes to layers assembled onto colloids. *Macromolecules*, 32, 2317–2328.

Chiarelli, P.A., Johal, M.S., Casson, J.L., Roberts, J.B., Robinson, J.M. & Wang, H.-L. (2001) Controlled fabrication of polyelectrolyte multilayer thin films using spin-assembly. *Advanced Materials*, 13, 1167–1171.

Cho, J., Char, K., Hong, J.D. & Lee, K.B. (2001) Fabrication of highly ordered multilayer films using a spin self-assembly method. *Advanced Materials*, 13, 1076–1078.

Cho, K.L., Hill, A.J., Caruso, F. & Kentish, S.E. (2015) Chlorine resistant glutaraldehyde crosslinked polyelectrolyte multilayer membranes for desalination. *Advanced Materials*, 27, 2791–2796

Choi, J., Sung, H., Ko, Y., Lee, S., Choi, W., Bang, J. & Cho, J. (2014) Layer-by-layer assembly of inorganic nanosheets and polyelectrolytes for reverse osmosis composite membranes. *Journal of Chemical Engineering of Japan*, 47, 180–186.

Clark, S.L., Montague, M.F. & Hammond, P.T. (1997) Ionic effects of sodium chloride on the templated deposition of polyelectrolytes using layer-by-layer ionic assembly. *Macromolecules*, 30, 7237–7244.

Cui, Y., Wang, H., Wang, H. & Chung, T.S. (2013) Micro-morphology and formation of layer-by-layer membranes and their performance in osmotically driven processes. *Chemical Engineering Science*, 101, 13–26.

De Grooth, J., Haakmeester, B., Wever, C., Potreck, J., de Vos, W.M. & Nijmeijer, K. (2015) Long term physical and chemical stability of polyelectrolyte multilayer membranes. *Journal of Membrane Science*, 489, 153–159.

Dubas, S.T. & Schlenoff, J.B. (1999) Factors controlling the growth of polyelectrolyte multilayers. *Macromolecules*, 32, 8153–8160.

Dubas, S.T. & Schlenoff, J.B. (2001) Swelling and smoothing of polyelectrolyte multilayers by salt. *Langmuir*, 17, 7725–7727.

Duong, P.H.H., Zuo, J. & Chung, T.-S. (2013) Highly crosslinked layer-by-layer polyelectrolyte FO membranes: understanding effects of salt concentration and deposition time on FO performance. *Journal of Membrane Science*, 427, 411–421.

El-Hashani, A., Toutianoush, A. & Tieke, B. (2008) Use of layer-by-layer assembled ultrathin membranes of dicopper-[18]azacrown-N-6 complex and polyvinylsulfate for water desalination under nanofiltration conditions. *Journal of Membrane Science*, 318, 65–70.

Erel-Unal, I. & Sukhishvili, S.A. (2008) Hydrogen-bonded hybrid multilayers: film architecture controls release of macromolecules. *Macromolecules*, 41, 8737–8744.

Fadhillah, F., Zaidi, S.M.J., Khan, Z., Khaled, M., Rahman, F. & Hammond, P. (2012) Development of multilayer polyelectrolyte thin-film membranes fabricated by spin assisted layer-by-layer assembly. *Journal of Applied Polymer Science*, 126, 1468–1474.

Fadhillah, F., Zaidi, S.M.J., Khan, Z., Khaled, M.M., Rahman, F. & Hammond, P.T. (2013) Development of polyelectrolyte multilayer thin film composite membrane for water desalination application *Desalination*, 318, 19–24.

Fu, H. & Kobayashi, T. (2010) Multilayer composite surfaces prepared by electrostatic self-assembled technique for desalination. *Desalination*, 264, 115–122.

Glinel, K., Moussa, A., Jonas, A.M. & Laschewsky, A. (2002) Influence of polyelectrolyte charge density on the formation of multilayers of strong polyelectrolytes at low ionic strength. *Langmuir*, 18, 1408–1412.

Gong, L., Zhang, L., Wang, N., Li, J., Ji, S., Guo, H., Zhang, G. & Zhang, Z. (2014) In situ ultraviolet-light-induced TiO_2 nanohybrid superhydrophilic membrane for pervaporation dehydration. *Separation and Purification Technology*, 122, 32–40.

Guillaume-Gentil, O., Zahn, R., Lindhoud, S., Graf, N., Voros, J. & Zambelli, T. (2011) From nanodroplets to continuous films: how the morphology of polyelectrolyte multilayers depends on the dielectric permittivity and the surface charge of the supporting substrate. *Soft Matter*, 7, 3861–3871.

Harris, J.J. & Bruening, M.L. (2000) Electrochemical and in situ ellipsometric investigation of the permeability and stability of layered polyelectrolyte films. *Langmuir*, 16, 2006–2013.

Harris, J.J., DeRose, P.M. & Bruening, M.L. (1999) Synthesis of passivating, nylon-like coatings through cross-linking of ultrathin polyelectrolyte films. *Journal of the American Chemical Society*, 121, 1978–1979.

Hong, S.U. & Bruening, M.L. (2006) Separation of amino acid mixtures using multilayer polyelectrolyte nanofiltration membranes. *Journal of Membrane Science*, 280, 1–5.

Hong, S.U., Malaisamy, R. & Bruening, M.L. (2006a) Optimization of flux and selectivity in Cl-/SO_4^{2-} separations with multilayer polyelectrolyte membranes. *Journal of Membrane Science*, 283, 366–372.

Hong, S.U., Malaisamy, R. & Bruening, M.L. (2007) Separation of fluoride from other monovalent anions using multilayer polyelectrolyte nanofiltration membranes. *Langmuir*, 23, 1716–1722.

Hong, S.U., Miller, M.D. & Bruening, M.L. (2006b) Removal of dyes, sugars, and amino acids from NaCl solutions using multilayer polyelectrolyte nanofiltration membranes. *Industrial & Engineering Chemistry Research*, 45, 6284–6288.

Hong, S.U., Ouyang, L. & Bruening, M.L. (2009) Recovery of phosphate using multilayer polyelectrolyte nanofiltration membranes. *Journal of Membrane Science*, 327, 2–5.

Hoogeveen, N.G., Cohen Stuart, M.A. & Fleer, G.J. (1996a) Polyelectrolyte adsorption on oxides. 1. Kinetics and adsorbed amounts. *Journal of Colloid and Interface Science*, 182, 133–145.

Hoogeveen, N.G., Cohen Stuart, M.A. & Fleer, G.J. (1996b) Polyelectrolyte adsorption on oxides. 2. Reversibility and exchange. *Journal of Colloid and Interface Science*, 182, 146–157.

Ishigami, T., Amano, K., Fujii, A., Ohmukai, Y., Kamio, E., Maruyama, T. & Matsuyama, H. (2012) Fouling reduction of reverse osmosis membrane by surface modification via layer-by-layer assembly. *Separation and Purification Technology*, 99, 1–7.

Jin, H., An, Q.-F., Zhao, Q., Qian, J.-W. & Zhu, M. (2010) Pervaporation dehydration of ethanol by using polyelectrolyte complex membranes based on poly (N-ethyl-4-vinylpyridinium bromide) and sodium carboxymethyl cellulose. *Journal of Membrane Science*, 347, 183–192.

Jin, W., Toutianoush, A. & Tieke, B. (2003) Use of polyelectrolyte layer-by-layer assemblies as nanofiltration and reverse osmosis membranes. *Langmuir*, 19, 2550–2553.

Kaner, P., Johnson, D.J., Seker, E., Hilal, N. & Altinkaya, S.A. (2015) Layer-by-layer surface modification of polyethersulfone membranes using polyelectrolytes and AgCl/TiO$_2$ xerogels. *Journal of Membrane Science*, 493, 807–819.

Klitzing, R.v. (2006) Internal structure of polyelectrolyte multilayer assemblies. *Physical Chemistry Chemical Physics*, 8, 5012–5033.

Kolarik, L., Furlong, D.F., Joy, H., Struijk, C. & Rowe, R. (1999) Building assemblies from high molecular weight polyelectrolytes. *Langmuir*, 15, 8265–8275.

Kotov, N.A. (1999) Layer-by-layer self-assembly: the contribution of hydrophobic interactions. *Nanostructured Materials*, 12, 789–796.

Krasemann, L. & Tieke, B. (1998) Ultrathin self-assembled polyelectrolyte membranes for pervaporation. *Journal of Membrane Science*, 150, 23–30.

Krasemann, L. & Tieke, B. (2000) Selective ion transport across self-assembled alternating multilayers of cationic and anionic polyelectrolytes. *Langmuir*, 16, 287–290.

Krasemann, L., Toutianoush, A. & Tieke, B. (2001) Self-assembled polyelectrolyte multilayer membranes with highly improved pervaporation separation of ethanol/water mixtures. *Journal of Membrane Science*, 181, 221–228.

Kwon, S.-B., Lee, J.S., Kwon, S.J., Yun, S.-T., Lee, S. & Lee, J.-H. (2015) Molecular layer-by-layer assembled forward osmosis membranes. *Journal of Membrane Science*, 488, 111–120.

Ladam, G., Schaad, P., Voegel, J.C., Schaaf, P., Decher, G. & Cuisinier, F. (2000) In situ determination of the structural properties of initially deposited polyelectrolyte multilayers. *Langmuir*, 16, 1249–1255.

Lajimi, R.H., Abdallah, A.B., Ferjani, E., Roudesli, M.S. & Deratani, A. (2004) Change of the performance properties of nanofiltration cellulose acetate membranes by surface adsorption of polyelectrolyte multilayers. *Desalination*, 163, 193–202.

Laurent, D. & Schlenoff, J.B. (1997) Multilayer assemblies of redox polyelectrolytes. *Langmuir*, 13, 1552–1557.

Lavalle, P., Gergely, C., Cuisinier, F.J.G., Decher, G., Schaaf, P., Voegel, J.C. & Picart, C. (2002) Comparison of the structure of polyelectrolyte multilayer films exhibiting a linear and an exponential growth regime: an in situ atomic force microscopy study. *Macromolecules*, 35, 4458–4465.

Lehr, B., Seufert, M., Wenz, G. & Decher, G. (1995) Fabrication of poly(p-phenylene vinylene) (PPV) nanoheterocomposite films via layer-by-layer adsorption. *Supramolecular Science*, 2, 199–207.

Li, X., De Feyter, S., Chen, D., Aldea Vandezande, S.P., Du Prez, P. & Vankelecom, I.F.J. (2008) Solvent-resistant nanofiltration membranes based on multilayered polyelectrolyte complexes. *Chemistry of Materials*, 20, 3876–3883.

Li, X., Goyens, W., Ahmadiannamini, P., Vanderlinden, W., De Feyter, S. & Vankelecom, I.F.J. (2010) Morphology and performance of solvent-resistant nanofiltration membranes based on multilayered polyelectrolytes: study of preparation conditions. *Journal of Membrane Science*, 358, 150–157.

Li, Y., Wang, X. & Sun, J. (2012) Layer-by-layer assembly for rapid fabrication of thick polymeric films. *Chemical Society Reviews*, 41, 5998–6009.

Linford, M.R., Auch, M. & Möhwald, H. (1998) Nonmonotonic effect of ionic strength on surface dye extraction during dye-polyelectrolyte multilayer formation. *Journal of the American Chemical Society*, 120, 178–182.

Liu, C., Shi, L. & Wang, R. (2015) Enhanced hollow fiber membrane performance via semi-dynamic layer-by-layer polyelectrolyte inner surface deposition for nanofiltration and forward osmosis applications. *Reactive and Functional Polymers*, 86, 154–160.

Lvov, Y., Ariga, K., Onda, M., Ichinose, I. & Kunitake, T. (1997) Alternate assembly of ordered multilayers of SiO$_2$ and other nanoparticles and polyions. *Langmuir*, 13, 6195–6203.

Lvov, Y., Ariga, K., Onda, M., Ichinose, I. & Kunitake, T. (1999) A careful examination of the adsorption step in the alternate layer-by-layer assembly of linear polyanion and polycation. *Colloids and Surfaces* A, 146, 337–346.

Lvov, Y., Decher, G. & Möhwald, H. (1993) Assembly, structural characterization, and thermal-behavior of layer-by-layer deposited ultrathin films of poly(vinyl sulfate) and poly(allylamine). *Langmuir*, 9, 481–486.

Lvov, Y., Onda, M., Ariga, K. & Kunitake, T. (1998) Ultrathin films of charged polysaccharides assembled alternately with linear polyions. *Journal of Biomaterials Science, Polymer Edition*, 9, 345–355.

Malaisamy, R. & Bruening, M.L. (2005). High-flux nanofiltration membranes prepared by adsorption of multilayer polyelectrolyte membranes on polymeric supports. *Langmuir*, 21, 10,587–10,592.

Miller, M.D. & Bruening, M.L. (2004) Controlling the nanofiltration properties of multilayer polyelectrolyte membranes through variation of film composition. *Langmuir*, 20, 11,545–11,551.

Miller, M.D. & Bruening, M.L. (2005) Correlation of the swelling and permeability of polyelectrolyte multilayer films. *Chemistry of Materials*, 17, 5375–5381.

Ouyang, L., Malaisamy, R. & Bruening, M.L. (2008) Multilayer polyelectrolyte films as nanofiltration membranes for separating monovalent and divalent cations. *Journal of Membrane Science*, 310, 76–84.

Park, J., Park, J., Kim, S.H., Cho, J. & Bang, J. (2010) Desalination membranes from pH-controlled and thermally-crosslinked layer-by-layer assembled multilayers. *Journal of Materials Chemistry*, 20, 2085–2091.

Porcel, C., Lavalle, P., Ball, V., Decher, G., Senger, B., Voegel, J.C. & Schaff, P. (2006) From exponential to linear growth in polyelectrolyte multilayers. *Langmuir*, 22, 4376–4383.

Porcel, C., Lavalle, P., Decher, G., Senger, B., Voegel, J.C. & Schaff, P. (2007) Influence of the polyelectrolyte molecular weight on exponentially growing multilayer films in the linear regime. *Langmuir*, 23, 1898–1904.

Qi, S., Li, W., Zhao, Y., Ma, N., Wei, J., Chin, T.W. & Tang, C.Y. (2012) Influence of the properties of layer-by-layer active layers on forward osmosis performance. *Journal of Membrane Science*, 423-424, 536–542.

Qiu, C., Qi, S. & Tang, C.Y. (2011) Synthesis of high flux forward osmosis membranes by chemically crosslinked layer-by-layer polyelectrolytes. *Journal of Membrane Science*, 381, 74–80.

Quinn, J.F., Johnston, A.P.R., Such, G.K., Zelikin, A.N. & Caruso, F. (2007) Next generation, sequentially assembled ultrathin films: beyond electrostatics. *Chemical Society Reviews*, 36, 707–718.

Ronge, J., Bosserez, T., Martel, D., Nervi, C., Boarino, L., Taulelle, F., Decher, G., Bordiga, S. & Martens, J.A. (2014) Monolithic cells for solar fuels. *Chemical Society Reviews*, 43, 7963–7981.

Saeki, D., Imanishi, M., Ohmukai, Y., Maruyama, T. & Matsuyama, H. (2013) Stabilization of layer-by-layer assembled nanofiltration membranes by crosslinking via amide bond formation and siloxane bond formation. *Journal of Membrane Science*, 447, 128–133.

Salomäki, M., Vinokurov, I.A. & Kankare, J. (2005) Effect of temperature on the buildup of polyelectrolyte multilayers. *Langmuir*, 21, 11,232–11,240.

Saren, Q., Qiu, C.Q. & Tang, C.Y. (2011) Synthesis and characterization of novel forward osmosis membranes based on layer-by-layer assembly. *Environmental Science and Technology*, 45, 5201–5208.

Schlenoff, J.B., Ly, H. & Li, M. (1998) Charge and mass balance in polyelectrolyte multilayers. *Journal of the American Chemical Society*, 120, 7626–7634.

Schlenoff, J.B., Rmaile, A.H. & Bucur, C.B. (2008) Hydration contributions to association in polyelectrolyte multilayers and complexes: visualizing hydrophobicity. *Journal of the American Chemical Society*, 130, 13,589–13,597.

Schoeler, B., Kumaraswamy, G. & Caruso, F. (2002) Investigation of the influence of polyelectrolyte charge density on the growth of multilayer thin films prepared by the layer-by-layer technique. *Macromolecules*, 35, 889–897.

Schonhoff, M. (2003) Self-assembled polyelectrolyte multilayers. *Current Opinion in Colloid & Interface Science*, 8, 86–95.

Setiawan, L., Wang, R., Tan, S., Shi, L. & Fane, A.G. (2013) Fabrication of poly(amide-imide)-polyethersulfone dual layer hollow fiber membranes applied in forward osmosis by combined polyelectrolyte cross-linking and deposition. *Desalination*, 312, 99–106.

Shan, W., Bacchin, P., Aimar, P., Bruening, M.L. & Tarabara, V.V. (2010) Polyelectrolyte multilayer films as backflushable nanofiltration membranes with tunable hydrophilicity and surface charge. *Journal of Membrane Science*, 349, 268–278.

Steitz, R., Jaeger, W. & Klitzing, R.v. (2001) Influence of charge density and ionic strength on the multilayer formation of strong polyelectrolytes. *Langmuir*, 17, 4471–4474.

Stockton, W.B. & Rubner, M.F. (1997) Molecular-level processing of conjugated polymers .4. Layer-by-layer manipulation of polyaniline via hydrogen-bonding interactions. *Macromolecules*, 30, 2717–2725.

Sukhishvili, S.A. & Granick, S. (2000) Layered, erasable, ultrathin polymer films. *Journal of the American Chemical Society*, 122, 9550–9551.

Sullivan, D.M. & Bruening, M.L. (2005) Ultrathin, cross-linked polyimide pervaporation membranes prepared from polyelectrolyte multilayers. *Journal of Membrane Science*, 248, 161–170.

Tan, H.L., McMurdo, M.J., Pan, G. & Van Patten, P.G. (2003) Temperature dependence of polyelectrolyte multilayer assembly. *Langmuir*, 19, 9311–9314.

Tieke, B. (2011) Coordinative supramolecular assembly of electrochromic thin films. *Current Opinion in Colloid & Interface Science*, 16, 499–507.

Tieke, B., Ackern, F.V., Krasemann, L. & Toutianoush, A. (2001) Ultrathin self-assembled polyelectrolyte multilayer membranes. *European Physical Journal* E: *Soft Matter and Biological Physics*, 5, 29–39.

Toutianoush, A., Jin, W., Deligoz, H. & Tieke, B. (2005) Polyelectrolyte multilayer membranes for desalination of aqueous salt solutions and seawater under reverse osmosis conditions. *Applied Surface Science*, 246, 437–443.

Voigt, U., Jaeger, W., Findenegg, G.H. & von Klitzing, R (2003) Charge effects on the formation of multilayers containing strong polyelectrolytes. *The Journal of Physical Chemistry* B, 107, 5273–5280.

Wang, J., Yao, Y., Yue, Z. & Economy, J. (2009) Preparation of polyelectrolyte multilayer films consisting of sulfonated poly (ether ether ketone) alternating with selected anionic layers. *Journal of Membrane Science*, 337, 200–207.

Wang, N., Zhang, G., Ji, S., Qin, Z. & Liu, Z. (2010) The salt-, pH- and oxidant-responsive pervaporation behaviors of weak polyelectrolyte multilayer membranes. *Journal of Membrane Science*, 354, 14–22.

Xu, J., Feng, X. & Gao, C. (2011) Surface modification of thin-film-composite polyamide membranes for improved reverse osmosis performance *Journal of Membrane Science*, 370, 116–123.

Xu, J., Gao, C. & Feng, X. (2010) Thin-film-composite membranes comprising of self-assembled polyelectrolytes for separation of water from ethylene glycol by pervaporation. *Journal of Membrane Science*, 352, 197–204.

Yin, M., Qian, J., An, Q., Zhao, Q., Gui, Z. & Li, J. (2010) Polyelectrolyte layer-by-layer self-assembly at vibration condition and the pervaporation performance of assembly multilayer films in dehydration of isopropanol. *Journal of Membrane Science*, 358, 43–50.

Yoo, D., Shiratori, S.S. & Rubner, M.F. (1998) Controlling bilayer composition and surface wettability of sequentially adsorbed multilayers of weak polyelectrolytes. *Macromolecules*, 31, 4309–4318.

Zhang, G., Yan, H., Ji, S. & Liu, Z. (2007) Self-assembly of polyelectrolyte multilayer pervaporation membranes by a dynamic layer-by-layer technique on a hydrolyzed polyacrylonitrile ultrafiltration membrane. *Journal of Membrane Science*, 292, 1–8.

Zhang, P., Qian, J., Yang, Y., An, Q., Liu, X. & Gui, Z. (2008) Polyelectrolyte layer-by-layer self-assembly enhanced by electric field and their multilayer membranes for separating isopropanol-water mixtures. *Journal of Membrane Science*, 320, 73–77.

Zhao, Q., Qian, J.W., An, Q.F., Yang, Q. & Zhang, P. (2008) A facile route for fabricating novel polyelectrolyte complex membrane with high pervaporation performance in isopropanol dehydration. *Journal of Membrane Science*, 320, 8–12.

Zhao, Q., Qian, J., An, Q. & Sun, Z. (2010) Layer-by-layer self-assembly of polyelectrolyte complexes and their multilayer films for pervaporation dehydration of isopropanol. *Journal of Membrane Science*, 346, 335–343.

Zhou, Y., Yu, S., Gao, C. & Feng, X. (2009) Surface modification of thin film composite polyamide membranes by electrostatic self deposition of polycations for improved fouling resistance. *Separation and Purification Technology*, 66, 287–294.

Zhu, Z.Q., Feng, X.S. & Penlidis, A. (2006) Self-assembled nano-structured polyelectrolyte composite membranes for pervaporation. *Materials Science and Engineering* C: *Materials for Biological Applications*, 26, 1–8.

Zhu, Z., Feng, X. & Penlidis, A. (2007) Layer-by-layer self-assembled polyelectrolyte membranes for solvent dehydration by pervaporation. *Materials Science and Engineering* C: *Materials for Biological Applications*, 27, 612–619.

CHAPTER 3

Fabrication and application areas of mixed matrix flat-sheet membranes

Derya Y. Koseoglu-Imer & Ismail Koyuncu

3.1 INTRODUCTION

Polymer-nanoparticle composite materials have unique characteristics, such as high mechanical strength, good electrical conductivity, optical and thermal properties. Nanoparticles can gain high functionality and therefore they increase the overall performance of conventional materials, such as membranes, used in environmental applications. Membrane separation properties can be controlled for each specific application by the proper choice of fabrication components (main polymer, solvent, additives like nanoparticles, pore forming agents, etc.) and parameters (evaporation time and temperature, coagulation bath temperature, etc.). Moreover, especially in recent years, the membrane fouling problem can be avoided with different membrane fabrication components. Membrane fouling can be defined as the uncontrolled deposition of particles, colloids, macromolecules or salt ions from feed solution at the membrane surface or inside the membrane pores. It is a severe problem for membrane materials used in pressure-driven processes such as reverse osmosis (RO), nanofiltration (NF), ultrafiltration (UF), and microfiltration (MF) and also for other membrane processes. Different polymers (Polyethersulfone (PES), polysulfone (PS), cellulose acetate (CA), polyvinylidene fluoride (PVDF), etc.) could be selected as membrane materials for these membrane systems according to their physical and chemical characteristics, such as good chemical, heat, mechanical and cleaning resistance and environmental endurance, as well as easy processing and manufacturing. For instance, the intrinsic hydrophobicities of some polymers are high and so this results in hydrophobic membrane materials and leads to a low membrane flux, poor anti-fouling properties and low application and useful life. The fouling causes a decrease in membrane performance, either temporarily or permanently. The fouling mechanism includes the interaction between the membrane surface and the foulants (inorganic, organic, and biological substances in many different forms). The foulant molecules not only physically interact with the membrane surface but also chemically degrade the membrane material. Most of the latest membrane fouling studies focused on the physical or chemical modification of membrane material for low fouling properties. These studies can be summarized in three main areas; (i) the modification of the membrane surface with *in-situ* physical and chemical treatments, (ii) the coating of the membrane with special materials that have low fouling properties and (iii) the preparation of the membrane by adding nanomaterials (mixed matrix membranes, MMMs). MMMs are formed by the addition of inorganic or metal oxide particles, having micrometer or nanometer sizes, to the polymeric casting solution or by *in-situ* generation. Over the past few years, the rapid growth in nanotechnology has aroused significant interest in the use of nanomaterials in membrane applications. So far, the membrane modification studies using nanotechnological methods in particular have achieved useful results. Nanoparticles (NPs) are defined as particles having the size of 1–100 nm and they have unique magnetic, electrical, optical, mechanical and structural properties. More recently, several natural and engineered nanomaterials have also proved to have exceptional properties, including chitosan, silver nanoparticles (nAg), photocatalytic TiO_2, fullerol, aqueous fullerene nanoparticles (nC60), and carbon nanotubes (CNT). Moreover, some nanoparticles, such as silver (Ag), copper (Cu), zinc oxide (ZnO), titanium oxide (TiO_2), etc., have antibacterial properties and thus show high toxicity to a broad spectrum of microorganisms, including bacteria, fungi, viruses and yeasts, and have been studied as antibacterial agents in different areas. The combination of membrane chemistry and the antibacterial properties of NPs may particularly solve the biofouling problem in membrane systems. To prevent fouling problems, NPs can be applied by directly coating on to the membrane surface or by blending in the polymer matrix of the membrane during the membrane fabrication process.

This chapter is focused on identifying the fabrication and application areas of mixed matrix flat-sheet membranes.

3.2   FABRICATION OF MIXED MATRIX FLAT-SHEET MEMBRANES

3.2.1   *Self-assembled mixed matrix membranes (MMMs)*

Self-assembly provides a feasible and low-cost method for preparing the membranes including nanoparticles (Yang *et al.*, 2014). This method is a bottom-up approach but it is limited to the fabrication of single-sided coatings on planar substrates. This technique does not require complicated equipment and can be easily applied to a large variety of materials (Askar *et al.*, 2013). A solution-based bottom-up assembly process represents the layer-by-layer (LbL) self-assembly, it provides the sequential adsorption of nanometer-thick monolayers of oppositely charged polyelectrolytes and nanoparticles to form a molecular-level controlled membrane. It is a very successful process with which to form cross-linked films with interfacial interactions between the carbon nanotube and polymers through dense covalent bonding and the stiffening of polymer chains (Zhang and Cui, 2012). Kim *et al.* (2003) designed self-assembled TiO_2 membranes to create membrane surfaces that have a low biofouling performance. They assembled the anatase TiO_2 nanoparticles on thin film composite membranes through bonding between hydroxyl groups of TiO_2 nanoparticles and carboxyl functional groups of aromatic polyamide thin film layer (Kim *et al.*, 2003). Since TiO_2 nanoparticles have a high affinity to water molecules and generate hydrogen bonding with groups in the polymer segment, it is also a good hydrophilic agent for modification of the membrane surface by self-assembly (Li X. *et al.*, 2014).

Bae *et al.* (2006) used the self-assembly procedure for membranes with TiO_2 nanoparticles. They modified the sulfonated PES membranes as fouling-resistant mixed matrix membranes and used the electrostatic self-assembly method between TiO_2 nanoparticles and sulfonic acid groups on the membrane surface. In this study, the sulfonated polyethersulfone (SPES) ultrafiltration membranes were fabricated with the phase inversion method and then the fabricated membranes were dipped into the TiO_2 nanoparticle solution including the controlled hydrolysis of titanium tetraisopropoxide. The fouling mitigation property of the mixed matrix membrane was examined using membrane bioreactor (MBR) sludge, which contains a great number of various foulants. The fouling mitigation ratios were evaluated quantitatively; the filtration resistances were calculated according to the resistance in a series model. The results showed that both the amount and rate of membrane fouling decreased in the mixed matrix membrane compared to the pristine polymeric membrane. The mixed matrix membrane had a higher stabilized relative flux (36% of the initial flux) than the pristine membrane (20% of the initial flux). The cake layer resistance of the mixed matrix membrane was decreased when compared to the pristine PES membrane (from 58.70×10^{11} to 33.27×10^{11} m^{-1}) and also the pore resistance of the mixed matrix membrane decreased from 4.31×10^{11} to 2.23×10^{11} (Bae *et al.*, 2006).

Mansourpanah *et al.* (2009) assembled TiO_2 nanoparticles on the surface of polyethersulfone (PES)/polyimide (PI) and –OH functionalized PES/PI blend membranes. They exposed the membranes to UV light after membrane fabrication using different concentrations of TiO_2. UV light gained a photocatalytic property to the membrane surface and this property increased the membrane hydrophilicity. With UV irradiated TiO_2 deposited blend membranes, –OH groups provided excellent adhesion of TiO_2 nanoparticles on the membrane surface and this increased the reversible deposition onto the membrane surface and diminished irreversible fouling into the membrane pores. Li X. *et al.* (2014) prepared a polyethersulfone (PES) hybrid ultrafiltration (UF) membrane with TiO_2 nanoparticles using a novel method. This method was to combine the self-assembly of TiO_2 nanoparticles with the formation of PES membranes *via* the non-solvent induced phase separation (NIPS) process. TiO_2 nanoparticles were self-assembled around the membrane pores during the NIPS process by controlling the hydrolysis and condensation of precursor (tetrabutyltitanate-TBT) confirmed by TEM and SEM-EDX mapping. The water permeability of the hybrid PES membranes was four times higher than that of the pristine PES membranes, $50.8 \, L \, m^{-2} \, h^{-1}$ for pristine PES and $235.9 \, L \, m^{-2} \, h^{-1}$ for hybrid membranes. An anti-fouling experiment, using BSA and humic acid (HA) as model foulants, showed that the PES hybrid membrane had good organic anti-fouling properties. Importantly, the leaching of self-assembly TiO_2 nanoparticles was moderate and this provided a stable and sustainable anti-fouling activity for the hybrid membranes (Li X. *et al.*, 2014).

3.2.2 *Surface modified mixed matrix membranes (MMMs)*

Surface treatment is one of the modification methods used for fabricated pristine membranes to increase the surface hydrophilicity and to reduce the membrane fouling. The surface modification of the membranes may be performed by coating, grafting, covalent bond, cold plasma treatment and incorporating hydrophilic fillers, including inorganic nanoparticles (Moghimifar *et al.*, 2014).

Coating is a physical modification method and includes the coating material(s) forming a thin layer that non-covalently adheres to the membrane surface. Coating methods can be divided into five techniques: coating by physical adsorption, heat curing, coating with a monolayer using Langmuir-Blodgett or analogous techniques, deposition from a glow discharge plasma, and casting or extrusion of two solutions by simultaneous spinning (Nady *et al.*, 2011). With these methods, nanoparticles can be coated as a layer on the membrane surface by physical or chemical bonds. It might provide a simple way to place nanoparticles onto the membrane surface. Although locating nanoparticles in a direct position with contact to the filtering solution improves the efficiency of nanoparticles, this procedure could lead to early detachment of the nanoparticles from the membrane surface with time. Nevertheless, with the physical coating method, a release of nanoparticles from the membrane material could be observed due to the difficulty of immobilizing them on the membranes without using binding mediums to form covalent bonds between the nanoparticles and the membrane. Despite many attempts being made to find appropriate organic binders for nanoparticles, the release of nanoparticles from the membrane still creates important operational problems over a long filtration period (Mericq *et al.*, 2015).

3.2.3 *Thin film nanocomposite (TFN) membranes*

Thin film nanocomposite membranes (TFN) were recently introduced as a new field in membrane technology. It is believed that the concept of dispersing nanophase materials into a polymeric matrix has brought a new perspective to advanced membrane materials. TFN membranes are fabricated by embedding porous nanoparticle materials, such as pure metals, metal oxides, silicon nanoparticles and carbon nanomaterials, into a polyamide matrix layer (Shen *et al.*, 2013). Nanomaterials give benefits to membranes, such as unique optical, thermal and physiochemical activities, so organic-inorganic membranes with nanoparticles have a high separation efficiency, low cost and easy operation. However, these membranes acquire some specific properties, like high permeability and selectivity, good chemical, thermal and mechanical stability, crystallinity, catalytic activity and antibacterial characteristics (Mollahosseini and Rahimpour, 2014).

Thin film membranes used for nanofiltration or reverse osmosis are polyamide thin film composites (TFC), which are prepared by an interfacial polymerization technique. Polyamide membranes have several disadvantages, such as relatively low water flux, low chlorine resistance and poor anti-fouling properties. Therefore, the fabrication of polyamide membranes that have a high water permeability and simultaneously high salt rejection is still a challenge and a strong research topic for the development of new and improved polyamide membranes. One effective

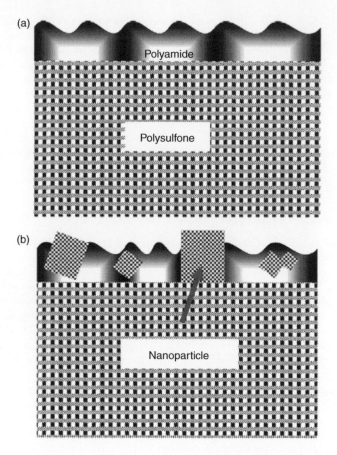

Figure 3.1.   Conceptual illustration of (a) TFC and (b) TFN membrane structures (Jeong *et al.*, 2007).

strategy is to introduce inorganic nanoparticles into the polyamide skin layer to produce inorganic-polyamide membranes. Incorporation of inorganic materials into the polymer matrix generally increases hydrophilicity and typically contributes to fouling reduction (Xu *et al.*, 2013).

Different nanoparticles have been used to modify polyamide (PA) TFC membranes, such as zeolite, TiO_2, silver, silicate or alumina. Zeolite nanoparticles are particularly used as inorganic materials to be embedded into the active layer to modify the PA-TFC RO membranes. Incorporation of zeolite nanoparticles into the PA active layer of TFC membranes can significantly improve the water flux without a large loss of salt rejection under high pressure in the RO process. Dong *et al.* (2016) found that TFN membranes with zeolite nanoparticles had a much higher water permeability than the TFC membranes (15.7 *vs.* $6.4 \times 10^{-6}\,\mathrm{m\,s^{-1}\,MPa^{-1}}$). These authors explained that this trend was due to the internal pores of zeolite nanoparticles increasing the membrane surface roughness. Jeong *et al.* (2007) fabricated a TFN polyamide RO membrane including 0.004–0.4% (w/v) of NaA zeolite nanoparticles dispersed within 50–200 nm thick polyamide films (Fig. 3.1). Pure polyamide membranes exhibited surface morphologies characteristic of commercial polyamide RO membranes, whereas TFN membranes had measurably smoother and more hydrophilic, negatively charged surfaces. At the highest nanoparticle ratio, TFN morphology was visibly different and pure water permeability ($3.8 \times 10^{-12}\,\mathrm{m\,Pa^{-1}s^{-1}}$) was nearly double that of the polyamide membrane ($2.1 \times 10^{-12}\,\mathrm{m\,Pa^{-1}s^{-1}}$) with equivalent solute rejections. It was suggested that the zeolite nanoparticle pores played an active role in water permeation and

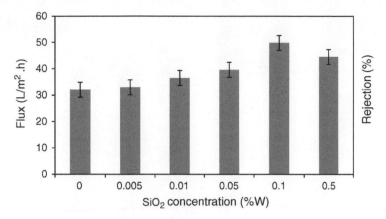

Figure 3.2.   Fluxes of polyamide TFC-RO membranes by using silica (SiO_2) nanoparticles at RO tests with aqueous NaCl salt solutions (11,000 mg L^{-1}) (Peyki *et al.*, 2015).

salt rejection. As a concept, TFN membrane technology may offer new options in tailoring RO membrane separation performance and material properties (Jeong *et al.*, 2007).

Kim *et al.* (2012) synthesized a new type of TFN membrane having antibacterial properties by interfacial polymerization. They used two different nanomaterials, which were acid modified multi-walled carbon nanotubes (MW-NTs) for a support layer and nanosilver (nAg) particles for a thin film layer. MW-NTs at 5.0 wt% in the support layer and nAg particles at 10 wt% in the thin film layer enhanced the pure water permeability of the TFN membrane by 23 and 20%, respectively, compared to 0 wt% of these components in their respective layers. The water permeability and hydrophilicity was enhanced in the TFN membrane that had a CNT support layer compared to a non-CNT matrix through the diffusive tunnel effects of nanopores in MW-NTs. nAg particles in the thin film particularly increased the anti-biofouling properties of the TFN membranes and did not change the salt rejections of the membranes. *Pseudomonas aeruginosa* PAO1 batch tests indicated greater anti-adhesive and antibacterial properties of the TFN membrane compared to similar membranes without nAg particles (Kim *et al.*, 2012).

Peyki *et al.* (2015) modified the surface of polyamide TFC-RO membranes by using silica (SiO_2) nanoparticles. The SiO_2 nanoparticles were added to amine solution with different concentrations of 0.005 to 0.5 wt%. FT-IR spectroscopy and SEM images confirmed the incorporation of SiO_2 nanoparticles on the polyamide active layer. The surface hydrophilicity and roughness of the membranes increased with the increasing SiO_2 concentration in the amine solution. The salt rejection and flux of the membranes were tested with NaCl aqueous solution. The flux values of the membranes increased gradually at lower levels of nanoparticles (0.005–0.1 wt%) and then decreased at 0.5 wt% (as seen in Fig. 3.2). Moreover, the rejection values increased at lower concentrations of SiO_2 (0.005 and 0.01 wt%) and then decreased with an increasing content of SiO_2. Long-term experiments showed that the modified RO membranes had lower flux decline compared to neat thin film composite RO membranes. The flux decline for unmodified RO membranes was 31%, while the flux declines for 0.005. 0.01, 0.05 and 0.1 wt% SiO_2 modified membranes were about 10%, 16%, 13% and 15%, respectively (Peyki *et al.*, 2015).

3.2.4   Blending method

The most conventional and simple method for the synthesis of polymer/inorganic mixed matrix membrane (MMM) is direct mixing of the nanoparticles into the polymer solution. The mixing can generally be done by melt blending or solution blending. Solution blending is a simple way to fabricate polymer-inorganic MMM (Fig. 3.3). In this method, a polymer is dissolved in a proper solvent to form an homogenous polymer solution, and then inorganic nanoparticles are added

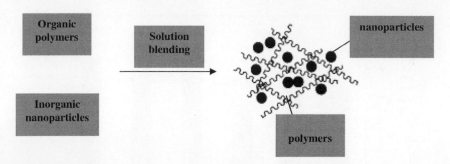

Figure 3.3.   Polymer and inorganic phases connected by Van der Waals force or hydrogen bonds at blending method (Cong *et al.*, 2007).

into the solution and dispersed by mixing. The polymer solution is cast by removing the solvent and the membrane is obtained (Cong *et al.*, 2007). The main difficulty in the mixing process is to achieve an effective dispersion of the nanoparticles in the polymer matrix because they have a strong tendency to form agglomerates (Kango *et al.*, 2013). Firsty, the nanoparticles should be added to solvent and stirred with sonication for preventing their agglomeration in polymer solution when the solubility of nanoparticle is low in solvent. The aggregation of nanoparticles in the membranes and the easy leaching of nanoparticles from the membranes are the main concerns of the solution blending method. However, commercial polymeric membranes can be modified by blending with hydrophilic materials during the industrial membrane production. It is a common practical technique and no additional processing steps are needed during or after the phase inversion process (Wu *et al.*, 2013).

Wang *et al.* (2012) fabricated the mixed matrix (polyvinylidene fluoride (PVDF) and graphene oxide (GO)) blended ultrafiltration membranes using the immersion phase inversion process. GO-blended PVDF membranes appeared to have a more hydrophilic surface and have a higher pure water fluxes recovery ratio than the unblended PVDF membranes due to the hydrophilic nature of GO. The permeation properties and mechanical structure of the blended membranes can obviously be improved. The permeability of the blended membrane increased by 96.4%; the tensile strength increased by 123%. Moreover, the contact angle decreased from 79.2° to 60.7°, which implies that the anti-fouling ability of the membrane was improved. Similarly, Silva *et al.* (2015) blended polyvinylidene fluoride (PVDF) with multi-walled carbon nanotubes (MW-CNTs) for the fabrication of flat-sheet direct contact membrane distillation (DCMD) membranes by the phase inversion method. The synthesis parameters, such as MW-CNT loading, polyvinylpyrrolidone (PVP) addition and MW-CNT surface chemistry, affected markedly the membrane properties and performances. MW-CNT/PVDF membranes prepared with functionalized MW-CNTs had a smaller pore size and lower contact angles (more hydrophilic), thus the functionalization of MW-CNTs is not recommended for this application. Sponge-like pores and smaller thickness (i.e. membranes prepared without PVP) allowed complete salt rejection (i.e. 100%). In contrast with larger thickness and elongated finger-like pores (resulting from PVP addition) salt exclusion ranged from 88.8 to 98.6%. Overall, the MW-CNT/PVDF blended membrane prepared with 0.2 wt% MW-CNTs (without adding PVP) exhibited the best performance in DCMD, presenting total salt rejection and a higher permeate flux (9.5×10^{-3} kg m^{-2} s^{-1}) than that obtained with a commercial PVDF membrane (7.8×10^{-3} kg m^{-2} s^{-1}).

Arsuaga *et al.* (2013) used different metal oxides nanoparticles (TiO_2, Al_2O_3 and ZrO_2) for the fabrication of pressure-driven membranes. The nanoparticles were blended in PES polymer solution. The membranes were characterized with hydrophilicity (contact angle), morphology (SEM), porosity, and thermal analysis (TGA). Membrane fouling was studied with BSA and humic acids as model organic foulants. Entrapped metal oxides caused a more open and porous structure in membrane morphology and significantly enhanced the anti-fouling property and long-term flux stability of the membranes. PES-Al$_2$O$_3$ membranes exhibited the highest value of 208.9 L m^{-2} h^{-1} while the permeability of pristine PES membranes was 180 L m^{-2} h^{-1}. Strong correlations were observed between physico-chemical properties such as porosity, hydrophilicity and permeability of modified membranes with the spatial particle distribution in the membrane structure. The fouling resistance of modified membranes was significantly reduced. The relative fouling resistances of the membranes were found to be 7.56 × 10^{12} m^{-1} for pristine PES and 5.52 × 10^{12}, 4.79 × 10^{12} and 5.46 × 10^{12} m^{-1} for PES-TiO$_2$, PES-Al$_2$O$_3$ and PES-ZrO$_2$, respectively.

3.3 NANOMATERIAL TYPES FOR FABRICATION OF MIXED MATRIX MEMBRANES (MMMs)

3.3.1 *Polymeric mixed matrix membranes (MMMs) with silver nanoparticles (AgNPs)*

In recent years, several studies have demonstrated the potential to use silver nanoparticles (AgNPs) as antibacterial agents in membrane separation processes. AgNPs have three different antibacterial mechanisms: (i) attachment to the cell membrane drastically disturbing the permeability and respiration of the membrane, (ii) penetration inside the bacteria cytoplasm and interaction with sulfur- and phosphorus-containing compounds, such as DNA, causing further damage and (iii) release of silver ions, which also have a known bactericidal effect as reported by Feng *et al.* (2000). The release of silver ions is also involved in mechanisms (i) and (ii) and it is the definitive molecular toxicant under aerobic conditions (Cruz *et al.*, 2015).

AgNPs have strong antibacterial activity against a different type of bacteria, but the use of AgNPs as biocide is mainly limited by its relatively high cost. Localized loading of small amounts of silver adjacent to the membrane surface, where biofilms develop, is viable for protecting the membrane from biofouling. Therefore, the anti-biofouling potential of AgNPs is well established in the literature, but the loading method of AgNPs on the membrane material is one of the main obstacles (Ben-Sasson *et al.*, 2014). During the fabrication of membranes with AgNP, the different pathways can be used to incorporate silver nanoparticles in the membrane structure, such as *ex-situ* synthesis (subsequent addition to the casting solution), or *in-situ* reduction of ionic silver during the phase inversion process (Taurozzi *et al.*, 2008). Furthermore, the silver can be applied by coating AgNP directly on to the surface of the membranes or by embedment in the polymer matrix of the membrane itself (Yang *et al.*, 2009; Zodrow *et al.*, 2009). Zodrow *et al.* (2009) prepared AgNP incorporating membranes having antimicrobial properties by using the phase inversion technique, in which the dope solution containing the polymer and AgNP was casted on a support and then immersed in a water bath to provide polymer precipitation. They found that the addition of AgNP prevented the bacterial attachment and biofilm formation on the membrane surface. Moreover, the authors suggested that the release of ionic silver can be the main mechanism.

The combination of polymer and AgNPs enhanced the separation properties due to the prevention of particle deposition (Babu *et al.*, 2010). AgNPs have a high antibacterial activity and this property may provide the long-term stability and efficient operation for mixed matrix membranes in filtration, especially for prolonged operational time periods. In the literature, there have been a number of studies to improve the distribution of AgNPs and its stability. Cellulose acetate (CA), chitosan, polyacrylonitrile (PAN) and polysulfone (PS) were the most popular polymeric materials studied with AgNPs (Basri *et al.*, 2011). However, CA, chitosan and PAN polymers have disadvantages such as poor solubility in common organic solvent and high cost. Therefore, PES appeared to be a more promising polymer because it is an easy handling material with good solubility in most of the organic solvents and is resistant to chemical attack (Susanto and Ulbricht, 2009; Wang *et al.*, 2005).

Liu *et al.* (2013) fabricated novel nanofiltration and forward osmosis membranes, including AgNP, using the layer-by-layer (LbL) assembly method. The incorporation of AgNPs in the membranes did not adversely affect the membrane separation performance in NF and forward osmosis

(FO) processes at low AgNPs incorporation levels (0.22–1.19 wt% as silver). The membranes with silver exhibited excellent antibacterial properties against both Gram-positive *Bacillus subtilis* and Gram-negative *Escherichia coli*. They showed that the performances of the membranes, including AgNPs, are highly dependent on silver incorporation in the membranes, which could be controlled by using different membrane synthesis methods and the addition of AgNPs.

3.3.2   *Polymeric mixed matrix membranes (MMMs) with silica nanoparticles (SiNPs)*

Silica nanoparticles (SiNPs) have high surface areas, narrow pore size distributions, high porosity and low density (Zargar *et al.*, 2016). They have been studied extensively for various applications. At polymeric membrane fabrication, SiNPs have been widely used to fabricate the gas separation membranes to enhance both gas permeability and selectivity. They increased the permeability of mixed matrix membranes without changing their selectivity due to their good compatibility with the polymer matrix. Furthermore, the functional –OH group on the silica should improve the hydrophilicity of the membrane and have a positive effect on the fouling resistance of the membrane (Huang *et al.*, 2012).

The use of SiNPs in membrane fabrication has not yet been studied to a large extent, despite the wide availability of different types with a wide range of particle sizes. Jadav and Singh (2009) incorporated two types of SiNPs *in-situ* into polyamide films. One type of SiNP was commercial Ludox® HS 40 (particle size of 16 nm); the other type of SiNP was a lab-synthesized colloid prepared from controlled hydrolysis of tetraethylortho silicate (TEOS) (0.4 wt% suspension in water). SiNPs were trapped into the polymer film with high temperature curing. It was observed that the effective pore radius of the mixed matrix membranes increased when the silica concentration in the casting solution was increased. Depending on the SiNPs content, the average pore radius of the membranes varied from a smaller radius of 0.34 nm to a larger radius of 0.73 nm. Furthermore, the number of pores in the membrane also increased. Silica has the advantage of improving the mechanical stability and, as a result of partial leaching, introduced additional pores in the membranes. Silica nanoparticles with silanol groups on the surface could act as active sites for polymer adsorption or reaction, thus providing a platform for anchoring anti-fouling ligands (Wu *et al.*, 2013).

3.3.3   *Polymeric mixed matrix membranes (MMMs) with carbon nanotubes (CNT)*

Carbon nanotube (CNT) was discovered in 1991 and classified as single-walled (SW-CNT) and multi-walled (MW-CNT). CNTs have attracted significant research areas because of their unique structure, physical, electronic, and thermal properties. MW-CNTs are more often preferred because of their relatively low cost and more advanced stage of commercial production. In recent times, CNT/polymer composites are prepared with the addition of CNTs into the intermediate and final polymer products, and these composites significantly increase the separation performance of the hybrid membranes (Wang *et al.*, 2014). MW-CNTs are naturally hydrophobic but they can be easily modified with different treatment methods and attached to different functional groups, such as hydrogen, nitrenes, fluorine, carbenes, radicals and aryl radicals. Acid treatment develops some hydrophilic part (–COOH, and –OH) on the MW-CNT substrate. Carbon part on one side of the MW-CNT makes it hydrophobic and creates sites for the attachment of hydrophobic polymer by hydrophobic-hydrophobic interaction and $\pi-\pi$ stacking. On the other side, acid and hydroxyl groups are created after acid treatment and this brings the hydrophilic part by hydrogen bonding, dipole-dipole interaction and dispersion forces (Irfan *et al.*, 2014).

CNTs are most widely used in membrane applications in water and wastewater treatment and they are also effective adsorbents for contaminants in water and wastewater treatment processes. Membrane roughness, surface functional groups, hydrophilicity, and fouling can influence permeate fluxes of CNT composite membranes (Kim *et al.*, 2013).

Recently, carbon nanotube (CNT) membranes have been investigated by many researchers as a novel membrane technology for water treatment. When CNTs are used as pores of membranes, they appear to allow fast fluid flow due to their strong hydrophobicity (Park *et al.*, 2014). In polymeric membranes that have CNTs, the membrane fabrication processes may not be so easy. The size and uniformity of CNTs are very important parameters because they may cause stability problems in membranes.

3.3.4 *Polymeric mixed matrix membranes (MMMs) with alumina nanoparticles (Al_2O_3NP)*

Alumina nanoparticles (Al_2O_3NP) are one of the nanoparticles used in polymeric mixed matrix membranes. Al_2O_3NP can increase the mechanical resistance and hydrophilicty and also modify the membrane morphology during the membrane formation in the phase inversion process (Homayoonfal *et al.*, 2015). Homayoonfal *et al.* (2015) synthesized a polysulfone (PSf)/alumina (Al_2O_3NP) mixed matrix membrane with the principal aim of reducing biofouling in the membrane bioreactors. The filtration experiments indicated that the addition of Al_2O_3NP increased the membrane hydrophilicity and so the pure water flux was enhanced from $450 \, L \, m^{-2} \, h^{-1}$ to $1600 \, L \, m^{-2} \, h^{-1}$ for 0.02 Al_2O_3/PSf and $2075 \, L \, m^{-2} \, h^{-1}$ for 0.03 Al_2O_3/PSf. The separation efficiency was maintained because of decreasing porosity. The presence of alumina nanoparticles up to the polymer concentration of 0.03 wt% resulted in an 83% reduction in biofouling (Homayoonfal *et al.*, 2015). The comprehensive study concerning the fabrication of mixed matrix membranes with Al_2O_3NP was carried out by Yan *et al.* (2005). They fabricated PVDF membranes with incorporated Al_2O_3NP. Dimethylacetamide (DMAC) was used as the solvent and the dope solution consisted of 19 wt% PVDF polymer, 0–4 wt% Al_2O_3NP and additional additives (hexad-sodium phosphate, and polyvinylpyrrolidone (PVP)). They found a remarkable change in the contact angle values. The contact angle of the bare PVDF membrane was 84° and decreased to 57–59° for membranes with 2–4% of Al_2O_3NP. Only the membrane with the lowest concentration of Al_2O_3NP still had a higher contact angle (68°). The porosity and rejection values remained unchanged in all experiments. Interestingly, SEM images showed that the addition of Al_2O_3NP did not affect the structure of the surface and cross-section. Furthermore, the mechanical properties of the membranes had improved; the tensile strength had increased remarkably, whereas the break elongate ratio reached a maximum at 2% Al_2O_3NP added. From these studies, it can be said that Al_2O_3NP increased the membrane hydrophilicty.

3.3.5 *Polymeric mixed matrix membranes (MMMs) with TiO_2 nanoparticles*

In the membrane fabrication studies, titanium dioxide (TiO_2) was one of the most preferred nanoparticles. Titanium dioxide nanoparticles (TiO_2NP) loaded membranes have received much attention because of their high hydrothermal nature, chemical stability and photocatalytic activity (Li D. *et al.*, 2014). Mixed matrix membranes with TiO_2NP can be manufactured in two ways, with assembling or blending. In the assembling method, TiO_2NP is coated to the surface of porous membranes with self-organization. The cross-linking degree of polymeric matrix and the types of bonds between the polymer and inorganic molecules can control the assembled membrane structure. TiO_2NP is directly added to the polymeric casting solution during the blending method and the homogeneity is the main problem. Some mechanical and chemical approaches (introduction of organic functional groups on the TiO_2 surface) have been applied for the modification of TiO_2NP to increase homogenous dispersivity, reduce agglomeration, improve the stability of TiO_2NP in polymer and enhance the nanofiller-polymer interaction (Vatanpour *et al.*, 2012).

Mericq *et al.* (2015) prepared low fouling PVDF membranes with TiO_2NP by NIPS (Non-solvent Induced Phase Separation) wet-process. They obtained a typical asymmetric membrane structure. The membrane structure, hydrophilic properties and permeability were improved in comparison with the PVDF neat membrane when the TiO_2NP concentration was increased up to an optimum concentration of 25 wt%. Under UV irradiation the phenomena of super-hydrophilicity due to the presence of TiO_2NP in the composite membrane permits the suppression of pure water permeate flux decline and the achievement of higher fluxes. Fouled composite membranes after

BSA filtration were successfully cleaned using water and UV irradiation. Permeate flux was totally recovered after this cleaning (Mericq *et al.*, 2015).

Under UV light TiO$_2$NPs produces some radicals such as OH$^\bullet$, O$_2^{-\bullet}$, and HO$_2^\bullet$ in water and these radicals are assigned as bactericidal action. The adhesion of the bacterial cells to TiO$_2$NP is controlled by hydrophobic and charge interactions and the active oxygen reaches and damages the bacterial cell wall. Due to the strong bactericidal properties as photocatalytic under UV light, the mixed matrix membranes having TiO$_2$NP inhibit the growth of microorganisms on the membrane surface, and thus membrane biofouling is reduced (Kochkodan and Hilal, 2015). This reduction in biofouling was demonstrated when TiO$_2$ modified membranes were used for water filtration, which showed that the fluxes of the modified membranes were 1.7–2.3 times higher compared with those for the control samples. The flux values of TiO$_2$ modified membranes reached to approximately 0.3 L h^{-1} after 4 days, while it was 0.15 L h^{-1} for the control membrane (Kochkodan *et al.*, 2008).

3.4   APPLICATION OF MIXED MATRIX MEMBRANES (MMMs)

3.4.1   *Antimicrobial activity*

Microbial growth and uncontrolled biofilm formation cause the membrane biofouling and it is a serious operational issue in filtration for water and wastewater treatment. It decreases membrane permeability and permeates quality, and so increases the energy costs of the separation process. Antimicrobial chemicals can be used for the prevention of biofouling problems in membranes, especially at the membrane fabrication step. Membranes having antimicrobial chemicals inhibit bacterial growth on the membrane surface and help to provide pathogen-free clean water. Development of antimicrobial or biofouling resistant membranes has focused on membrane surface modification to reduce the adhesion of bacteria or biopolymers, or the inhibition of bacterial growth. Several studies have incorporated antimicrobial nanoparticles in membranes. A commonly used method is to disperse the nanoparticles in the polymer solution during the phase inversion process (Wu *et al.*, 2015). Among antimicrobial nanoparticles, AgNPs are the most popular antimicrobial agent used in different areas. There are different methods to integrate AgNPs into the matrix, such as electroless plating and vacuum deposition. Immobilization of silver element or ions onto the modified polymeric surface through metal-ligand interaction was found to be the most efficient way (Zhang *et al.*, 2013).

Zhang *et al.* (2014) prepared polyethersulfone membranes with biogenic AgNP synthesized from *Lactobacillus fermentum*. Biogenic AgNP was introduced into the membrane by blending at different concentrations. The fabricated membranes were tested for physical properties such as water permeability, MWCO (molecular weight cut-off), contact angle, SEM and AFM. Cross-section micrographs demonstrated that the biogenic silver nanoparticles dispersed well into the PES matrix without aggregation. Well dispersed biogenic AgNP contained the hydroxyl and amino groups, which are known as hydrophilic groups, and the hydrophilicity of the PES membranes slightly increased and so the protein adsorption on the membrane surface decreased significantly. Moreover, biogenic AgNP improved the water permeability and did not sacrifice the selectivity of protein. The evaluation of silver release from the mixed matrix membranes indicated the good stability of biogenic silver nanoparticles in the membrane matrix. The results of the disk diffusion test revealed the excellent antibacterial activity of the membranes, including biogenic AgNP. In addition, the sludge immersion and the bacterial suspension filtration experiments showed that the membranes with biogenic AgNP not only prevented the bacteria attachment on the membrane surface but also inhibited the reproduction and development of biofilms. As seen in Figure 3.4, the biogenic AgNP was an effective material for decreasing the biofouling in the membrane process.

Similarly, carbon based nanomaterials, like single-walled carbon nanotubes (SW-NTs), significantly reduce the bacterial and viral suspension in water due to their antimicrobial properties (Kang *et al.*, 2007). Ahmed *et al.* (2013) coated nitrocellulose membranes with a nanomaterial

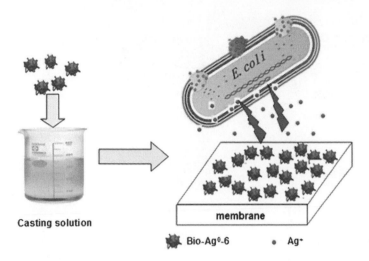

Figure 3.4.   The antibacterial mechanism of biogenic AgNP on the membrane surface (Zhang *et al.*, 2014).

solution containing 97 wt% of polyvinyl-N-carbazole (PVK) and 3 wt% of single-walled carbon nanotubes (SW-NTs) (97:3 wt% ratio PVK:SW-NT). Membranes coated with the nanomaterial solution exhibited significant antimicrobial activity towards Gram-positive and Gram-negative bacteria (80–90%); and presented a virus removal efficiency of ~2.5 logs. The possible antibacterial mechanism was explained as cellular inactivation with cell membrane damage, because the higher efflux of intracellular material (deoxyribonucleic acid, DNA) was analyzed in the permeate of membranes with SW-CNTs rather than in the filtrate of control membranes.

Rahimpour *et al.* (2011) fabricated novel poly(vinylidene fluoride) (PVDF)/sulfonated polyethersulfone (SPES) blend membranes with the immersion precipitation technique and then modified these membranes with TiO_2NP for antibacterial activity. The results of this antibacterial study indicated that the membranes with TiO_2NP efficiently eliminated *E. coli* growth after UV treatment due to the photocatalytic bactericidal effect of the TiO_2 catalyst. The bactericidal effect of UV/TiO_2 photocatalysis is due to the presence of reactive oxygen species like O_2^-, H_2O_2 and HO generated by TiO_2 or the direct UV illumination of the cells. The bactericidal effect of TiO_2 photocatalytic on the death of *E. coli* cells was explained with HO radical attack and lipid peroxidation reaction.

Huang *et al.* (2014) deposited the silver nanoparticles on the silica sphere surface to form Ag-SiO_2. They improved the anti-fouling performance and enhanced the dispersion of silver in the membrane. Ag-SiO_2 mixed matrix polyethersulfone (PES) membranes were fabricated by the phase inversion method. The effect of Ag-SiO_2 content on the membrane performance was investigated. The antibacterial and anti-formation properties of the membranes were systematically studied with pure cultures of both *Escherichia coli* and *Pseudomonas* sp. and also with a mixed bacteria culture in an activated sludge bioreactor. The results showed that the prepared Ag-SiO_2/PES membrane exhibited improved filtration performance and excellent antibacterial and anti-biofouling properties.

3.4.2   *Desalination*

Desalination is a unique process that removes the salts and minerals from water medium and is a key solution in solving the challenges of global water scarcity. Along with conventional desalination technologies like thermal distillation, the reverse osmosis (RO) process contributed greatly to the success of sea/brackish water desalination, despite high energy consumption (Emadzadeh

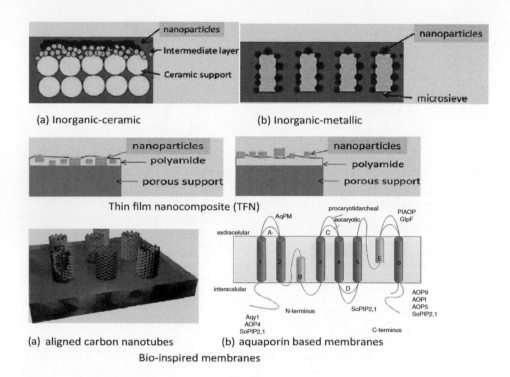

(a) Inorganic-ceramic   (b) Inorganic-metallic

Thin film nanocomposite (TFN)

(a) aligned carbon nanotubes   (b) aquaporin based membranes

Bio-inspired membranes

Figure 3.5.   Classification of novel desalination membranes (Buonomenna, 2013).

et al., 2014). Besides this, membrane fouling is the other problem for the RO process. Therefore, recent studies have focused on developing RO membranes with consistent high water flux by improving the anti-fouling property of the membrane surface. To prevent membrane fouling, some inorganic nanomaterials were introduced into RO membranes during the fabrication of thin film composite (TFC) membranes. Inorganic nanomaterials can improve film formation by offering the following benefits: (i) increasing the diffusion rate of monomers to the interface, (ii) expanding the wet zone on the top of the support layer and (iii) capturing by-products and controlling reaction pH (buffer agent) (Emadzadeh *et al.*, 2015a). Moreover, inorganic nano-additives (e.g. titanium dioxide (TiO_2), silica, silver, and zeolite nanoparticles) improve the diffusion features of the formed membrane and also enhance their fouling resistance. The nanotechnology concept has led to new desalination membranes having high permeability, catalytic reactivity, and fouling resistance. Among the numerous concepts proposed, the most promising to date include zeolitic and catalytic nanoparticle coated membranes, mixed matrix and bio-inspired membranes, such as hybrid protein-polymer biomimetic membranes, aligned nanotube membranes, and isoporous block copolymer membranes. Figure 3.5 shows the classification of novel desalination membranes (Buonomenna, 2013).

Among the various nanoparticles, zeolite nanoparticles are the most popular nano-fillers in RO membranes. The physical and chemical properties of zeolite nanoparticles affect markedly the TFN membrane performance and it leads to more permeable polyamide active layers for water desalination. Generally, it can be said that zeolite nanoparticles provide preferential flow paths for water transport when they have a tight pore distribution less than the diameter of a hydrated salt ion. Besides this, it is also believed that the presence of zeolite nanoparticles in the polyamide layer may change the structure of thin film layer at the RO membrane surface by the formation of nano-gaps at the organic-inorganic interfaces, which can reduce the cross-linking density of the polyamide layer (Dong *et al.*, 2015).

Over the past several years, the forward osmosis (FO) process using membrane-based technology has attracted considerable attention among membrane scientists as a potential desalination process to compete with RO in the future. Despite the distinct advantages offered by FO membranes in the water desalination process, the research on how to further improve the properties of thin film composite (TFC) FO membranes, particularly at the top active skin layer, still remains as the main research focus of researchers (Emadzadeh *et al.*, 2014). Tian *et al.* (2015) fabricated a novel TFN-FO membrane on nanofibrous substrate with functionalized MW-CNT. MW-CNT increased the tensile modulus by 53% and substrate porosity by 18%. TFN membranes achieved a significantly higher water flux, i.e. $61 \, L \, m^{-2} \, h^{-1}$ in active layer-draw solution orientation. Emadzadeh *et al.* (2015b) prepared a self-synthesized TFN-FO membrane incorporated with hydrophilic functionalized titanate nanotubes (TNTs) and tested these membranes for FO desalination. TNTs improved the surface morphology and separation performance. The water flux of the FO membrane was significantly improved from $21.0 \, L \, m^{-2} \, h^{-1}$ of TFC (control) to $31.5 \, L \, m^{-2} \, h^{-1}$ in PRO mode by adding only 0.05 wt% of TNTs. It was concluded that modifying PA layer properties by adding TNTs was an effective approach to enhancing FO performance of typical TFC membranes because the large specific surface area and high pore volumes of TNTs might provide abundant adsorption sites and diffusion channels for water.

3.4.3 *Fuel cell*

Fuel cells generate the electricity by a chemical reaction. They are becoming increasingly popular due to the contribution of conventional fossil fuels to environmental pollution and the reduction of the overall oil supply. Among the various types of fuel cells, direct methanol fuel cells (DMFCs) have attracted extensive attention because of advantages like their low weight, simple system design, high energy density at low operating temperature, low emission and the ease of handling their liquid fuel. The key part of DMFCs is a proton exchange membrane (PEM), which is used to segregate both sides of the electrodes in order to prevent an internal electric current between the two electrodes and provide a charge carrier for the protons (Hasanabadi *et al.*, 2013). Nafion® (DuPont, USA), Flemion® (Asahi Glass, Japan) and Aciplex® (Asahi Kasei, Japan) are mostly used for membrane material, but these materials are expensive and are major contributors to the system cost. The development of the membranes with a lower cost and higher performance is necessary (Peighambardoust *et al.*, 2010). Therefore, full cell researchers focused on making new PEMs with high proton conductivity, durability, thermal stability, maximum power density and low fuel crossover, and low cost. The hybrid organic-inorganic composite or mixed matrix membranes have risen as an attractive option. The combination of organic and inorganic properties may overcome the limitations of the pure polymeric membranes. The utilization of novel polymers and the incorporation of inorganic materials as fillers in the pure polymer are several approaches for the development of new membrane material. Hydrophilic inorganic materials, such as silicon dioxide (SiO_2), titanium dioxide (TiO_2), and zirconium dioxide (ZrO_2), and heteropolyacids, such as phosphotungstic acid and silicotungstic acid, have been widely studied. Inorganic materials enhanced the mechanical properties, and also contributed to the blocking of the fuels, such as methanol, by increasing the transport pathway tortuousness and improving the proton conductivity, which resulted in better cell performance (Kim *et al.*, 2014).

Aslan and Bozkurt (2014) prepared proton conducting nano-titania composite membranes. They also discussed the production and characterization of proton conducting super acid membranes. During membrane fabrication, sulfated nano-titania was firstly synthesized by hydrolysis and precipitation of titanyl sulfate (TS) and was then blended with sulfonated polysulfone (SPSU). The maximum proton conductivity of the prepared membrane was obtained as $0.002 \, S \, cm^{-1}$ at 150°C.

Aviles-Barreto and Suleiman (2015) studied the synthesis and characterization of mixed matrix membranes based on the sulfonated poly(styrene-isobutylene-styrene) (SIBS) and SW-CNT loading and functional groups' substitution with the aim of evaluating their transport properties for direct methanol fuel cells (DMFC). The addition of SW-CNT limited the passage of methanol through the membranes due to the partial blockage of the free volume by the SW-CNT. The high water content and transport results showed two different types of water inside the polymer membranes (bound and bulk water). Large amounts of bulk water seemed to inhibit the transport of protons, decreasing the proton conductivity of the membranes. For DMFC applications, selectivity values suggested that 0.1 wt% is the optimum SW-CNT loading. Using carboxylic groups as terminal functionalization in the SW-CNT had a zero impact on the overall transport properties (e.g. m selectivity) for DMFC applications.

3.5   CONCLUSION AND OUTLOOK

Mixed matrix membranes with nano-additives can be fabricated with different methods and have numerous application areas. These membranes are named as high-performance membranes due to relatively high flux, easy cleaning, good mechanical strength and dimensional stability and they are also suitable for different areas such as water filtration or fuel cell application. Modification of classical polymeric membranes with nanomaterials has become an expanding field of research, as the introduction of specific groups can improve the structure and performance of membranes. The key steps for this membrane fabrication are as follows: (i) choice of suitable inorganic nano-additives for homogenous dispersity in the membrane matrix, (ii) understanding of the relationship between polymer and additive for the ideal membrane structure, (iii) improving the stability of nano-additives in the membrane structure after fabrication for a long service life and (iv) regeneration or cleaning of mixed matrix membranes for reuse. Much research and development is still needed in order to understand the performance criteria of these membranes and to fabricate the ideal polymer-inorganic mixed matrix membrane.

ACKNOWLEDGMENTS

The authors are grateful to TUBITAK (The Scientific and Technological Research Council of Turkey) for their financial support under grant (Project No: 111Y095).

REFERENCES

Ahmed, F., Santos, C.M., Mangadlao, J., Advincula, R. & Rodrigues, D.F. (2013) Antimicrobial PVK:SW-NT nanocomposite coated membrane for water purification: performance and toxicity testing. *Water Research*, 47, 3966–3975.

Arsuaga, J.M., Sotto, A., Rosario, G., Martinez, A., Molina, S., Teli, S.B. & Abajo, J. (2013) Influence of the type, size, and distribution of metal oxide particles on the properties of nanocomposite ultrafiltration membranes. *Journal of Membrane Science*, 428, 131–141.

Askar, K., Phillips, B.M., Fanga, Y., Choi, B., Gozubenli, N., Jiang, P. & Jiang, B. (2013) Self-assembled self-cleaning broadband anti-reflection coatings. *Colloids and Surfaces* A: *Physicochemical and Engineering Aspects*, 439, 84–100.

Aslan, A. & Bozkurt, A. (2014) Nanocomposite membranes based on sulfonated polysulfone and sulfated nano-titania/NMPA for proton exchange membrane fuel cells. *Solid State Ionics*, 255, 89–95.

Aviles-Barreto, S. & Suleiman, D. (2015) Effect of single-walled carbon nanotubes on the transport properties of sulfonated poly(styrene-isobutylene-styrene) membranes. *Journal of Membrane Science*, 474, 92–102.

Babu, V.R., Kim, C., Kim, S., Ahn, C. & Lee, Y.I. (2010) Development of semi-interpenetrating carbohydrate polymeric hydrogels embedded silver nanoparticles and its facile studies on *E. coli*. *Carbohydrate Polymer*, 81, 196–202.

Bae, T., Kim, I. & Tak, T. (2006) Preparation and characterization of fouling-resistant TiO_2 self-assembled nanocomposite membranes. *Journal of Membrane Science*, 275, 1–5.

Basri, H., Ismail, A.F. & Aziz, M. (2011) Polyethersulfone (PES)-silver composite UF membrane: effect of silver loading and PVP molecular weight on membrane morphology and antibacterial activity. *Desalination*, 273, 72–80.

Ben-Sasson, M., Lu, X., Bar-Zeev, E., Zodrow, K.R., Nejati, S., Qi, G., Giannelis, E.P. & Elimelech, M. (2014) *In-situ* formation of silver nanoparticles on thin-film composite reverse osmosis membranes for biofouling mitigation. *Water Research*, 62, 260–270.

Buonomenna, M.G. (2013) Nano-enhanced reverse osmosis membranes. *Desalination*, 314, 73–88.

Cong, H., Radosz, M., Towler, B. F. & Shen, Y. (2007) Polymer-inorganic nanocomposite membranes for gas separation. *Separation and Purification Technology*, 55, 281–291.

Cruz, M.C., Ruano, G., Wolf, M., Hecker, D., Vidaurre, E.C., Schmittgens, R. & Rajal, V.B. (2015) Plasma deposition of silver nanoparticles on ultrafiltration membranes: antibacterial and anti-biofouling properties. *Chemical Engineering Research and Design*, 94, 524–537.

Dong, H., Zhao, L., Zhang, L., Chen, H., Gao, C. & Ho, W.S.W. (2015) High-flux reverse osmosis membranes incorporated with NaY zeolite nanoparticles for brackish water desalination. *Journal of Membrane Science*, 476, 373–383.

Dong, L., Huang, X., Wang, Z., Yang, Z., Wang, X. & Tang C.Y. (2016) A thin-film nanocomposite nanofiltration membrane prepared on a support with *in-situ* embedded zeolite nanoparticles. *Separation and Purification Technology*, 166, 230–239.

Emadzadeh, D., Lau, W.J., Matsuura, T., Rahbari-Sisakht, M. & Ismail, A.F. (2014) A novel thin film composite forward osmosis membrane prepared from PSf-TiO$_2$ nanocomposite substrate for water desalination. *Chemical Engineering Journal*, 237, 70–80.

Emadzadeh, D., Lau, W.J., Rahbari-Sisakht, M., Daneshfar, A., Ghanbari, M., Mayahi, A., Matsuura, T. & Ismail, A.F. (2015a) A novel thin film nanocomposite reverse osmosis membrane with superior anti-organic fouling affinity for water desalination. *Desalination*, 368, 106–113.

Emadzadeh, D., Lau, W.J., Rahbari-Sisakht, M., Ilbeygi, H., Rana, D., Matsuura, T. & Ismail, A.F. (2015b) Synthesis, modification and optimization of titanate nanotubes-polyamide thin film nanocomposite (TFN) membrane for forward osmosis (FO) application. *Chemical Engineering Journal*, 281, 243–251.

Feng, Q.L., Wu, J., Chen, G.Q., Cui, F.Z., Kim, T.N. & Kim, J.O. (2000) A mechanistic study of the antibacterial effect of silver ions on *Escherichia coli* and *Staphylococcus aureus*. *Journal of Biomedical Materials Research*, 52, 662–668.

Hasanabadi, N., Ghaffarian, S.R. & Sadrabadi, M.M.H. (2013) Nafion-based magnetically aligned nanocomposite proton exchange membranes for direct methanol fuel cells. *Solid State Ionics*, 232, 58–67.

Homayoonfal, M., Mehrnia, M.R., Rahmani, S. & Mojttahedi, Y.M. (2015) Fabrication of alumina/ polysulfone nanocomposite membranes with biofouling mitigation approach in membrane bioreactors. *Journal of Industrial and Engineering Chemistry*, 22, 257–367.

Huang, J., Zhang, K., Wang, K., Xie, Z., Ladewig, B. & Wang, H. (2012) Fabrication of polyethersulfone-mesoporous silica nanocomposite ultrafiltration membranes with antifouling properties. *Journal of Membrane Science*, 423–424, 362–370.

Huang, J., Wang, H. & Zhang, K. (2014) Modification of PES membrane with Ag-SiO$_2$: reduction of biofouling and improvement of filtration performance. *Desalination*, 336, 8–17.

Irfan, M., Ildris, A., Yusof, N.M., Khairuddin, N.F.M. & Akhmal, H. (2014) Surface modification and performance enhancement of nano-hybrid f-MW-CNT/PVP90/PES hemodialysis membranes. *Journal of Membrane Science*, 467, 73–84.

Jadav, G.L. & Singh, P.S. (2009) Synthesis of novel silica-polyamide nanocomposite membrane with enhanced properties. *Journal of Membrane Science*, 328, 257–267.

Jeong, B.H., Hoek, E.M.V., Yan, Y., Subramani, A., Huang, X., Hurwitz, G., Ghosh, A.K. & Jawor, A. (2007) Interfacial polymerization of thin film nanocomposites: a new concept for reverse osmosis membranes. *Journal of Membrane Science*, 294, 1–7.

Kang, S., Pinault, M., Pfefferle, L.D. & Elimelech, M. (2007) Single-walled carbon nanotubes exhibit strong antimicrobial activity. *Langmuir*, 23, 8670–8673.

Kango, S., Kalia, S., Celli, A., Njuguna, J., Habibi, Y. & Kumar, R. (2013) Surface modification of inorganic nanoparticles for development of organic-inorganic nanocomposites: a review. *Progress in Polymer Science*, 38, 1232–1261.

Kim, S., Kwak, S., Sohn, S. & Park, T. (2003) Design of TiO$_2$ nanoparticle self-assembled aromatic polyamide thin-film-composite (TFC) membrane as an approach to solve biofouling problem. *Journal of Membrane Science*, 211, 157–165.

Kim, E.S., Hwang, G., El-Din, M.G. & Liu, Y. (2012) Development of nanosilver and multi-walled carbon nanotubes thin-film nanocomposite membrane for enhanced water treatment. *Journal of Membrane Science*, 394–395, 37–48.

Kim, E.S., Liu, Y., El-Din, M.G. (2013) An *in-situ* integrated system of carbon nanotubes nanocomposite membrane for oil sands process-affected water treatment. *Journal of Membrane Science*, 429, 418–427.

Kim D.J., Jo, M.J. & Nam, S.Y. (2014) A review of polymer-nanocomposite electrolyte membranes for fuel cell application. *Journal of Industrial and Engineering Chemistry*, 21, 36–52.

Kochkodan, V. & Hilal, N. (2015) A comprehensive review on surface modified polymer membranes for biofouling mitigation. *Desalination*, 356, 187–207.

Kochkodan, V., Tsarenko, S., Potapchenko, N., Kosinova, V. & Goncharuk, V. (2008) Adhesion of microorganisms to polymer membranes: a photobactericidal effect of surface treatment with TiO_2. *Desalination*, 220, 380–385.

Li, D., Wang, H., Jing, W., Fan, Y. & Xing, W. (2014) Fabrication of mesoporous TiO_2 membranes by a nanoparticle-modified polymeric sol process. *Journal of Colloid and Interface Science*, 433, 43–48.

Li, X., Fang, X., Pang, R., Li, J., Sun, X., Shen, J., Han, W. & Wang, L. (2014) Self-assembly of TiO_2 nanoparticles around the pores of PES ultrafiltration membrane for mitigating organic fouling. *Journal of Membrane Science*, 467, 226–235.

Liu, X., Qi, S., Li, Y., Yang, L., Cao, B. & Tang, C.Y. (2013) Synthesis and characterization of novel antibacterial silver nanocomposite nanofiltration and forward osmosis membranes based on layer-by-layer assembly. *Water Research*, 47, 3081–3092.

Mansourpanah, Y., Rahimpour, A., Farhadian, A. & Taheri, A.H. (2009) Formation of appropriate sites on nanofiltration membrane surface for binding TiO_2 photo-catalyst: performance, characterization and fouling-resistant capability. *Journal of Membrane Science*, 330, 297–306.

Mericq, J.P., Mendret, J., Brosillon, S. & Faur, C. (2015) High performance PVDF-TiO_2 membranes for water treatment. *Chemical Engineering Science*, 123, 283–291.

Moghimifar, V., Raisi, A. & Aroujalian, A. (2014) Surface modification of polyethersulfone ultrafiltration membranes by corona plasma-assisted coating TiO_2 nanoparticles. *Journal of Membrane Science*, 461, 69–80.

Mollahosseini, A. & Rahimpour, A. (2014) Interfacially polymerized thin film nanofiltration membranes on TiO_2 coated polysulfone substrate. *Journal of Industrial and Engineering Chemistry*, 20, 1261–1268.

Nady, N., Franssen, M.C.R., Zuilhof, H., Eldin, M.S.M., Boom, R. & Schroen, K. (2011) Modification methods for poly(arylsulfone) membranes: a mini-review focusing on surface modification. *Desalination*, 275, 1–9.

Park, S.M., Jung, J., Lee, S., Baek, Y., Yoon, J., Seo, D.K. & Kim, Y.H. (2014) Fouling and rejection behavior of carbon nanotube membranes. *Desalination*, 343, 180–186.

Peighambardoust, S.J., Rowshanzamir, S. & Amjadi, M. (2010) Review of the proton exchange membranes for fuel cell applications. *International Journal of Hydrogen Energy*, 35, 9349–9384.

Peyki, A., Rahimpour, A. & Jahanshahi, M. (2015) Preparation and characterization of thin film composite reverse osmosis membranes incorporated with hydrophilic SiO_2 nanoparticles. *Desalination*, 368, 152–158.

Rahimpour, A., Jahanshahi, M., Rajaeian, B. & Rahimnejad, M. (2011) TiO_2 entrapped nano-composite PVDF/SPES membranes: preparation, characterization, antifouling and antibacterial properties. *Desalination*, 278, 343–353.

Shen, J., Yu, C., Ruan, H., Gao, C. & Bruggen, B.V. (2013) Preparation and characterization of thin-film nanocomposite membranes embedded with poly(methyl methacrylate) hydrophobic modified multiwalled carbon nanotubes by interfacial polymerization. *Journal of Membrane Science*, 442, 18–26.

Silva, T.L.S., Morales-Torres, S., Figueiredo, J.L. & Silva, A.M.T. (2015) Multi-walled carbon nanotube/PVDF blended membranes with sponge- and finger-like pores for direct contact membrane distillation. *Desalination*, 357, 233–245.

Susanto, H. & Ulbricht, M. (2009) Characteristics, performance and stability of polyethersulfone ultrafiltration membranes prepared by phase separation method using different macromolecular additives. *Journal of Membrane Science*, 327, 125–135.

Taurozzi, J.S., Arul, H., Bosak, V.Z., Burban, A.F., Voice, T.C., Bruening, M.L. & Tarabara, V.V. (2008) Effect of filler incorporation route on the properties of polysulfone-silver nanocomposite membranes of different porosities. *Journal of Membrane Science*, 325, 58–68.

Tian, M., Wang, Yi-N. & Wang, R. (2015) Synthesis and characterization of novel high-performance thin film nanocomposite (TFN) FO membranes with nanofibrous substrate reinforced by functionalized carbon nanotubes. *Desalination*, 370, 79–86.

Wang, Y., Yang, Q., Shan, G., Wang, C., Du, J., Wang, S., Li, Y., Chen, X., Jing, X. & Wei, Y. (2005) Preparation of silver nanoparticles dispersed in polyacrylonitrile nanofiber film spun by electrospinning. *Materials Letter*, 59, 3046–3049.

Wang, T., NanShen, J., GuangWu, L. & Bruggen, B.V. (2014) Improvement in the permeation performance of hybrid membranes by the incorporation of functional multi-walled carbon nanotubes. *Journal of Membrane Science*, 466, 338–347.

Wang, Z., Yu, H., Xia, J., Zhang, F., Li, F., Xia, Y. & Li, Y. (2012) Novel GO-blended PVDF ultrafiltration membranes. *Desalination*, 299, 50–54.

Wu, H., Mansouri, J. & Chen, V. (2013) Silica nanoparticles as carriers of antifouling ligands for PVDF ultrafiltration membranes. *Journal of Membrane Science*, 433, 135–151.

Wu, J., Yu, C. & Li, Q. (2015) Regenerable antimicrobial activity in polyamide thin film nanocomposite membranes. *Journal of Membrane Science*, 476, 119–127.

Vatanpour, V., Madaeni, S.S., Khataee, A.R., Salehi, E., Zinadini, S. & Monfared, H.A. (2012) TiO$_2$ embedded mixed matrix PES nanocomposite membranes: influence of different sizes and types of nanoparticles on antifouling and performance. *Desalination*, 292, 19–29.

Xu, G.R., Wang, J.N. & Li, C.J. (2013) Strategies for improving the performance of the polyamide thin film composite (PA-TFC) reverse osmosis (RO) membranes: surface modifications and nanoparticles incorporations. *Desalination*, 328, 83–100.

Yan, L., Li, Y.S. & Xiang, C.B. (2005) Preparation of poly (vinylidene fluoride) (PVDF) ultrafiltration membrane modified by nano-sized alumina (Al$_2$O$_3$) and its antifouling research. *Polymer*, 46, 7701–7706.

Yang, H.L., Lin, J.C.T. & Huang, C. (2009) Application of nanosilver surface modification to RO membrane and spacer for mitigating biofouling in seawater desalination. *Water Research*, 43, 3777–3786.

Yang, Y., Wang, S., Wang, Y., Wang, Y., Wang, X., Wang, Q. & Chen, M. (2014) Advances in self-assembled chitosan nanomaterials for drug delivery. *Biotechnology Advances*, 32(7), 1301–1316.

Zargar, M., Hartanto, Y., Jin, B. & Dai, S. (2016) Hollow mesoporous silica nanoparticles: A peculiar structure for thin film nanocomposite membranes. *Journal of Membrane Science*, 519, 1–10.

Zhang, D. & Cui, T. (2012) Tunable mechanical properties of layer-by-layer self-assembled carbon nanotube/polymer nanocomposite membranes for M/NEMS. *Sensors and Actuators* A, 185, 101–108.

Zhang, M., Field, R.W. & Zhang, K. (2014) Biogenic silver nanocomposite polyethersulfone UF membranes with antifouling properties. *Journal of Membrane Science*, 471, 274–284.

Zhang, S., Qiu, G., Ting, Y.P. & Chung, T.S. (2013) Silver-PEGylated dendrimer nanocomposite coating for anti-fouling thin film composite membranes for water treatment. *Colloids and Surfaces* A: *Physicochemical and Engineering Aspects*, 436, 207–214.

Zodrow, K., Brunet, L., Mahendra, S., Li, D., Zhang, A., Li, Q.L. & Alvarez, P.J.J. (2009) Polysulfone ultrafiltration membranes impregnated with silver nanoparticles show improved biofouling resistance and virus removal. *Water Research*, 43(3), 715–723.

CHAPTER 4

Stimulus-responsive nano-structured bio-hybrid membranes reactors in water purification

Abaynesh Yihdego Gebreyohannes & Lidietta Giorno

4.1 INTRODUCTION

4.1.1 *Bio-hybrid membrane processes in industrial biocatalysis*

Over the past two decades membrane technology has shown great potential towards replacing conventional processes owing to the numerous advantages that this technology holds (Brunetti *et al.*, 2010; Cassano *et al.*, 2003; Charpentier, 2005; Gugliuzza and Drioli, 2014; Judd and Judd, 2006). It permits working under mild temperatures and pressure while avoiding the use of additives or solvents (Drioli and Giorno, 2009). They practically have the potential to intensify multi-stage unit operations into a single-step process that eventually comes with a significantly reduced footprint (Drioli and Romano, 2001).

However, membrane processes have a range of inherent limitations, e.g. a membrane system designed to treat wastewater may be limited by the water's high concentration of suspended solids, viscosity, osmotic pressure, and temperature when trying to attain a target quality (Gebreyohannes *et al.*, 2015b). Therefore, the optimal separation process in many cases may be a membrane-based hybrid process.

Here, hybrid membrane processes refer to the integration of membranes with biocatalysis in membrane bioreactors (MBR). The biocatalyst used in MBR can be of microorganisms, algal biomass, or purified enzymes. Among these, purified enzymes have been playing a substantial role in many industrial products, such as the preparation of functional foods from cheap and renewable raw agricultural materials, wastewater treatment, animal feed, in research, bakeries, pharmaceuticals, neutraceuticals, winery, breweries, fruit juice industries, diagnostics and bioenergy generation. In 2012 and 2013, the global market for industrial enzymes was nearly U\$4.5 and U\$4.8 billion respectively, with an expected market rise to around U\$7.1 billion by 2018 (Dewan, 2014). The immobilized enzyme sales in 2010, excluding industries that use their own immobilized enzymes, were U\$130 million with an expected rise to U\$230 million by 2015.

From an engineering standpoint, process intensification through the integration of multifunctional systems makes MBRs a major part of the future solution to chemical production that may be limited by material resources and energy availability. MBRs with different biocatalysts, e.g. pectinase, laccase, lipase, amylase, acylase, etc., have been applied in different sectors to foster overall productivity, e.g. in the fermentation of bio-renewables (Machsun *et al.*, 2010), in waste valorization (Gebreyohannes *et al.*, 2013), in the pharmaceutical industry (Xiao-Ming and Wainer, 1993), and degradation of emerging hospital contaminants (Abejón *et al.*, 2015).

MBRs have also been used to make membrane filtration economically feasible by reducing membrane fouling caused by the presence of huge amounts of contaminants (e.g. pectins ~6.5% ton^{-1} washed coffee beans) (Beyene *et al.*, 2013). MBRs have also been used to hydrolyze highly concentrated polysaccharides, thus rendering wastewaters suitable for bioethanol production. MBRs can also be used to treat emerging contaminants, which are resistant to microbial degradation or advanced oxidation. In general, they form an important strategy to increase the sustainability of processes by offering new and better possibilities in bio-commodity manufacturing, environmental protection or bioenergy generation.

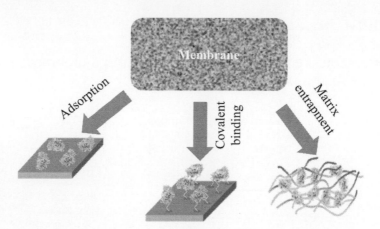

Figure 4.1. The different mechanisms employed to immobilize enzymes on membranes: physical entrapment, physicochemical interactions, chemical bond, and molecular recognition.

For instance, in wastewater treatment the integration brings performance improvements depending on feed characteristics or desired product quality (Singh, 2005). So it is a strategy designed to benefit from the synergy between biodegradation of organic substrate and membrane-based separation in a single unit (Drioli and Romano, 2001). The bio-hybridization (integrated bioreactor with membrane separation) may also permit the rationalization of improvements in the produced water quality. It also comes with reduced capital cost, equipment size, environmental pollution, footprint and ease of fractionation and direct and indirect energy savings (Drioli and Romano, 2001). So it can be claimed to be one of the key factors that have contributed to the successful widespread use of membrane processes in water treatment (Singh, 2005).

4.1.2 *MBR configurations*

The biocatalysts in MBR can be used either free suspended in the reaction mixture or they can be immobilized on the membrane physically or chemically (Mazzei *et al.*, 2010). The choice depends on the properties of the reaction system. For example, bioconversions for which the homogeneous catalyst distribution is particularly important are optimally performed in a reactor with the biocatalyst compartmentalized by the membrane in the reaction vessel. The membrane is used to retain large components, such as the enzyme and the substrate, while allowing small molecules (e.g. the reaction product) to pass through. Since enzymes are biological catalysts that are not consumed in the reactions in which they participate, they may be used repeatedly for as long as they remain active. However, in most applications enzymes are mixed in a solution with substrates and cannot be economically recovered. This single use is obviously uneconomical when the cost of the enzymes as well as the required downstream processing are considered. Immobilization of the biocatalyst in a biochemical reactor configuration may help to separate them from the reaction mixture for use over multiple cycles. Immobilized enzymes permit an increase in stability and prevent reaction inhibition by products (Giorno and Drioli, 2000). Moreover, in batch reactors, both substrate and enzyme have equal residence time. In contrast, in continuous flow reactor configurations, the average substrate residence time within the reactor is far shorter than that of the immobilized enzyme catalyst. Hence for a fixed amount of enzyme, the continuous flow reactor has a better productivity; assuming that that all substrate that come in contact with the immobilized enzyme are converted (Chaplin and Bucke, 1990).

Numerous methods exist for enzyme immobilization to create bio-hybrid matrices. The principal strategies for enzyme immobilization on membranes are entrapment, adsorption, chemical grafting or molecular recognition (Fig. 4.1). Chemical grafting has so far been the best when compared to entrapment or adsorption in terms of better immobilization stability. It is mainly based on ionic binding, cross-linking or covalent linking. The sites involved in this chemistry are generally carboxylic acid, hydroxyls, amino or quaternary ammonium groups, which are created on the membrane by various methods such as direct chemical surface modification or plasma or UV activation. The reactive sites thus created allow the immobilization of the enzyme by using coupling reagents such as glutaraldehyde.

4.1.3   *Challenges of enzyme immobilization direct onto membrane*

Although the immobilization of enzymes onto the membrane generally enhances their stability, it is also mostly accompanied by a significant loss in enzyme activity relative to the free enzyme activity (Jochems *et al.*, 2011). This may be due to the active site being blocked by substrate accessibility, occurrence of multiple-point binding, or enzyme denaturation. This inverse relationship has been widely observed in the literature. However, Giorno *et al.* (2007) and Mazzei *et al.* (2009) demonstrated that this is not a general rule. Indeed, by properly governing the microenvironment and transport properties, immobilized enzymes (lipase and β-glucosidase) can have improved stability as well as native catalytic activity and selectivity. The amount of immobilized enzyme is an important parameter and strongly affects the reactor performance. A highly cross-linked enzyme may eventually reduce the intrinsic membrane permeability.

Another very critical limitation of direct immobilization onto the membrane is the limited membrane life cycle due to the enzyme activity. For example, when a covalently linked enzyme is denatured, removing the denatured enzyme to reuse the membrane is difficult in practice. In addition to deactivation, the enzyme may get covered with excess substrate, particularly when mass transfer rate is faster than reaction rate, which is the case for most biochemical membrane reactors. Thus, even in the absence of a loss in apparent enzyme activity, there is a need to apply membrane cleaning to remove the over-accumulated substrates without affecting the enzyme activity. This oversaturated membrane first of all causes reduced membrane water productivity. At the same time, the diffusion resistance encountered by the product molecules can sometimes cause the product to accumulate near the immobilized enzyme to an undesirably high level, leading to product inhibition. Moreover, cleaning may pose another difficulty, since enzymes are highly sensitive to the change in the microenvironment that may be induced by the applied cleaning strategy. In addition, because an immobilized enzyme preparation is intended for a prolonged period of operation, there is also a gradual, but noticeable, loss in the apparent enzyme activity. This will eventually raise the demand for fresh enzyme replacement or the need for the replacement of the entire immobilized enzyme layer.

Overall, the direct integration of enzymes into the membrane matrix limits the possibility to replace a denaturized enzyme by a fresh one or to clean a membrane that is oversaturated with substrate, forcing premature disposal of the membrane module.

4.1.4   *Alternative enzyme immobilization technique*

Alternatively, large sized retainable carrier particles are available to perform important biocatalysis in MBRs. These include alginate beads, silica particles, diatomaceous earth or surface functionalized biopolymers. These particles, once activated with enzyme, can be deposited on the membrane or suspended with the bulk stream to affect the membrane biocatalysis. They give a better degree of freedom for reutilization compared to the direct integration of enzyme onto membrane. But the problem with such carrier particles is, firstly, that the retention of the biocatalytic particles on the membrane is based on size exclusion. This will imply that one has to always take the molecular weight cut-off of the membrane into consideration in order to select the enzyme carrier or vice versa. However, the main drawback to the extensive use of these particles as industrial biocatalysts is that they exhibit similar particle sizes to other retainable components of the feed stream by the membrane. As a result, separation based on size exclusion from the rest of the mixture for reutilizing the immobilized enzyme is a great challenge. Therefore, the successful application of biochemical membrane reactors on a larger scale should overcome all the aforementioned limitations.

Below is a summary of immobilized enzyme MBR challenges:

- Loss of enzyme activity in continuous reaction.
- Over accumulation of substrates and difficulty in membrane cleaning without degrading the enzyme activity.
- Hindered transport of reaction products due to fouling.
- Design of a reactor for exothermic reactions.
- The problem of separating immobilized enzymes from unreacted substrate and other retained components (when using enzymes immobilized on particles of the same size as retained components).
- Decrease in apparent activity of immobilized enzymes and ease of removing deactivated immobilized enzymes.

4.2   PROGRESS TOWARDS STIMULUS-RESPONSIVE ENZYME IMMOBILIZATION TECHNIQUES

To overcome these major hurdles, one could emphasize the reversible enzyme immobilization or formation of dynamic biocatalytic layers. The next generation enzyme immobilization techniques may thus focus on the formation of bio-nanocomposite layers on the membrane by inducing various smart responses to direct alterations in the membranes environment, which provides ease of reversible enzyme immobilization without sacrificing performances. The different environmental changes include:

- pH/ionic strength response.
- Temperature response.
- Photo irradiation response.
- Magnetic/electric field response.

4.2.1   *Temperature, pH or ionic strength signal-responsive enzyme immobilization*

Temperature sensitive polymers are based on polymer-water interactions, particularly specific hydrophobic-hydrophilic balancing effects, and the configuration of side groups (Wandera *et al.*, 2010). For example, Zhu *et al.* (2013) developed a temperature responsive smart enzyme-Pluronic polymer nanoconjugate that can be readily dissolved in organic solvents for homogenous catalysis at 40°C with enhanced apparent catalytic activities. Temperature-induced precipitation at 4°C was later employed to recover the soluble enzyme-polymer nanoconjugates from the reaction mixture.

pH-sensitive enzyme immobilization can be achieved by the immersion precipitation of graft copolymers with pH-sensitive side chains or layer-by-layer assembly of side groups and enzymes (Wandera *et al.*, 2010). Enzymes are multi-charged molecules, hence, multilayer assemblies of polyelectrolytes and enzymes through electrostatic interaction of positive and negative charges represents the simplest way to perform reversible enzyme immobilization within the membrane pore domain. The simplicity and versatility of this method after immobilizing glucose oxidase and alkaline phosphatase was tested in terms of loading efficiency, biocatalytic effect, storage stability and ease of enzyme/membrane regeneration (Smuleac *et al.*, 2006). Avidin-biotin affinity interaction tests indicated that minor conformational changes occurred due to protein-polyelectrolyte chain interactions, which otherwise is a major challenge during enzyme immobilization such as by covalent binding. Moreover, 72% of the enzyme was recovered; hence the membrane was regenerated after it was washed with deionized water containing 0.5 M NaCl. The overall effectiveness of this method depends on (i) the overall charge on the protein and polyelectrolyte multilayers and the nature of the polyelectrolytes (weak or strong), chain length, (ii) use of suitable membrane material, pore size and (iii) operational parameters (transmembrane flux, pH, temperature, ionic strength).

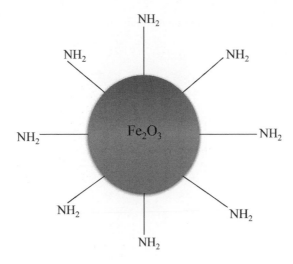

Figure 4.2.   Ferric based magnetic nanoparticles with functional side groups.

For example, Smuleac *et al.* (2006) reported that the amount of immobilized enzyme on an oppositely charged membrane under convective flow was 25 times higher than the amount obtained when both the membrane and protein exhibited similar charges.

Therefore, these classes of either temperature or pH responsive enzyme immobilization could represent a pertinent solution in the future strategy of reversible enzyme immobilization.

4.2.2   *Leading role of nanosized magnetic responsive carriers for the enzyme immobilization*

Superparamagnetic nanoparticles (NP^{SP}) are most often superparamagnetic iron oxide Fe_3O_4 or γ-Fe_2O_3, or "soft" metallic iron and "hard" magnetic materials, e.g. Co, Ni, FeN, FePt, FePd, etc. (Thevenot *et al.*, 2013). Since they have zero memory of their magnetic property in the absence of an external magnetic field, they can be well dispersed in a reaction mixture (Yeon *et al.*, 2009). Coating with polymers is an approach aimed at creating biocompatible environments for these nanoparticles (Miguel-Sancho *et al.*, 2011). In addition, introducing surface functional groups as shown in Figure 4.2 to assist the chemical immobilization of enzymes on the surface of these nanosized particles is a well demonstrated science (Brullot *et al.*, 2012; Miguel-Sancho *et al.*, 2012; Xiao-Ming and Wainer, 1993).

The resulting materials represent an important tool for biotechnology because they can be used in a large variety of processes, e.g. the food industry, waste treatment, production of chemicals, drug delivery, cell transplantation or cell immobilization. In particular, the biocompatibility of magnetic nanoparticles and the high surface to volume ratio that they can provide makes them interesting hosts for enzymes. Enzymes can be easily incorporated into or grafted on to ferric based magnetic nanoparticles to form biohybrids. These biohybrids are indeed bionanocomposites that hold versatile interesting properties like mechanical, optical, electrical, ionic, bio-sensors, and catalyst properties. These properties can also be modified *via* control of their microstructure. Since the first introduction of enzyme immobilization on iron oxide particles support in the 1970s (Robinson *et al.*, 1973), NP^{SP}s have gained numerous applications as magnetically separable high-performance supporters for biocatalysis when they are dispersed in a reaction mixture (Lee *et al.*, 2008). Compared to micro-scale supports, nanobiocatalysts could achieve higher enzyme loading capacity and better mass transfer efficiency.

According to Figure 4.3a–b, enzyme encapsulation on paramagnetic beads, though started as early as the 1970s, has seen an exponential growth in the last ten years. Especially in the last three years (2011–2014), there has been a threefold increase in the overall research activities, as shown by both the number of publications and citations compared to those seen in 2007.

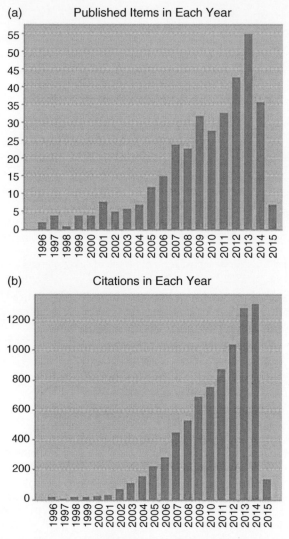

Figure 4.3.   Protein immobilization on magnetic particles: trend over the last twenty years: (a) published articles in English and (b) number of citation for these articles. Source: http://apps.webofknowledge.com.

The highest number of publications so far occurred in 2013 (Fig. 4.3a). As shown in Figure 4.4, 80% of the enzymatically activated magnetic particles were employed in different biocatalysis, e.g. lignoceluloisic biodegradation, while 13% were in immunoassay. By far the largest number of the investigations utilized the enzyme lipase for various applications, followed by laccase. In the following section, details about the potential applications of bionanocomposites in membrane biocatalysis for water treatment will be illustrated taking different case studies.

4.2.2.1   *Micro-structured membrane reactors with nano-designed biocatalysis*
4.2.2.1.1   *Magnetic carrier with immobilized enzyme for effective biofouling control
in membrane bioreactor*
In MBR, membrane biofouling arising from biofilm growth and subsequent membrane permeability loss is the main issue for the widespread application of MBRs in wastewater treatment.

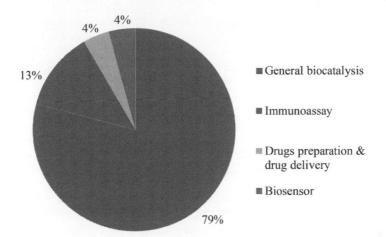

Figure 4.4. Application based categorization of enzyme immobilization on ferromagnetic particles based on publications registered in the year 2013. Source: http://apps.webofknowledge.com.

Bacteria regulates their group behaviors, such as biofilm formation using signal molecules called quorum sensing. Inactivation of N-acyl homoserine lactone, autoinducer of Gram-negative bacteria through enzymatic quenching using enzyme acylase, have been demonstrated as an effective intrinsic biofouling reduction mechanism. However, the use of free acylase in continuously operated MBR is not feasible due to: (i) enzyme activity loss as a result of microorganisms and other contaminant materials in the bioreactor, (ii) leaching of free enzymes through the membrane pores into the permeate and (iii) enzymatic loss along with excess sludge withdrawn from the bioreactor.

To avert this problem, an interesting work by Yeon *et al*. (2009) reported on the use of magnetic particles to immobilize the quorum quenching acylase. Magnetic ion exchange resin with a net positive charge made from γ-Fe_2O_3, divinyl benzene and glycidal methacrylate was used as a magnetic core. Prior to enzyme immobilization, the magnetic core was coated with anionic polyelectrolyte (polystyrene sulfonate) and cationic polyelectrolyte (chitosan) through layer-by-layer assembly. Subsequently, acylase was immobilized on the aminated magnetic particle, either by physical adsorption or cross-linking using glutaraldehyde.

In a batch MBR, the biocatalytic magnetic particles were added to a flask containing the wastewater and the activated sludge made from ten bacterial consortiums, with a working volume of 150 mL. By placing the reservoir flask on a magnetic stirrer, it was possible to retain the biocatalytic particles within the reservoir. Hence it was possible to avoid membrane fouling control that might arise from the shearing effect of the biocatalytic particles.

Alternatively, the magnetic responsiveness of the biocatalytic particles was used to recover and recycle the enzyme back to the reactor from the sludge withdrawn every day in order to adjust the sludge retention time to 50 days in a lab-scale MBR. In this configuration, a Polyvinylidene fluoride PVDF hollow fiber membrane was immersed into a 1 L reservoir containing the feed wastewater, activated sludge and the biocatalytic particles for a long-term continuous operation. The acylase activated magnetic particles were retained within the bioreactor due to their larger particle size as compared to the membrane pore size. Downstream of the MBR, a permanent magnet installed along the sludge disposal line was able to recover the biocatalytic magnetic particles withdrawn along with the excess sludge, and returned it back to the MBR by a reversed pumping flow.

The biofouling controlling efficiency of the developed system was evaluated by following the transmembrane pressure (*TMP*) change over time as well as analyzing the permeate quality. Results were compared with an identical parallel control experiment, except with no addition of the biocatalytic magnetic particles.

In the control MBR, the *TMP* surpassed 30 kPa after 48 h of operation in both the first and second cycle operations, resulting in a premature membrane replacement. On the contrary, in the MBR integrated with magnetic biocatalytic particles, the initial 10 kPa *TMP* was maintained well in both the first and second cycle operations. Biofilm mitigation by the acylase activated magnetic particles in the MBR further helped in elongating the backwashing or relaxation interval. From analysis of the permeate quality, the chemical oxygen demand COD of the control and biocatalytic MBR were 19–22 mg mL^{-1} and 6–19 mg mL^{-1}, respectively, meaning the presence of biocatalytic magnetic particles in the MBR did not alter or deteriorate the organic degrading capacity of the bacterial community in the MBR. It is worth noting that the application of the magnetic biocatalytic particles in the MBR retarded production of extracellular polymeric substances (EPS) in the biocake by about nine factors (polysaccharide from 84 to 10 mg g^{-1} biocake; protein from 114 to 15 mg g^{-1} biocake).

Overall, the developed MBR that integrated acylase activated magnetic particles benefited from stable enzyme immobilization, easy recycling and multiple use due to easy magnetic isolation, no enzyme loss during sludge withdrawal, well prevented membrane biofouling due to the reduction of soluble microbial products (SMP) in the mixed liquor and extracellular polymeric substances (EPS) in the membrane biocake by the immobilized acylase.

4.2.2.1.2   *Magnetic-stimulus-responsive layer for enhanced membrane biocatalysis*
Nano-designed membranes can be fabricated by blending engineered nanoparticles with polymeric or inorganic membranes or by assembling them into porous membranes (Kim and Van der Bruggen, 2010). Hybridization of polymeric membranes with inorganic fillers in a mixed matrix membrane exhibits a positive mutual influence on both phases. The inclusion of metal oxide nanoparticles, for example, induced improved permselectivity, fouling resistance, improved permeates quality, inactivation of viruses, and bacteria. In particular, mitigation of membrane fouling due to the catalytic properties of some metal oxide nanoparticles has provided nano-designed membranes with a built-in organic foulants oxidative property.

Producing smart magnetic responsive materials by mixing polymeric materials with superparamagnetic nanoparticles (NP^{SP}) has also become a more interesting and efficient strategy (Daraei *et al.*, 2013). Indeed, the magnetic moment of those membranes containing a small amount of the NP^{SP} is much larger than those of molecular magnets (Thevenot *et al.*, 2013). This allows the membrane to respond to the weak stimuli of the permanent or alternating magnetic field. The incorporation of NP^{SP} in the polymers offers not only new possibilities but also the chance for the improvement of already established systems with respect to better control (Huang *et al.*, 2012).

4.2.2.1.3   *Magnetic-stimulus-responsive layer for enhanced membrane biocatalysis:*
case study direct attachment of biocatalytic film on electrode bearing
a cylinder magnet without O-ring
The earliest paper ever treated in this chapter, by Calvot *et al.* (1975), deals with the use of a magnetic enzyme membrane that contains enzyme decarboxylases and magnetic particles as the active part of the specific protein sensors. 5 mg of lysine decarboxylase enzyme molecules are cross-linked with 2.5 mL of a solution containing 6% plasma albumin (inert protein) using 0.7% glutaraldehyde as a cross-linker. 3 g of commercial magnetic iron oxide particles are embedded in the membrane structure. The mixture was casted on flat glass plate in order to obtain a membrane of homogeneous thickness. Subsequent to 2 h of cross-linking, the plate with the magnetic protein layer was immersed in a coagulation bath containing distilled water. Integration of the magnetic beads allowed direct attachment of the film on a pC0 electrode bearing a cylinder magnet without O-ring.

4.2.2.1.4   *Magnetic-stimulus-responsive layer for enhanced membrane biocatalysis:*
case study of pectin and lignoceluloisic (wheat arabinoxylan) hydrolysis
An alternative to the direct integration of the enzyme and magnetic particles within the polymer matrix as shown in the previous section, stimulus-responsive programmable layers on membranes can be formed *via* attraction through reversible physical forces. In this case, magnetic guidance

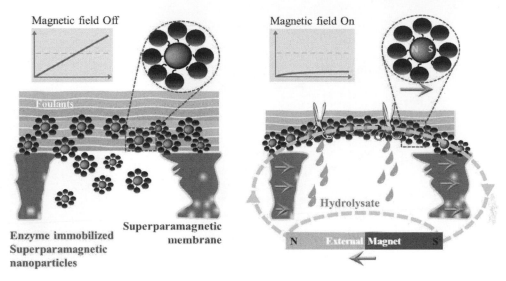

Figure 4.5. Working principle of magnetic responsive biocatalytic system for *in-situ* foulant degradation. Reprinted from Gebreyohannes *et al.* (2015a).

of the bionanocomposites and the hybrid membranes, provided by an external magnetic field, may offer the key strategy. For example, in an attempt to fill the gap of permanent loss of both enzyme and membrane in the cases of enzyme deactivation and membrane oversaturation during direct enzyme immobilization on the membrane, one could envisage a spatial and temporal control of physical immobilization of the enzyme triggered by reversible magnetic forces. More particularly, the use of stimulus-responsive particles may hold a great potential to enable the dispersal and recovery of biocatalysts used in facilitating reaction occurring at the membrane-solution interface. Tuning the reversible immobilization of enzyme on membrane, using the concept of superparamagnetism, can come with a number of benefits:

- It may help in the formation of a dynamic layer that may shield the membrane from direct contact with pollutants in the feed.
- Temporary pore size reduction of microporous membranes.
- Activation of the monolayer with biocatalyst to achieve biocatalysis at the membrane-solution interface.

In a recent work, nanoscale tuning of enzyme immobilization on the surface of a polymeric membrane to enhance surface biocatalysis in a magnetic responsive biocatalytic membrane reactor has been demonstrated (Gebreyohannes *et al.*, 2015a). The novelty of the process lies in the use of NP^{SP}, both as enzyme carriers to form bionanocomposites and as nanofillers to form organic/inorganic (O/I) magnetic responsive hybrid membranes. The embedded NP^{SP} acted as a magnetic field actuator when an external magnetic field was applied.

This combination triggered magnetically guided reversible enzyme immobilization on the surface of the membrane. Hence it has potentially mimicked the micro-nano architecture of the covalent immobilization of the enzyme on the membrane (Fig. 4.5).

This multidisciplinary approach gave an efficient performance in *in-situ* degradation of simulated wastewater rich in pectin, one of the most important foulants in food based wastewater membrane filtration. By depositing 1 to $2 \, \mathrm{g \, m^{-2}}$ of pectinase activated magnetic nanoparticles with an average diameter of 8 nm, between a 40 to 100% reduction in filtration resistance during microfiltration of $0.3 \, \mathrm{g \, L^{-1}}$ pectin solution at $17 \, \mathrm{L \, m^{-2} \, h^{-1}}$ constant flux has been observed (Fig. 4.6a). An excellent performance ($\sim$75% filtration resistance reduction) is also demonstrated

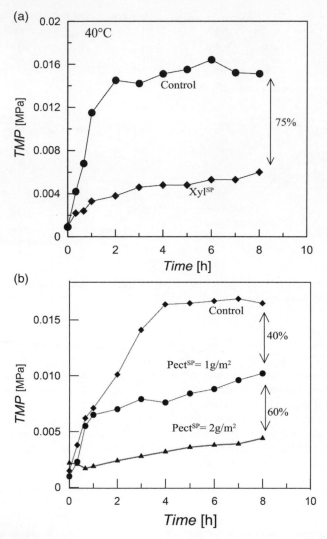

Figure 4.6.   (a) Effect of the presence of a dynamic layer of xylanase activated magnetic nanoparticles (Xyl^{SP}) on *TMP* during the filtration of $0.3\,g\,L^{-1}$ wheat arabinoxylan at a constant flux of $17\,L\,m^{-2}\,h^{-1}$ compared to a system containing the same amount of dynamic layer of neutral magnetic nanoparticles at 40°C; (b) transmembrane pressure evolution during the filtration of $0.3\,g\,L^{-1}$ pectin through a layer of $1\,g\,m^{-2}$ neutral NP^{SP} layer as a control and a BMR^{SP} containing $1\,g\,m^{-2}$ or $2\,g\,m^{-2}$ Pect^{SP} layer at 40°C, pH 4.5 and a fixed flux of $17\,L\,m^{-2}\,h^{-1}$. Reprinted from Gebreyohannes *et al.* (2015a).

in the hydrolysis of arabinoxylan (Fig. 4.6b), a typical polysaccharide in streams used for bio-ethanol production or present in brewery and bakery wastewaters. Moreover, when forming a dynamic layer of both pectinase and xylanase activated magnetic nanoparticles, it was possible to intensify the hydrolysis of multi-stage bioreactor requiring substrate in a single stage (Fig. 4.7).

The developed system has also become a suitable platform for multiple uses of both enzyme and membrane. Most interestingly, authors' demonstrated possible uses of the membrane even after the enzyme had been deactivated, since the system gives the degree of freedom to remove the damaged nano-layer from the surface of the membrane. This has resolved a major missing gap in the traditional biocatalytic membrane reactor that normally would have left the fate of the

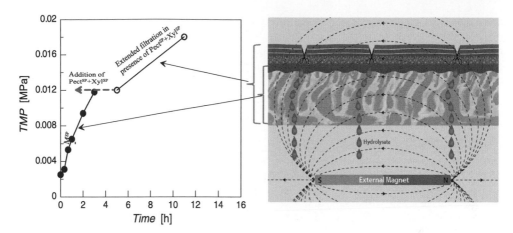

Figure 4.7. Effectiveness of using enzyme blend (Xyl^{SP} and $Pect^{SP}$ mixtures) on degrading mixed pectin and arabinoxylan fouling layers (feed containing $0.3\,mg\,mL^{-1}$ wheat arabinoxylan and $0.3\,mg\,mL^{-1}$ pectin mixed in equal ratio) $17\,L\,m^{-2}\,h^{-1}$ flux at $40°C$: two step filtration, involving a first step on bare M^{SP} (during first 3 h), followed by dispersion of a mixture of Xyl^{SP} and $Pect^{SP}$ on the pre-fouled M^{SP} and subsequent filtration (3–11 h). Reprinted from Gebreyohannes *et al.* (2015a).

biocatalytic membrane module to be disposal, since detaching an immobilized enzyme from the membrane is practically impossible. Apart from membrane fouling control, the system is believed to play an important role in the development of more efficient (micro)-reactors and separation technologies with numerous potential applications that span from environmental remediation to large-scale production of bio-commodities.

4.3 CONCLUDING REMARKS AND FUTURE PERSPECTIVES

Combinatorial biocatalysis and membrane separations with either physically or covalently immobilized enzymes in biocatalytic membrane reactors has become an important avenue in the development of more efficient (micro)reactors and separation technologies. However, the inevitable enzyme denaturation over time and the loss of system performance has significantly limited the industrial applications of immobilized enzymes. As a result, only a minor fraction of the total industrial enzyme market is covered by immobilized enzyme sales.

In recent decades, highly adaptive membranes or membrane processes that can respond well to changes in the micro/macro environment have been developed. Tailoring reversible enzyme immobilization through combinatorial stimulus-responsive enzyme immobilization techniques holds a great potential in the widespread use of MBRs. In this chapter, we have focused on how the different signal-responsive interactions within the membrane or its surroundings can be tuned and monitored in controlled environments in order to modulate reversible enzyme immobilization. These stimuli-responsive membranes have a strong potential for future applications in bio-separations, anti-fouling surfaces and creating self-cleaning or self-refreshing membranes. Switchable membrane surface properties may also open up a new strategy towards *in-situ* process control, online monitoring or the easy process automation of many bio-technological processes. Therefore, stimulus-responsive enzyme immobilization as a whole can be employed to "re-engineer" membrane biocatalysis.

Nano-designed magnetic responsive MBRs have been demonstrated to hold a bigger future prospect for designing the next generation of responsive membranes towards advanced functions. Retrofitting this approach in an existing process would involve the replacement of the current membranes by membranes containing small amounts of magnetic nanoparticles. Inclusion of the magnetic nanoparticles hardly needs the re-designing of established industrial membrane manufacturing processes and will eventually come with the additional benefit of improved membrane permselectivity. This method could also provide new possibilities for consolidating multi-substrate reactions by using a mixture of magnetic nanoparticles activated with different enzymes.

The future innovations on the basis of stimulus-responsive enzyme immobilization will lead to the design of more complex membrane systems that are capable of mimicking nature.

REFERENCES

Abejón, R., De Cazes, M., Belleville, M.P. & Sanchez-Marcano, J. (2015) Large-scale enzymatic membrane reactors for tetracycline degradation in WWTP effluents. *Water Research*, 73, 118–131.

Beyene, A., Yemane, D., Addis, T., Assayie, A.A. & Triest, L. (2013) Experimental evaluation of anaerobic digestion for coffee wastewater treatment and its biomethane recovery potential. *International Journal of Environmental Science and Technology*, 11(7), 1881–1886.

Brullot, W., Reddy, N.K., Wouters, J., Valev, V.K., Goderis, B., Vermant, J. & Verbiest, T. (2012) Versatile ferrofluids based on polyethylene glycol coated iron oxide nanoparticles. *Journal of Magnetism and Magnetic Materials*, 324(11), 1919–1925.

Brunetti, A., Scura, F., Barbieri, G. & Drioli, E. (2010) Membrane technologies for CO_2 separation. *Journal of Membrane Science*, 359(1–2), 115–125.

Calvot, C., Berjonneau, A.-M., Gellf, G. & Thomas, D. (1975) Magnetic enzyme membranes as active elements of electrochemical sensors: Specific amino acid enzyme electrodes. *FEBS Letters*, 59(2), 258–262.

Cassano, A., Drioli, E., Galaverna, G., Marchelli, R., Di Silvestro, G. & Cagnasso, P. (2003) Clarification and concentration of citrus and carrot juices by integrated membrane processes. *Journal of Food Engineering*, 57(2), 153–163.

Chaplin, M.F. & Bucke, C. (1990) *Enzyme Technology*. Cambridge University Press, Cambridge, UK.

Charpentier, J.C. (2005) Process intensification by miniaturization. *Chemical Engineering & Technology*, 28(3), 255–258.

Daraei, P., Madaeni, S.S., Ghaemi, N., Khadivi, M.A., Astinchap, B. & Moradian, R. (2013) Fouling resistant mixed matrix polyethersulfone membranes blended with magnetic nanoparticles: Study of magnetic field induced casting. *Separation and Purification Technology*, 109(0), 111–121.

Dewan, S.S. (2014) Global markets for enzymes in industrial applications. Report, BCC Research, BIO030H.

Drioli, E. & Giorno, L. (eds) (2009) *Membrane Operations: Innovative Separations and Transformations*. Wiley-VCH, Weinheim, Germany.

Drioli, E. & Romano, M. (2001) Progress and new perspectives on integrated membrane operations for sustainable industrial growth. *Industrial & Engineering Chemistry Research*, 40(5), 1277–1300.

Gebreyohannes, A.Y., Mazzei, R., Curcio, E., Poerio, T., Drioli, E. & Giorno, L. (2013) Study on the in situ enzymatic self-cleansing of microfiltration membrane for valorization of olive mill wastewater. *Industrial & Engineering Chemistry Research*, 52(31), 10,396–10,405.

Gebreyohannes, A.Y., Bilad, M.R., Verbiest, T., Courtin, C.M., Dornez, E., Giorno, L., Curcio, E. & Vankelecom, I.F.J. (2015a) Nanoscale tuning of enzyme localization for enhanced reactor performance in a novel magnetic-responsive biocatalytic membrane reactor. *Journal of Membrane Science*, 487(0), 209–220.

Gebreyohannes, A.Y., Curcio, E., Poerio, T., Mazzei, R., Di Profio, G., Drioli, E. & Giorno, L. (2015b) Treatment of olive mill wastewater by forward osmosis. *Separation and Purification Technology*, 147(0), 292–302.

Giorno, L. & Drioli, E. (2000) Biocatalytic membrane reactors: applications and perspectives. *Trends in Biotechnology*, 18(8), 339–349.

Giorno, L., D'Amore, E., Mazzei, R., Piacentini, E., Zhang, J., Drioli, E., Cassano, R. & Picci, N. (2007) An innovative approach to improve the performance of a two separate phase enzyme membrane reactor by immobilizing lipase in presence of emulsion. *Journal of Membrane Science*, 295(1–2), 95–101.

Gugliuzza, A. & Drioli, E. (2014) Membranes for ultra-smart textiles. In: Gugliuzza, A. (ed) *Smart Membranes and Sensors: Synthesis, Characterization, and Applications*. John Wiley & Sons, Inc. Hoboken, NJ, and Scrivener Publishing LLC, Salem, MA. pp. 401–419.

Huang, Z.-Q., Zheng, F., Zhang, Z., Xu, H.-T. & Zhou, K.-M. (2012) The performance of the PVDF-Fe$_3$O$_4$ ultrafiltration membrane and the effect of a parallel magnetic field used during the membrane formation. *Desalination*, 292, 64–72.

Jochems, P., Satyawali, Y., Diels, L. & Dejonghe, W. (2011) Enzyme immobilization on/in polymeric membranes: status, challenges and perspectives in biocatalytic membrane reactors (BMRs). *Green Chemistry*, 13(7), 1609–1623.

Judd, S. & Judd, C. (eds) (2006) *The MBR Book: Principles and Applications of Membrane Bioreactors in Water and Wastewater Treatment*. Elsevier Science.

Kim, J. & Van der Bruggen, B. (2010) The use of nanoparticles in polymeric and ceramic membrane structures: review of manufacturing procedures and performance improvement for water treatment. *Environmental Pollution*, 158(7), 2335–2349.

Lee, J., Lee, Y., Youn, J.K., Na, H.B., Yu, T., Kim, H., Lee, S.-M., Koo, Y.-M., Kwak, J.H., Park, H.G., Chang, H.N., Hwang, M., Park, J.-G., Kim, J. & Hyeon, T. (2008) Simple synthesis of functionalized superparamagnetic sagnetite/silica core/shell nanoparticles and their application as magnetically separable high-performance biocatalysts. *Small*, 4(1), 143–152.

Machsun, A., Gozan, M., Nasikin, M., Setyahadi, S. & Yoo, Y. (2010) Membrane microreactor in biocatalytic transesterification of triolein for biodiesel production. *Biotechnology and Bioprocess Engineering*, 15(6), 911–916.

Mazzei, R., Giorno, L., Piacentini, E., Mazzuca, S. & Drioli, E. (2009) Kinetic study of a biocatalytic membrane reactor containing immobilized β-glucosidase for the hydrolysis of oleuropein. *Journal of Membrane Science*, 339(1–2), 215–223.

Mazzei, R., Drioli, E. & Giorno, L. (2010) Biocatalytic membranes and membrane bioreactors. In: Enrico, E. & Lidietta, G. (eds) *Comprehensive Membrane Science and Engineering*. Elsevier, Oxford. pp. 195–212.

Miguel-Sancho, N., Bomatí-Miguel, O., Colom, G., Salvador, J.P., Marco, M.P. & Santamaría, J. (2011) Development of stable, water-dispersible, and biofunctionalizable superparamagnetic iron oxide nanoparticles. *Chemistry of Materials*, 23(11), 2795–2802.

Miguel-Sancho, N., Bomati-Miguel, O., Roca, A.G., Martinez, G., Arruebo, M. & Santamaria, J. (2012) Synthesis of magnetic nanocrystals by thermal decomposition in glycol media: effect of process variables and mechanistic study. *Industrial & Engineering Chemistry Research*, 51(25), 8348–8357.

Robinson, P.J., Dunnill, P. & Lilly, M.D. (1973) The properties of magnetic supports in relation to immobilized enzyme reactors. *Biotechnology and Bioengineering*, 15(3), 603–606.

Singh, R. (2005) Hybrid membrane systems – applications and case studies. In: Singh, R. (ed) *Hybrid Membrane Systems for Water Purification: Technology, Systems Design and Operations*. Elsevier Science. pp. 131–196.

Smuleac, V., Butterfield, D.A. & Bhattacharyya, D. (2006) Layer-by-layer-assembled microfiltration membranes for biomolecule immobilization and enzymatic catalysis. *Langmuir*, 22(24), 10,118–10,124.

Thevenot, J., Oliveira, H., Sandre, O. & Lecommandoux, S. (2013) Magnetic responsive polymer composite materials. *Chemical Society Reviews*, 42(17), 7099–7116.

Wandera, D., Wickramasinghe, S.R. & Husson, S.M. (2010) Stimuli-responsive membranes. *Journal of Membrane Science*, 357(1–2), 6–35.

Xiao-Ming, Z. & Wainer, I.W. (1993) On-line determination of lipase activity and enantioselectivity using an immobilized enzyme reactor coupled to a chiral stationary phase. *Tetrahedron Letters*, 34(30), 4731–4734.

Yeon, K.M., Lee, C.H. & Kim, J. (2009) Magnetic enzyme carrier for effective biofouling control in the membrane bioreactor based on enzymatic quorum quenching. *Environmental Science and Technology*, 43(19), 7403–7409.

Zhu, J., Zhang, Y., Lu, D., Zare, R.N., Ge, J. & Liu, Z. (2013) Temperature-responsive enzyme-polymer nanoconjugates with enhanced catalytic activities in organic media. *Chemical Communications*, 49(54), 6090–6092.

CHAPTER 5

Study of carbon nanotubes' embedment into porous polymeric membranes for wastewater treatment

John A. Anastasopoulos, Amaia Soto Beobide, Theodoros Karachalios,
Katerina Kouravelou & George A. Voyiatzis

5.1 INTRODUCTION

Over-exploitation of the Earth's finite fresh water resources has already impacted both ecosystems and global water reserves. The United Nations estimates that human consumption of water increased sixfold in the 20th century, while the population only tripled. Under those conditions, governments and industry worldwide have recognized that improving the water infrastructure is critical (Calvert Investments, 2013). Water contaminated by industry and agriculture with heavy metal ions, pesticides, organic compounds, endocrine disruptive compounds, and nutrients (phosphates, nitrates, nitrites) has to be efficiently treated to protect against human intoxication. Furthermore, incidental sludge from industrial wastewater treatment facilities is commonly highly contaminated with toxic compounds. Hence, the need of advanced technologies for water treatment results is huge.

Membrane bioreactor (MBR) technology is regarded as a key element of advanced wastewater reclamation and reuse schemes, and can considerably contribute to sustainable water management. MBRs, which were introduced in order to aim at the coupling of membrane separation properties with a biochemical reaction, have been implemented for wastewater treatment and reuse in municipal, agricultural and a variety of industrial sectors in Europe, the Middle East and in North African countries.

Polymeric materials are extensively used within the MBR systems as active membranes and this is justified since there is a high availability of different chemistries combined to low cost (Andre de Sousa, 2009). The pore diameters of the commonly used microfiltration (MF) or ultrafiltration (UF) membranes is in the range between 0.01 to 0.1 μm, so that particulates and bacteria can be kept out of permeate, and the membrane system replaces the traditional gravity sedimentation unit in the biological sludge process.

To date, MBR systems combined with nanofiltration (NF) membranes are used typically in ground and surface water to produce drinking water, for reverse osmosis (RO) pre-treatment and wastewater treatment (Fig. 5.1) (Hillal et al., 2004). However, this combination has several drawbacks, as schematized in Figure 5.2.

The two-step process requires additional technologies resulting in additional costs and maintenance efforts in order to achieve high quality recycling water; relatively high capital and operating costs due to requiring a high-pressure pump, since NF operating at over 0.10 MPa results in energy consumption; recycling and/or treatment of toxic NF concentrate is laborious and expensive; the efficiency within MBR-NF depends on the composition of the under-treatment water, substances creating stress for bacteria reduce the organic compound filtering ability, while the building up of a filter cake at the MBR and/or NF/RO membranes furthermore reduces the overall performance of the system (Jiang et al., 2003); and lastly, the high concentration of salts may decrease the biodegradation of the sludge resulting in toxic disposal.

With an increased tendency to fouling, the MBR filtration performance inevitably decreases with filtration time. Membrane fouling is the most serious problem affecting system performance,

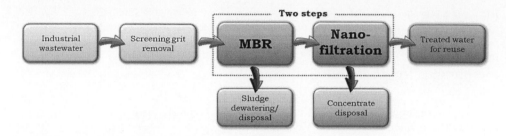

Figure 5.1.   Conventional membrane bioreactor process combined to nanofiltration.

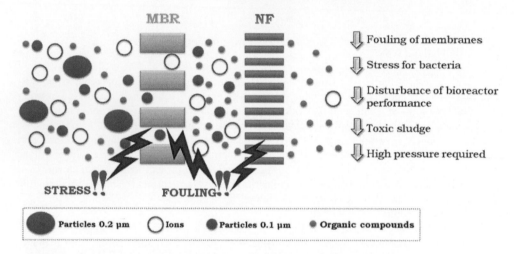

Figure 5.2.   2-step treatment process and threats of conventional MBR-NF systems.

as it leads to a significant increase in hydraulic resistance, resulting in a decline of permeates flux. Typical flux rates range from around $20\,\mathrm{L\,m^{-2}\,h^{-1}}$. Also, an increase of transmembrane pressure (*TMP*) occurs under constant flux conditions.

As an overall result, the commonly employed MBR-NF systems have a high operating efficiency with respect to cost and quality of treatment of wastewater containing high biodegradable organic compounds. However, they are inefficient with respect to water containing a high amount of stress-inducing substances, high coloring and low biodegradable organic compounds. Therefore, these systems can only be applied to wastewater with high biodegradable organic pollutants and thus they are considered as only one to one solution, an issue that was addressed by the EC funded project BioNexGen (www.bionexgen.eu).

To this end, a new class of functional low fouling membranes for MBR technology, while increasing both water permeability/flux and the rejection rate of low-molecular weight organics, was considered to be provided by the inclusion of carbon nanotubes (CNTs) into the pores of polymeric membranes.

5.2   WHY CARBON NANOTUBES?

Given the emphasis on the use of CNT in the membrane technology, it is manifested that well-aligned CNT can serve as competent pores in membranes for water desalination and decontamination applications (Elimelech and Phillip, 2011). The hollow CNT structure provides frictionless transport of water molecules, a feature that makes them suitable for the development of high flux separation systems. The channels of proper pore diameter CNT can constitute mass/volume barriers, rejecting even salt ions and allowing water transport through the nanotube hollows (Corry, 2008).

In addition to CNT-diameter synthesis control, the modification of CNT-cavities in order to selectively reject ions is also feasible (Khin *et al.*, 2012). Thus, CNT-membranes can be used as 'gate keepers' for size controlled separation of multiple pollutants. In addition, they show anti-fouling and self-cleaning properties and may be reusable (Das *et al.*, 2014). Both SW-CNT and MW-CNT have been used for water desalination studies (Dai *et al.*, 2006; Li and Zou, 2011; Nasrabadi and Foroutan, 2011). The CNT employment in advanced membrane technologies, since they provide a low energy solution for efficient water treatment and reuse, proves intriguing and challenging.

5.3   WATER TRANSPORT THROUGH CARBON NANOTUBES

Even though water is the most studied material on earth, and the web of science gives more than half a million articles with the word "water" in the title, its anomalous bulk properties are still surprising and the properties of highly confined water are fascinating. That is why the water flow through CNTs is an area of intense research activity and conflict.

The transport of all molecules, including charged molecules through nanochannels, is affected by the molecular size with respect to the lateral channel dimensions, and also by considerations of molecular entropy. Molecules can be excluded from channels by ion exclusion, by steric hindrance or because of the cost in internal entropy. On the basis of these three factors, nanochannels work as molecular sieves, much like membranes. Effects that happen at the channel walls become increasingly important when decreasing channel height. Typical examples of these effects are the occurrence of electrical double layers, slip and specific adsorption effects.

5.3.1   *Liquid slip*

It is generally agreed that the flow rates of water in CNTs depend strongly on the slip length that corresponds to the distance where the linear extrapolation of the velocity profile reaches zero (the tangential velocity component vanishes). Whereas no-slip represents a situation where the liquid in the first molecular layer is stagnant and all other molecules are sheared past the first molecular layer, the first molecular layer does move in slip flow, though with strong friction with the wall. The lower this friction with the wall, which is achieved, for example, by employing very hydrophobic walls, the less force is needed for a given flow velocity. Therefore, slip is very important in nanofluidics since it drastically reduces the required pressure in pressure-driven flows.

In Figure 5.3, the consequences of slip for the velocity of the liquid are clearly shown. The case on the right ($b = \infty$) is particularly informative, because it represents a situation where the liquid does not experience any friction with the wall and could theoretically be accelerated to the speed of light. Parameters that are known to influence slip in nanochannels are surface roughness and hydrophobicity. The slip length is the additional length from the wall at which the tangent to the fluid velocity at the wall is extrapolated to reach zero relative tangential velocity between fluid and solid. Generally, the slip length is increased for increasing hydrophobicity and decreasing surface roughness when the channel walls are hydrophilic. There is, however, convincing evidence that surface roughness in the case of hydrophobic channels leads to greater slip lengths (several tens of microns as compared to around 20 nm for smooth hydrophobic channels) because of air getting trapped in surface inhomogeneities (Sparreboom *et al.*, 2010).

Numerous experimental and simulation studies have been carried out in order to find the transport properties of water through CNTs. The flow enhancement results differ by 1–5 orders

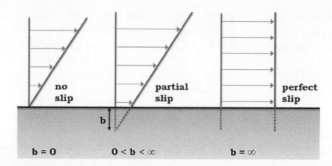

Figure 5.3.   Three cases of slip flow past a stationary surface. The slip length *b* that corresponds to the distance where the linear extrapolation of the velocity profile reaches zero is indicated (Sparreboom *et al.*, 2010).

of magnitude compared to the classical no-slip flow predictions. The slip length L_s and flow enhancement E for Hagen-Poiseuille flow are defined as follows:

$$u_s = u(R) = L_s \left| \left(\frac{\partial u\,(r)}{\partial u} \right)_{r=R} \right| \tag{5.1}$$

$$E = \frac{Q_{\text{slip}}}{Q_{\text{no-slip}}} = \left(1 + \frac{8L_s}{D} \right) \tag{5.2}$$

where u_s is the slip velocity (fluid velocity at the wall), Q_{slip} is the observed flow rate, $Q_{\text{no-slip}}$ is the expected flow rate using the no-slip boundary condition, D is the diameter of the tube and R is the radius of the tube. For a given fluid-solid combination the slip length is a useful property, which is commonly quoted in the nanofluidics literature. Above a certain channel width/tube diameter the slip length is independent of the channel size. Using Equation (5.2) one can estimate the flow enhancement if given the slip length (Kannam *et al.*, 2013).

Studies on CNTs have provided evidence for the flow of water through CNTs. Many studies have reported a higher than expected flow rate, when compared to the predictions of the classical Hagen-Poiseuille (HP) equation, for water flowing through CNT. Hummer *et al.* (2001) had done molecular dynamics computations to study the water conduction through a CNT and they pointed out that the inherent smoothness of the CNT, the tight hydrogen bonding, and the lower chemical potential were responsible for the pulse-like transmission of water through the nanotube. Molecular dynamics simulations performed by Joseph and Aluru (2008) have claimed that this enhancement in flow may be due to the existence of a hydrogen bond depleted area near the tube wall (Babu and Sathian, 2011).

Majumder *et al.* (2005) observed the frictionless movement of water molecules with high velocities from 95 to 430 cm s^{-1} MPa^{-1} through a 7 nm diameter CNT-pore studying the water passage through membranes consisting of an array of aligned CNT. They have claimed that the flow rates are four to five orders of magnitude faster than those of conventional fluid flow of between 0.0015 and 0.0057 cm s^{-1} MPa^{-1}.

However, recently Kannam *et al.* (2013) reviewed and compared all the experimental, computational/simulated and theoretical studies on the slip length of water in CNTs published since 2004; they suggested that there was only a $\sim$3 to 2 orders of magnitude water flow enhancement for CNTs of internal diameters of 1.66–6.5 nm, respectively. They also proposed that as the internal diameter of the tube increases above 10 nm, the flow rate slowly approaches the classical Navier-Stokes prediction with no significant enhancement. In the same context, Walther *et al.* (2013) performed large-scale molecular dynamics simulations, emulating for the first time the micrometer thick CNTs membranes used in the experiments previously mentioned. These simulations have been able to confirm only a 200-fold enhancement of the water flow for pure water and CNTs and have not confirmed the 100,000-fold enhancement experimentally claimed.

Molecular simulation dynamics have also shown that the water conductance of the (7,7) and (8,8) tubes are roughly double and quadruple that of the (6,6) tube respectively, given that the permeability of a single water chain forming in a tube (5,5) is a little under half of the one from a tube (6,6), due to the fact that water chains only form across the narrower pore half of the time (Corry, 2008). Accordingly, it is concluded that CNT type and structure play a critically important role in water transport and permeability through CNT.

Anastasiou *et al.* (2013) reported results from a detailed computer simulation study for the nano-sorption and mobility of four different small molecules, water included, inside smooth SW-CNTs. Especially for the narrowest or (6,6) of the CNTs considered, the results for the molecules of water were further confirmed through an additional Grand Canonical (μVT) Monte Carlo (GCMC) simulation; water molecules were found to form a 1D array (a string) of molecules with H bonds developing only axially. For large enough CNT-diameters (larger than about 13 Å), water diffusivity was found to be higher than the corresponding experimental value in the bulk by about 55%.

5.4   TYPE & FABRICATION OF CNT-MEMBRANES

The successful fusion of nanotechnology and membrane technology has been stated to lead to efficient next generation separation systems. The type and quality of CNT, the filling/host/substrate materials, the processing, and the fabrication methods used for the synthesis of CNT-membranes are the main factors influencing their performances. Currently, three types of CNT-membranes are available:

- Membranes made by arrays of vertically aligned CNT (*VA-CNT-membranes*).
- Mixed matrix CNT-membranes (*MM-CNT-membranes*).
- Membranes with CNT-infiltrated through their porosity (*CNT-infiltrated membranes*).

Illustrations showing the three types of CNT-membranes are depicted in Figure 5.4. Further discussion on the fabrication of the aforementioned CNT-membranes is stated in the following subsections.

5.4.1   *Vertically aligned-CNT-membranes (VA-CNT-membranes)*

The VA-CNT-membranes are synthesized by arranging perpendicular CNT with supportive filler contents between the tubes. These membranes are high molecular sieves with an intercalated filler matrix, such as polymer, between them. The fillers may also be epoxy, silicon nitride and others with no water permeability. The VA-CNT-membrane was first introduced by Hinds *et al.* (2004) with polystyrene as the filling material between CNT. The fabrication procedure was simple but the pore sizes were irregular. The membrane could not retain $Ru(NH_3)_6^{3+}$ ions initially, following H_2O plasma and HCl treatments, respectively. However, functionalization of CNT core with negatively charged carboxylate groups bounded the positively charged $Ru(NH_3)_6^{3+}$ ions. Biotin and streptavidin attachment onto the functionalized CNT-membranes reduced ion transport by a factor 5.5–15. Such functionalized membranes worked as 'gate keeper'-controlled chemical separators, demonstrating the ability to gate molecular transport through CNT cores for potential applications in chemical separations and sensing.

Holt *et al.* (2006) introduced a micro-electromechanical method for synthesizing VA-CNT-membranes with stimulated nano-fluidic functions. The growth of a dense, vertically aligned array of DWCNT- of sub 2-nm nanotube pore diameter on the surface of a silicon chip by catalytic chemical vapor deposition was followed by conformal encapsulation of the nanotubes by a hard, low-pressure chemical vapor deposited silicon nitride (Si_3N_4) matrix. The Si_3N_4 was used as a filler between the nanotube spaces in order to inhibit water flow through the nanotube gaps, and create stress to stimulate water flow through the tubes. The excess Si_3N_4 was removed from both sides of the membrane by ion milling, and the ends of the nanotubes were opened up by reactive ion etching. The membranes remained impermeable to the tested gases and liquids until the very last etching step. That was evidence for the fabrication of crack-free and void-free membranes. A similar procedure was also followed for the construction of MW-CNT-membrane in a preceding work in 2004 by Holt's group (Holt *et al.*, 2004). Finally, they observed increased (>3) folds over other no-slip water flux, and hydrodynamic flow, and enhanced ion selectivity compared to MW-CNT-membranes. The membrane transported $Ru^{2+}(bipyr)_3$ species with sizes up to 1.3 nm but blocked 2 nm Au particles, suggesting that their pore sizes were between 1.3 and 2 nm. The performances of those membranes generated promising results, such as high selectivity to multiple variants, high water fluxing and low energy consumption.

The key findings of the aforementioned works pointed out that the design of the CNT-pore sizes to match with the target molecules allows the molecular sieving effects to improve the membrane selectivity. This advancement was predicted to directly benefit the field of chemical separations and water purification processes.

More recently, Baek *et al.* (2014) fabricated VA-CNT-membranes from a Fe catalyst using a water-assisted thermal chemical vapor deposition onto a silicon (Si) wafer. After nanotube growth, the synthesized VA-CNTs were incorporated into epoxy resin for filling up the vacant areas. The performance of the fabricated membranes was evaluated in terms of various solvents' flux and diverse nanoparticles' rejection compared to commercial UF membranes with a similar pore size. The VA-CNT-membrane appeared to have a water flux approximately three times higher than the UF membrane and water transport approximately 70,000 times faster than conventional no-slip flow, while showing a similar rejection rate to the UF membrane.

5.4.2   *Mixed matrix-CNT-membranes (MM-CNT-membranes)*

The unique properties of CNT have spurred a great deal of interest in their use as nanofillers. Their introduction in polymer matrices represents a promising new direction for the development of composite materials. For separation purposes, most of the recent studies are mainly focused on the preparation of randomly dispersed and distributed CNT in polymer composites. In the majority of these studies, the nanocomposite membranes were prepared through physical solution mixing of CNT in the polymer dope, followed by the phase inversion technique to remove the solvent and allow the formation of the membrane. This fabrication approach poses a significant disadvantage since the randomly dispersed CNT in the polymer matrix mostly hinders the passage of fluid molecules through CNT tube channels. It is worth noting that, in the case of CNT mixed with a polymer matrix, CNT mainly acts as interlayer spacers to form gap abundant channels increasing the free volume of the composite. Therefore, high flux and permeability can only be realized when CNT are set as fast diffusion channels for the transport of fluid molecules (Goh *et al.*, 2014; Sharma *et al.*, 2010).

Nevertheless, a MM-CNT-membrane composed of polymeric blend or other composite materials with CNT (Fig. 5.4b) could be easily fabricated at a reduced cost and was introduced by Zimmerman *et al.* (1997) to overcome the disadvantages of polymeric membranes for gas purification. Generally, this type of membrane has significantly strengthened the water purification ability of the existing membranes in many ways. MM-CNT-membranes could be a cheap solution to overcoming certain hurdles of current conventional separation technologies, namely, the membrane fouling and pollutant precipitation. In addition to that, irregular pore size, deleterious micro pollutants, influent water quality and pH variations decrease membrane capacities. Additionally, fouling creates defects in membrane pores, causing pore blocking and complicating membrane regeneration. In this context, many studies have revealed the benefits of the use of MM-CNT-membranes in water purification. Choi *et al.* (2006) have used MM-CNT-membranes to improve the filtration capacity of UF membranes. It was observed to enhance water permeability, solutes retention and mechanical robustness of the membrane (Yang *et al.*, 2013).

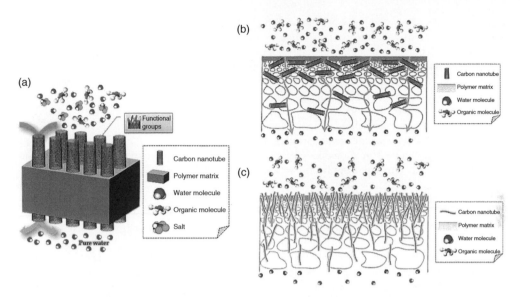

Figure 5.4. Schematic representation of different types of CNT-membranes: (a) membrane made by arrays of vertically aligned CNT; (b) an asymmetric membrane mixed with CNTs and (c) membrane by CNTs infiltrated through the porosity.

5.4.3 *Membranes with infiltrated CNT (infiltrated-CNT-membranes)*

The third approach of a CNT-membrane refers to infiltrated-CNT-membranes. Such a membrane is fabricated by the incorporation of CNT into a porous polymer matrix using filtration methods. Analytically, a polymeric porous membrane, usually microfiltration, is used as the main body of the CNT-membrane, and by filtering a very well dispersed CNT-suspension in organic or aquatic dispersant, using low vacuum during the infiltration process, CNT are embodied in the pores of the membrane active surface (selective layer). The incorporation of CNT alters the internal diameters of the polymeric pores, which are substituted by those of the CNT, either by SW, DW, or MW-CNT. Kim *et al.* (2007) have used polytetrafluoroethylene (PTFE) microfiltration (MF) membranes with a pore size of 0.2 µm as substrates and have studied the filtration of SW-CNT dispersed in THF. The membranes were developed for practical gas separation applications and exhibited high gas flux transport.

In the same context, Madaeni *et al.* (2013) have studied the embedment of MW-CNT of outer diameters 10–30 nm in polyvilylidene fluoride (PVDF) MF membrane with a pore size of 0.22 µm by means of filtering, and subsequently a polydimethylsiloxane (PDMS) coating was applied. The aim of that study was the preparation of a superhydrophobic nanofiltration (NF) membrane using a MF membrane as host substrate. The results revealed that the flux recovery ratio for membrane coated with 5 wt% PDMS was enhanced, showing that the fabricated superhydrophobic NF membrane possessed superior anti-biofouling property.

5.5 FUNCTIONALIZATION OF CARBON NANOTUBES FOR CNT-MEMBRANES

Depending on the intended purpose of a CNT-membrane, often the functionalization of CNT constitutes a precondition especially for membranes applied on water purification technologies. Pristine CNT is usually aggregate, resulting in a significant decrease of water flux and pollutant rejection capacities of the CNT-membranes. Generally, CNT appear contaminated with metal catalysts, impurities and physical heterogeneities (Mauter and Elimelech, 2008). Additionally, CNT are capped into hemisphere-like fullerene-type curvature during synthesis and purification

(Li *et al.*, 2003). These capped CNT get unzipped into open tips after special treatment and may be oxidized into specific functional groups to trap selective pollutants, as already mentioned. Different CNT-functionalization treatments can result in positive ($-NH^{3+}$), negative ($-COO^-$, sulfonic acids) and hydrophobic (aromatic rings) groups on CNT surfaces (Goh *et al.*, 2013; Kar *et al.*, 2012). These special chemical groups may lead to CNT-membranes with selectivity for particular pollutant retention and increased water flux through the nanotube hole. Commonly, functionalized CNT-membranes show good water permeability, mechanical and thermal stability, fouling resistance, pollutant degradation and self-cleaning properties (Qu *et al.*, 2013). CNT can also be decorated with various nanoparticles such as Cu, Ag, Au, Pt, Pd, TiO_2, polymers, and biomolecules (pollutant degradative enzymes, DNA and proteins) which have attractive membrane properties and thus broadened applications in water purification (Van Hooijdonk *et al.*, 2013).

5.6   FROM LAB-SCALE TO LARGE-SCALE POTENTIAL OF CNT-MEMBRANES

Figure 5.5 illustrates the fractionation capacity of RO, NF, UF and MF membranes. It is easily observable that the nano-porous cavities of CNT-membranes are suitable for rejecting micropollutants and ions in liquid phase. The hydrophobic hollow structures encourage frictionless movement of water molecules without the need of any energy-driven force to push water molecules through hollow tubes. The cytotoxic effects of CNT-membranes decrease biofouling and increase membrane life by killing and removing pathogens.

The fabrication process of CNT-membranes and functionalization of CNT are factors influencing the selective rejection of particular pollutants from water mixture (Goh *et al.*, 2013; Mauter and Elimelech, 2008). The dense porous architectures of RO and NF membranes require pressure to force the water molecules to pass through them. UF and MF membranes consume less energy

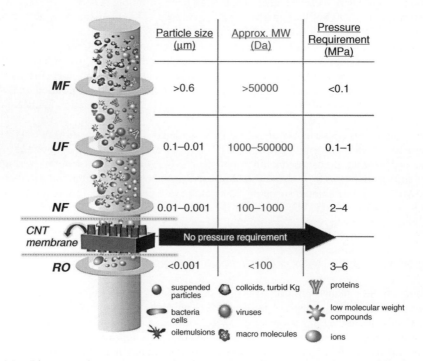

Figure 5.5.   Diagrammatic representation of major membrane filtration methods (www.alting.fr).

but retain only suspended particles. The hydrophobic hollow tubes of CNT provide a strong invitation to polar water molecules.

The nano-scale pore diameters reject salts and retain ions. In addition to that, CNT-membranes can be made highly reusable, less complex, durable, scalable, and eco-friendly without the need of complicated chemical transformation (Das *et al.*, 2014).

5.7 CNT-INFILTRATED COMMERCIAL MEMBRANES

The concept of the BioNexGen project was the fabrication of a new class of functional membranes with enhanced properties regarding permeability, selectivity and fouling resistance by the subsequent embedment of hollow CNTs in the thin selective layer of commercial UF polymeric membranes. In that case, the hollow CNTs would substitute the pores, namely the selective barriers of the selective layer; this is characteristically depicted in Figure 5.6. In an effort to fulfill the BioNexGen project concept many research groups, project partners, have contributed in different ways, i.e. theoretical aspects and computational modeling, commercial membrane supply, production and functionalization of appropriate CNTs, toxicological studies of CNTs, preparation of antimicrobial coatings, and preparation of hydrophilic coatings.

Our main objective was to study the experimental parameters influencing the efficient binding of CNTs in the thin selective layer of the ultrafiltration membrane (with pore diameters of ~40 nm) to transform/render it to a nanofiltration one with pores to be defined exclusively by the hollow CNT-internal diameters. In this way, membrane selectivity could be expanded to the ~200–1000 Da range enabling the rejection of a variety of organic pollutants of industrial wastewaters, presented in Figure 5.7.

5.7.1 *Materials and characterization*

5.7.1.1 *Commercial polymeric membranes*
The commercial polymeric membrane to be modified was supplied by Microdyn-Nadir GmbH. It was a hydrophilic, flat sheet ultrafiltration membrane, made of polyethersulfone (PES), manufactured by the phase separation process, on a non-woven polyethylene terephthalate (PET) substrate

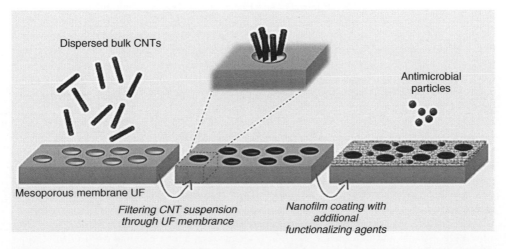

Figure 5.6.   Schematic representation of the BioNexGen project concept (www.bionexgen.eu). A suspension of dispersed CNTs is filtered through a UF membrane with pore diameters ~40 nm. CNTs are encapsulated in the pores of the thin selective layer of the membrane and finally, an antimicrobial coating is applied to stabilize CNTs.

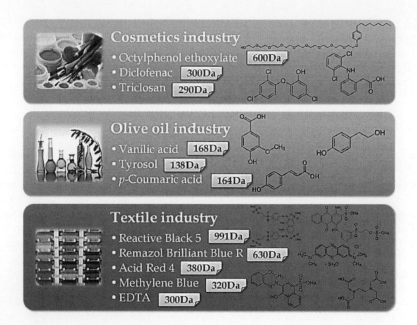

Figure 5.7.   Organic compounds to be rejected from CNT-membranes of industrial wastewaters.

for enhanced mechanical strength. The polymers usually used for membranes in MBR application are hydrophobic, which makes these membranes susceptible to fouling in the bioreactor liquors they are filtering. This normally necessitates either modification of the base polymers or further modification of the membranes to produce a hydrophilic surface (Ansorge *et al.*, 2014; Ohlrogge and Ebert, 2006). To receive an initially hydrophilic membrane, hydrophilic additives can be added to the polymer solution, or the base polymer can be modified. An example for the modification is sulfonation of polysulfone. The other possibility, in order to get a more hydrophilic surface, is the additional modification by chemical oxidation, organic chemical reaction, plasma treatment or grafting methods (Judd, 2006; Ohlrogge and Ebert, 2006). These techniques feature the possibility to functionalize the membrane surface to meet the requirements for special applications, but they always imply an additional manufacturing step that increases the manufacturing costs.

The membrane porous structure was conserved with the pores filled up with a filling agent; as such, in order to treat the membrane, it was rinsed, however, this needed to be done with caution because the membrane should not be dried. Therefore, a certain process consisting of the following steps was being followed for the rinsing of glycerin, in order to possess an opened-pore functional membrane:

- Submersion of the membrane in 2-propanol solution 25% v/v for 1 h.
- Rinsing and sinking of the membrane in triply distilled water overnight.
- Final rinsing with 3D water.

A schematic representation and an image of the commercial PES membrane are given in Figure 5.8.

The selective layer is exposed to the feed solution of the commercial PES membrane bore pores of average ~40 nm and in terms of rejection is characterized by a nominal *MWCO* (molecular weight cut off) of 150 kDa. The membrane is coded as "UP150" indicating the process of separation is produced for, by U:ultrafiltration, the material made of, by P:polyethersulfone, and the MWCO, by 150 in kDa. In Figure 5.9, the specifications of the membrane derived from the Microdyn-Nadir product catalog are illustrated.

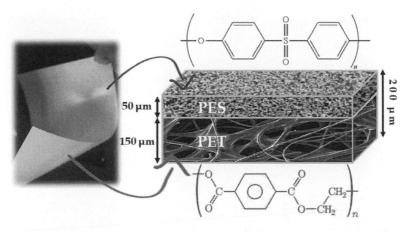

Figure 5.8. Schematic illustration and photograph of the commercial PES membrane structure.

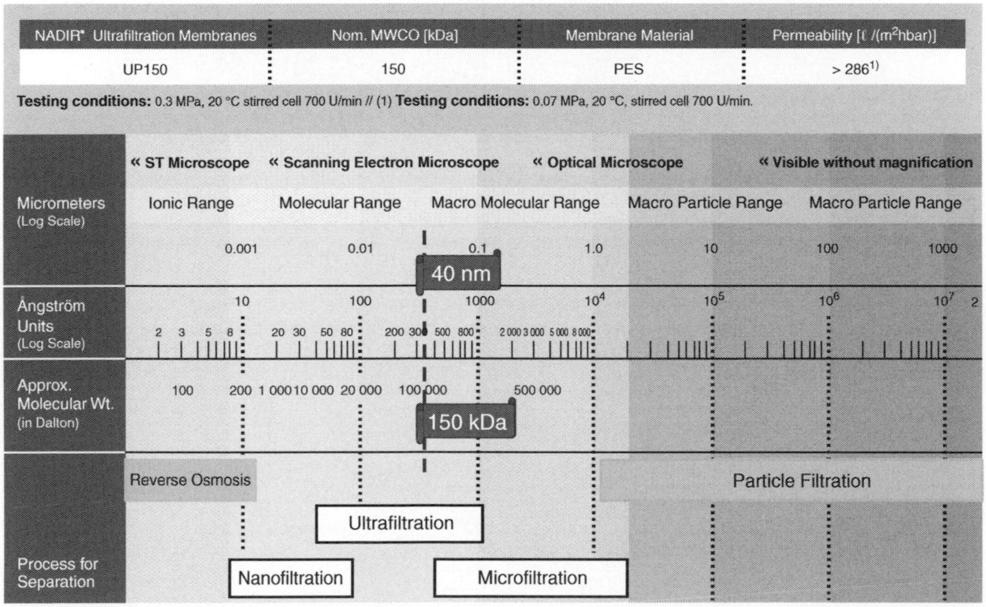

NADIR® Ultrafiltration Membranes	Nom. MWCO [kDa]	Membrane Material	Permeability [ℓ /(m²hbar)]
UP150	150	PES	> 286[1)]

Testing conditions: 0.3 MPa, 20 °C stirred cell 700 U/min // **(1) Testing conditions:** 0.07 MPa, 20 °C, stirred cell 700 U/min.

	« ST Microscope	« Scanning Electron Microscope	« Optical Microscope	« Visible without magnification	
Micrometers (Log Scale)	Ionic Range	Molecular Range	Macro Molecular Range	Macro Particle Range	Macro Particle Range
	0.001	0.01	0.1 **40 nm**	1.0	10	100	1000
Ångström Units (Log Scale)	10	100	1000	10⁴	10⁵	10⁶	10⁷ ₂
Approx. Molecular Wt. (in Dalton)	100	200 1 000 10 000 20 000	100 000 **150 kDa** 500 000			
	Reverse Osmosis				Particle Filtration
Process for Separation		Ultrafiltration			
	Nanofiltration		Microfiltration		

Figure 5.9. Specifications of UP150 membrane (data derived from Microdyn-Nadir product catalog).

5.7.1.2 *Characterization of the commercial PES membrane*

In order to understand the membrane structure, scanning electron microscopy (SEM) was utilized and micrographs of the membrane surface and cross-section were collected. In order not to destroy the structure of the membrane for the cross-section imaging, cryogenic cuts of the samples were carried out by sinking small specimens of membranes in liquid N_2 and carefully breaking the samples trying to avoid any eventual traumas. Representative SEM micrographs of the surface and the whole cross-section of UP150 membrane are shown in Figure 5.10.

The examination with SEM of the commercial PES membrane resulted in information-rich images regarding the membrane structure and morphology as described in the caption for Figure 5.10. In a nutshell, the membrane structure is characterized by the gradual increase of the pores from the "skin layer" of around 40 nm to a broad PES support and an even largely broadened PET

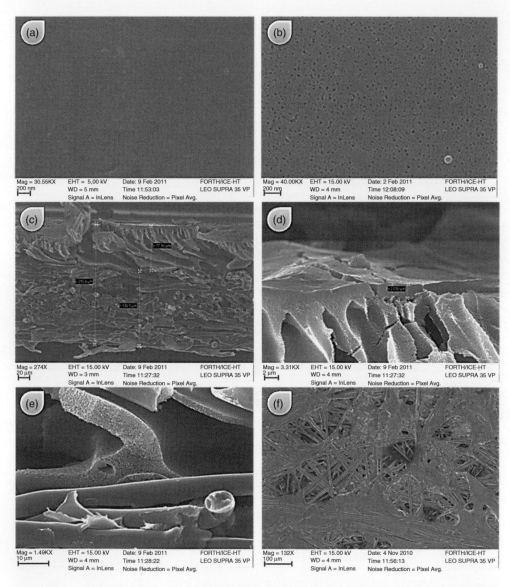

Figure 5.10.   SEM micrographs of the surfaces of UP150 membrane: (a) pristine PES surface (before filling agent withdrawal), (b) PES active surface after filling agent removal allowing the observation of external pores and pore distribution, (c) total cross area of PES membrane including the substrate PET layer allowing the determination of the different layers' thickness, (d) cross-section of membrane part indicating the selective PES layer ("skin layer") and the finger-type membrane support structure, (e) and (f) cross-section and surface respectively, of non-woven PET (substrate) allowing the observation of the broad network upon which PES layer is casted.

network. This "finger-type" structure seemed to favor the extrusion of CNTs from the bottom of the membrane (i.e. the PET substrate) directly to the "skin layer", targeting the encapsulation of CNTs in the narrow pores of the UF membrane that was the challenging main concept.

5.7.1.3   *Carbon nanotubes*
A variety of carbon nanotubes, SW-CNTs, DW-CNTs, Thin-MW-CNTs and MW-CNTs, with different features concerning the diameters, the length and the existence (or not) of chemical

Table 5.1. Time of the mechanical treatment process that CNTs underwent by the use of the tip sonication for CNTs cutting and the resulting lengths.

Time of treatment process	Resulting length of Thin MW/MW-CNTs
2 min	$\geq 10\,\mu$m
15 min	800 nm–2 μm
30 min	450 nm–1 μm
1 h	350–900 nm
2 h	350–700 nm

functionalities, were utilized in the study aiming at the efficient binding of CNTs into the pores of the commercial PES membrane. The majority of the CNTs used were supplied by Nanothinx S.A., while pristine SW- and DW-CNTs were purchased from Cheap Tubes Inc. CNTs supplied by Nanothinx S.A. were prepared by the catalyzed chemical vapor deposition (CCVD) method under pressure equal to 1 atm ($\sim$101.3 kPa) and temperature at 600–900°C. The carbon sources were either hydrocarbons or alcohols, while the catalysts constituting the determinant parameters of the dimensions of the external or internal CNT-diameters are IP protected by patent (Sotirchos and Mitri, 2007).

5.7.1.3.1 *Uncapping/cutting CNTs*
Pristine carbon nanotubes generally appear in bundles of long arrays with cupped ends. Two main methods have been utilized for the CNT-cutting/uncapping: the chemical modification by the use of acids (Liu and Shen, 2000; Nagasawa and Yudasaka, 2000; Wiltshire and Khlobystov, 2004; Zhao and Song, 2002) and the mechanical modification by the use of tip sonication. Thin-MW-CNTs (purity 94%, length $\geq 10\,\mu$m, external (ex.) diameter: 6–15 nm) and MW-CNTs (purity 97%, length $\geq 10\,\mu$m, ex. diameter: 15–35 nm) were used for this study. SEM was employed for the characterization of the initial and the final length.

The mechanical modification was accomplished by the use of the tip sonicator (UP400S Hielscher Co.) operating at 400 W, 24 kHz. In liquid, the rapid vibration of the tip causes cavitation, the formation and violent collapse of microscopic bubbles. The collapse of thousands of cavitation bubbles releases tremendous energy in the cavitation field. Objects and surfaces within the cavitation field are "processed". By increasing the amplitude setting, cavitation intensity within the sample is also increased (Cheng *et al.*, 2010).

Aqueous suspensions of CNTs of concentration 10 μg mL^{-1} were prepared and tip sonication was employed performing at 35% of the device amplitude for 5 different time periods: 2 min, 15 min, 30 min, 1 h and 2 h. Thereafter, the suspensions were collected and filtered for the collection of dry CNTs. The CNTs were spin-coated on a mica substrate and were examined through SEM imaging. The results regarding the final length of the mechanically-treated CNTs are summarized in Table 5.1. SEM micrographs of the samples before and after suffering both mechanical and chemical treatments are presented in Figure 5.11.

The chemical treatment of the CNTs was carried out by the use of acids. Particularly, 2 mg of Thin-MW-CNTs in powdered form were added to a beaker containing 8 mL of acid mixture H_2SO_4/HNO_3 in the ratio 3:1 and was placed into a sonication water bath operating at 35–40°C for 6 h. The resulting suspension was diluted by water and CNTs were filtrated and rinsed with NaOH solution 10 mM, and finally collected. SEM micrographs were collected indicating that the chemical treatment process led to CNT-length $\sim$320 nm near the values that mechanical treatment leads to (average $\sim$500 nm).

5.7.1.3.2 *Functionalization-CNT dispersion*
CNTs are characteristically of small diameter with a high aspect ratio and thus possess a large surface area. Even if chemically or mechanically-treated, CNTs appear in heavily entangled

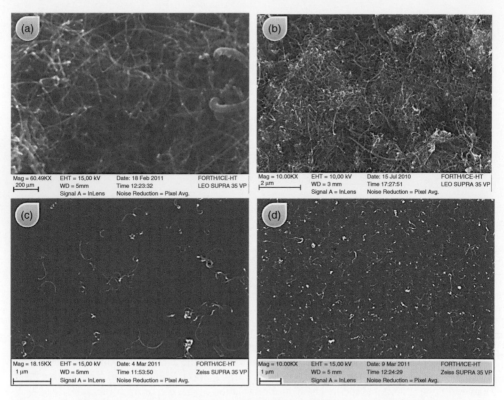

Figure 5.11.   SEM micrographs of: (a) Thin-MW-CNTs before treatment, (b) MW-CNTs before treatment (other magnification), (c) Thin-MW-CNTs after mechanical treatment for 2 h, and (d) Thin-MW-CNTs after chemical treatment.

bundles resulting in inherent difficulties in dispersion in polymer matrices, in organic solvents and in aqueous solutions. To resolve those problems, methods to modify the surface properties of CNTs were developed. These can be divided into chemical (covalent) and physical (non-covalent) functionalization as interactions between active materials and CNTs and were both employed to enhance the dispersion of CNTs for the sake of the CNT-embedment in polymer matrices.

Chemical functionalization is based on the covalent bond of functional groups onto the carbon form of CNTs. It can be performed at the end caps of nanotubes, which tend to be composed of highly reactive groups, as compared with the side walls. The side walls themselves contain defect sites, such as pentagon-heptagon pairs called Stone-Walls defects, sp^3-hybridized defects, and vacancies in the nanotube lattice. Direct covalent functionalization is associated with a change of hybridization from sp^2 to sp^3 and a simultaneous loss of p-conjugation system on graphene layer. This process can be made by reaction with some molecules of a high chemical reactivity.

Another method is defect functionalization of CNTs. This method was the one employed for the CNT-functionalization with –OH and –COOH groups of the CNTs used in the present study. The surface intrinsic defects are supplemented by oxidative damage to the nanotube framework by strong acids, which leave holes functionalized with oxygenated functional groups. In particular, treatment of CNTs with strong acid such as HNO_3, H_2SO_4 or a mixture of them, tends to uncap these tubes and to subsequently generate hydrophilic oxygenated functional groups that exhibit a higher affinity with water molecules.

In addition, surfactants were used to functionalize CNTs. The physical adsorption of surfactant on the CNT-surface lowers the surface tension of CNTs that effectively prevents their aggregation. Furthermore, the surfactant-treated CNTs overcome the van der Waals attraction by electrostatic/steric repulsive forces. Sodium dodecyl sulfate (SDS) is an anionic surfactant with

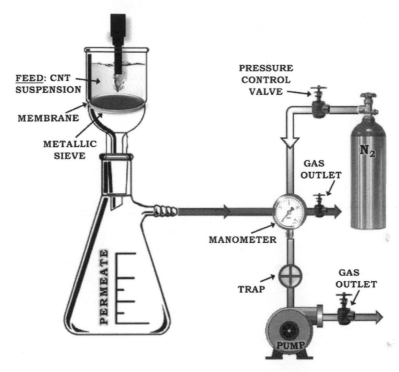

Figure 5.12.   Schematic representation of CNT-infiltration process through the PES membrane followed by a diagram describing the experimental setup.

the formula $CH_3(CH_2)_{11}OSO_3Na$. It consists of a 12-carbon tail attached to a sulfate group giving it amphiphilic properties and was extensively used for the dispersion of CNTs in the ratio 1:1 (SDS:Thin-MW/MW-CNTs) or 1.5:1 (SDS:SW-CNTs), in the aqueous media prepared. Another surfactant employed was the poloxamer 407, which is a non-ionic surfactant. It is a triblock copolymer consisting of a central hydrophobic block of polypropylene glycol (propylene glycol block: 56 repeat units) flanked by two hydrophilic blocks of polyethylene glycol (total PEG blocks: 101 repeat units). Poloxamer 407:CNTs ratio was the same as the SDS:CNTs.

5.7.2   *Infiltration of CNTs through commercial PES membranes*

The membrane cross area structure and morphology, as examined by SEM, seemed to favor the extrusion of CNTs from the wide porosity of the PET substrate towards the edges of the "finger-type" membrane channels, until their final encapsulation/protrusion at/from PES selective layer's-40-nm-pores using filtration methods. The PES membrane was placed in a dead-end filtration apparatus in such a manner that the PET substrate was exposed to the feed stream so that CNTs were able to enter the membrane porosity from the wide network of the membrane in order to encapsulate to the narrow-pore PES surface area. The whole procedure was assisted by additional means – actually at the stage of patentability – within the filtration process contributing to the alignment and the impulse of CNTs towards the membrane pores. Round-shaped specimens 4.7 cm in diameter cut from the flat sheet membrane were appropriate to match the cylinder funnel of the vacuum filtration apparatus (Supelco®) (300 mL volume); this allowed an active membrane surface 3.5 cm in diameter to be infiltrated. The membranes were left on a rigid stainless steel sieve and the apparatus was equipped with a pressure gage. Pressurized nitrogen gas was applied to limit the vacuum of the pump at 0.05 MPa for the proper operation of the membrane. A schematic representation of the experimental setup for the CNT-infiltration process is illustrated in Figure 5.12.

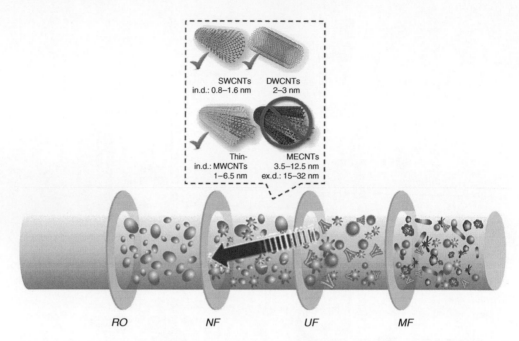

Figure 5.13.   Schematic illustration of the rejection capacity that the transformation of a UF membrane to a NF one would bring in with use of CNTs bearing appropriate internal diameters (Das *et al.*, 2014).

5.7.2.1   *Infiltration of CNTs dispersed in aqueous suspensions*

Preliminary filtration tests were performed using ethanol as CNT dispersant. However, due to the observation of swelling on PES membrane by the use of the solvent, all the infiltration processes for the encapsulation of the CNTs into the selective layer's pores presupposed the preparation of CNT-aqueous-suspensions. The 200 mL CNT-suspensions reached homogeneity by the use of the tip sonicator for 15 min in beakers of 300 mL prior to infiltration. Additionally, a variety of parameters were addressed in order to efficiently infiltrate the CNTs through the PES membrane in terms of water permeability and retention/rejection capacity/efficiency, which are listed as:

- *Type of CNTs*: Only certain types of CNTs with respect to their diameters were appropriate for the rejection of small organic molecules of molecular weight ranging from ~200–1000 Da. The transformation of the UF membrane to a NF one preconditioned the decrease of the pores from ~40 nm to less than 2 nm (Fig. 5.13). The successful transformation would be guaranteed exclusively by using SW-CNTs with internal diameters in the range 0.8–1.6 nm and probably DW-CNTs with 1–2 nm respectively. However, even with the use of Thin-MW-CNTs bearing internal diameters ranging from 1–6.5 nm, the rejection target could be approached, whereas MW-CNTs with external diameters 15–35 nm would not even be able to penetrate the ~40-nm-pores. Thereafter, there was a broad variety of available CNTs to be employed. The type and the features of the CNTs used for the infiltration studies are presented in Table 5.2.
- *Membrane conformation in the filtration apparatus*: As mentioned before, the membrane structure seems to favor the intrusion of CNTs from the broad formed substrate porosity to reach the pores of the skin layer. However, both the aforementioned configuration and the reverse one (i.e. the selective layer exposed to the feed during filtration) were tested to ensure the efficient CNT hooking into the membrane pores (Fig. 5.14). On occasion, both the processes were performed.
- *The volume of the feed solution (i.e. the CNT-suspension) and the concentration of the CNT-suspensions to be infiltrated through the polymeric membrane*: The capacity of the filtration

Table 5.2.   Type and features of CNTs tested for the infiltration study.

Type of CNT	Carbon purity [%]	External diameter [nm]	Internal diameter [nm]	Length [μm]	Price [€ g^{-1}]
SW (cheap tubes)	>90	1–2	0.8–1.6	5–30	90
SW-COOH	>90	1–2	0.8–1.6	5–30	90
DW (cheap tubes)	>90	2–4	1–2	5–15	60
Thin-MW (nanothinx)	94	6–15	1–6.5	≥10	16
Thin-MW-OH (nanothinx)	94	6–15	1–6.5	~0.6	36
Thin-MW-COOH (nanothinx)	94	6–15	1–6.5	≤1	31
MW (nanothinx)	97	15–35	3.5–12.5	≥10	14

The lengths of the CNTs given in the table are before CNTs suffer tip sonication.

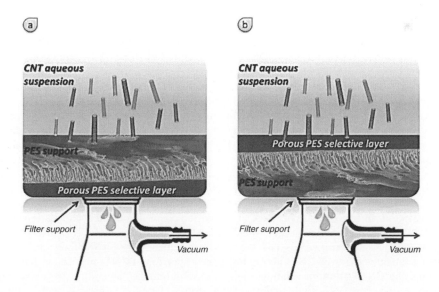

Figure 5.14.   Filtration processes were performed with either the substrate to be exposed to the CNT-suspension (a) or inversely, the selective layer to be exposed to the feed suspension (b).

cylinder determined the volume of the suspension to be infiltrated. The addition of the tip sonicator operating during the filtration process limited the volume of the suspensions due to the preconditions required in terms of the sonicator-probe position in order to operate properly. Thus, the volume of the feed suspension determined at 250 mL allowed the permeance of 50 mL for each infiltration. With respect to the concentration of the CNT-suspensions, a broad range of concentrations were studied, ranging from 0.078 μg mL^{-1} to 150 μg mL^{-1} in an effort to cover the whole active surface of the membrane with CNTs.

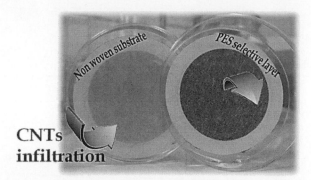

Figure 5.15.   The infiltration process carried out by filtering a CNT-suspension from the support side (left) to reach the selective layer (right). The black imprint of CNTs on the side of the selective layer is clearly indicating their existence to the surface pores, while the substrate is empty of CNTs, indicating that all CNTs were pushed towards the PES skin layer.

The determination of the parameters influencing the efficiency of the CNT-infiltration process was extracted through a vigorous effort to conceive the CNT-infiltration effect on the membrane by experimental observations in terms of visual and microscopic imaging. The black imprint of CNTs in the skin layer when imported from the substrate side was a clear indication that CNTs had followed the right path to reach the pores of the membrane. A representative assistant-illustration is depicted in Figure 5.15.

Concerning the membranes grafted with CNTs by filtration through the substrate, the parameter of the sonication-probe position, as well as of the concentration of the CNT-suspensions, seemed to favor the intrusion of CNTs towards the selective layer. Representative images are illustrated in Figures 5.16 and 5.17, noting the CNTs-imprint on the PES surface when the infiltrations are carried out either from the substrate or from the selective area.

Thereafter, groups of the prepared CNT-grafted-membranes were evaluated in terms of scanning electron microscopy (SEM), atomic force microscopy (AFM), Raman spectroscopy, water flux measurements and rejection tests. In the following subsection the characterization and evaluation of the membranes is presented.

5.7.2.2   *Treatment of carboxyl acid-functionalized CNTs*

CNTs bearing high van der Waals forces, high surface area and high aspect ratio inevitably cause self-aggregation. In aqueous suspensions, they tend to cohere due to their high surface energy and lack of chemical affinity with the dispersing medium. Even oxidized CNTs after carboxyl-functionalization show limited dispersion stability in aqueous media for long time periods. Further treatment of oxidized CNTs may improve the dispersion of CNTs in aqueous dispersants, facilitating the CNT-infiltrations through porous membranes. In this context, to obtain stable aqueous dispersions of CNTs for long time periods, deprotonation of Thin-MW-CNT-COOH and SW-CNT-COOH was carried out by treatment with an aqueous NaOH solution (Lee *et al.*, 2007), and their dispersion stability was examined. The investigation of the dispersion state of CNTs in liquids is quite difficult due to its opaque blackness even at very low concentrations. However, a distinguishable visible change of the darkness of the CNT-suspensions prepared took place only a few hours later and remained similar for several months (Fig. 5.18).

Two comparative studies between Thin-MW-CNTs and the same CNTs functionalized with carboxyl groups, and between SW-CNTs and carboxyl-functionalized CNTs, by the use or not of surfactant, were carried out. The observation of the dispersions of the CNT-suspensions was evaluated by means of UV-visible absorbance (Fig. 5.19) and zeta potential measurements (Table 5.3) that were carried out after five months in order to observe the eventual long-term stability.

Judging from both the photographs and the UV-vis absorption spectra, the CNTs, either Thin-MW-CNTs or SW-CNTs bearing carboxylic anions, showed much improved stability in water

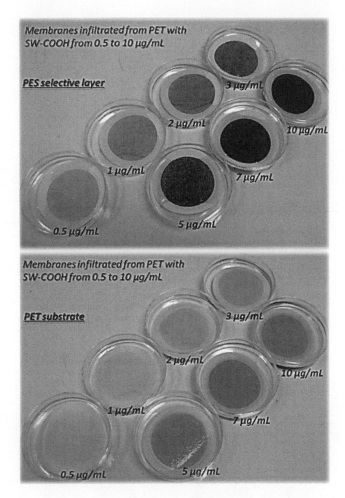

Figure 5.16.   Infiltrated membranes with SW-CNT-COOH in the concentration range 0.5–10 μg mL^{-1}. The black imprint on the selective layer appears to increase gradually with the suspension-concentration increase (left), whereas the infiltrations carried out from the PET substrates remain "clear".

due to the chemical affinity between the polar modified groups and water and the electrostatic repulsion.

The UV-vis absorption spectra indicated that the carboxylated CNTs with anion groups showed enhanced dispersion compared to the pristine or carboxyl-functionalized CNTs.

The origin of electrostatic repulsion can be found from the increased zeta potential (ζ) carboxylated CNTs presented. The ζ is an important indicative of ionically stabilized colloid systems. Greater magnitude (usually ± 25 mV) of the ζ endows the colloid system with an improved stability against coagulation (Lee *et al.*, 2007). In Table 5.3 the values of ζ of the CNT-suspensions are presented.

For the infiltration studies, carboxyl-functionalized with anion groups Thin-MW and SW-CNTs were also utilized.

5.7.2.3   *Characterization of the CNT-infiltrated membranes*

A Zeiss SUPRA 35VP scanning electron microscope was implemented for providing visual information of the top surface and cross-sectional morphology of the CNT-infiltrated membranes.

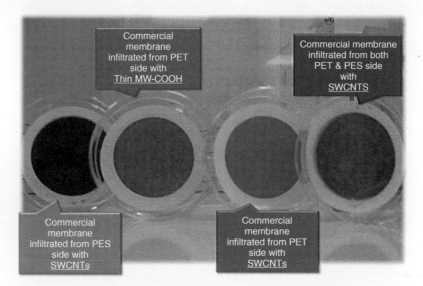

Figure 5.17.   Commercial membranes grafted with either Thin-MW-CNT-COOH or SW-CNTs from the substrate, the selective layer or both. The images show the selective surfaces in all cases.

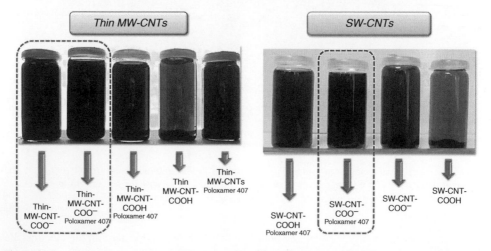

Figure 5.18.   Digital photographs of CNT-suspensions of pristine and modified CNTs five months after the treatment process. The most stable suspensions are presented in frames.

The membrane samples were cut into small pieces and left to dry. For cross-sectional images, the pieces were dipped in liquid nitrogen for 2 min, frozen and broken into two slices and kept in air for drying. The dried samples were glued on the support and gold sputtered to create electric conductivity. Representative SEM micrographs collected from the PES porous surfaces as well as from the membrane cross area are illustrated in Figures 5.20, 5.21 and 5.22.

For the membranes infiltrated through the support side, as one may notice through SEM micrographs, only a few CNTs seem to protrude the surface area (Fig. 5.20) while the majority of the CNTs appear to stack just before reaching the porous skin layer of the membrane (Figs. 5.20, 5.21). Therefore, the black imprint of CNTs on the surface of the membranes was attributed to their existence beneath the $\sim$2 μm skin layer. In Figure 5.21, the CNTs appear to lie on the PES surface, while some of them appear to have entered the membrane pores.

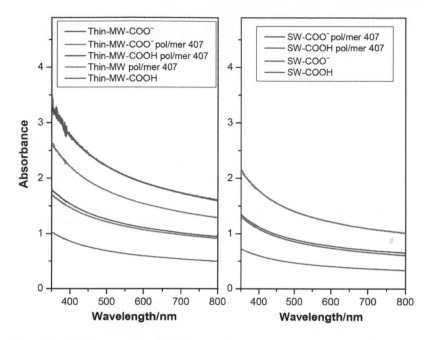

Figure 5.19.   UV-vis absorption spectra of CNT-suspensions treated and dispersed in different methods.

Table 5.3. Zeta potential values for the different CNT-suspensions.

Sample	Zeta potential [mV]
Thin MW-CNT $_{Poloxamer407}$	−8.51
Thin MW-CNT-COOH	−37.1
Thin MW-CNT-COOH $_{Poloxamer407}$	−28.4
Thin MW-CNT-COO$^-$	−41.1
Thin MW-CNT-COO$^-_{Poloxamer407}$	−30.6
SW-CNT-COOH	−31.7
SW-CNT-COOH $_{Poloxamer407}$	−25.6
SW-CNT-COO$^-$	−32.2
SW-CNT-COO$^-_{Poloxamer407}$	−30.1

To elucidate the influence of the addition of CNTs on membrane roughness, atomic force microscopy (AFM) was employed to analyze the surface morphology and roughness of the prepared membranes. The membrane surfaces in a few cases showed an enhancement of the root mean square roughness (RMS roughness) of the samples that had been grafted with CNTs. In Figure 5.23 the AFM topography of the PES surfaces of an unmodified (pristine) membrane and of a membrane infiltrated from the support side with SW-CNTs are presented. In the images, the brightest areas indicate the highest point of the membrane surface and the dark regions show valleys or pores. The enhancement of the RMS roughness of ∼5 nm was revealed through images, which is probably attributed to the CNT presence on the membrane surface.

In addition, laser Raman spectroscopy was implemented for the characterization of the CNT-membranes. Backscattering Raman spectra were collected from bulk SW-CNTs and Thin-MW-CNT-COOH in powdered form as well as from the pure UP150 membrane and from

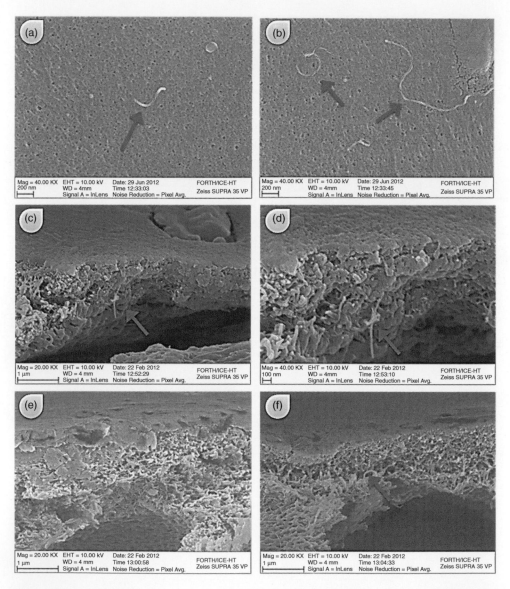

Figure 5.20.   (a), (b) PES surface of membrane grafted with Thin-MW-COOH (conc. $5\,\mu g\,mL^{-1}$), (c), (d), (e), (f) PES cross-section of membrane grafted with Thin-MW-COOH (conc. $5\,\mu g\,mL^{-1}$). The samples were CNT-infiltrated from the PET substrate.

membranes infiltrated with SW-CNTs or Thin-MW-CNT-COOH (from the PES selective layer). In Figure 5.24, the Raman spectra are presented.

The presence of D and G bands in the spectra of the CNT-infiltrated membranes is clear evidence for their existence near or on the skin layer. A note is made of the fact that Raman penetration depth is of the order of 1–2 μm, about the thickness of the selective layer.

5.7.2.4   *Performance of the CNT-infiltrated membranes*

Any eventual effect attributed to the infiltration of CNTs through the membranes was characterized by testing the performance of each sample of the prepared membranes in terms of pure water flux. The CNT-membranes were expected to exhibit enhanced water fluxes due to the contribution of

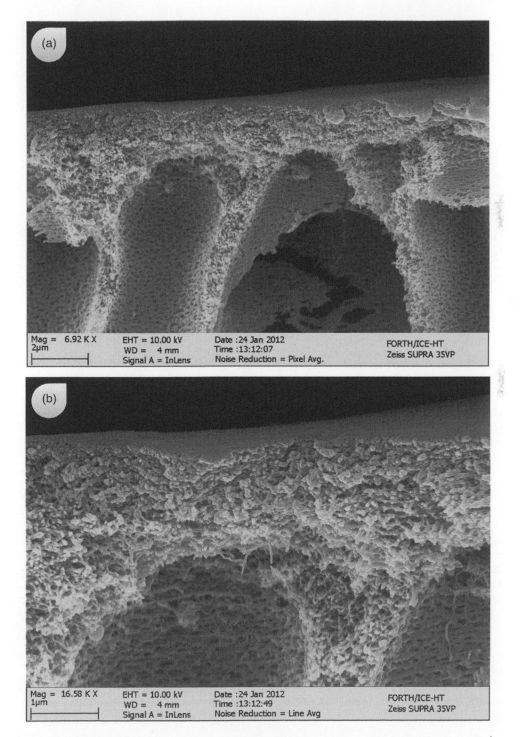

Figure 5.21.   (a), (b) PES cross-section of membrane grafted with Thin-MW-OH (conc. $1.25\,\mu g\,mL^{-1}$). The samples were CNT-infiltrated from the PET substrate.

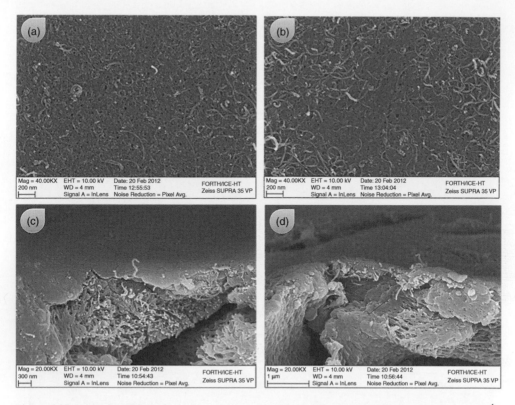

Figure 5.22.   (a), (b) PES surface of membrane grafted with Thin-MW-COOH (conc. 1.25 μg mL^{-1}), (c), (d) cross-sectional images from the same samples. The samples were CNT-infiltrated towards the PES selective layer.

the CNT-existence. The experiments were carried out in the dead-end filtration apparatus used for the infiltration of CNTs under 0.05 MPa transmembrane pressure at room temperature. The membrane surface area was 9.6 cm^2. The flux of each membrane was measured before and after the CNT-infiltration for reasons of comparison. For the flux determination, the filtration funnel was filled with 200 mL of triply distilled water, allowing for the permeance of 100 mL water the membrane should not get dry. The time needed for the 100 mL water to pass through the membrane was recorded by means of a chronometer. Thereafter, the membrane was placed in the opposite configuration and the infiltration with CNTs was carried out as previously described. After the CNT-infiltration, the membrane was turned again in the opposite configuration so as to allow the PES surface to be exposed to the feed solution and the time needed for another 50 mL of water to pass through the membranes was recorded. For the membranes infiltrated with CNTs directly through the PES selective layer, the membrane remained in the same place and the flow rate determination was carried out immediately after rinsing and cleaning the remaining CNT-suspension with water. The membrane pure water flux (J) was calculated by the following equation (Vatanpour *et al.*, 2011):

$$J = \frac{V}{A\Delta tP} \tag{5.3}$$

where V [L] was the volume of permeated water, A [m^2] was the membrane area, Δt [h] was the permeation time and P [Pa] was the pressure difference used.

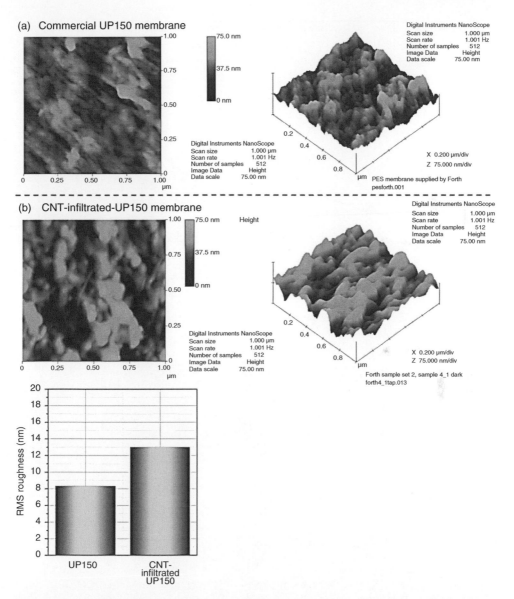

Figure 5.23.   AFM topography images: (a) pristine Commercial UP150 membrane and (b) UP150 membrane infiltrated with SW-CNTs (conc. $3\,\mu g\,mL^{-1}$). Down and left, RMS roughness of the pristine and the CNT-infiltrate-membrane are compared.

A note is made of the fact that in many cases the CNT-membranes showed enhanced flow rate, however, the results were not always reproducible.

Nevertheless, for those cases where the membranes exhibited an increase in the flow rate, the fluxes of the membranes were calculated using the Equation (5.3). The type and the concentration of the CNTs used for the infiltrations of those membranes, as well as the percentage of their flux increase, are presented in Figures 5.25 and 5.26.

For the retention performance study of the modified membranes, aqueous solutions of 50 mL of the model foulant consisted mainly of the organic dyes Remazol Brilliant Blue R (Mw: 630 Da) and Acid Red 4 (Mw: 380 Da) in equal concentration ($50\,\mu g\,mL^{-1}$) and were left passing through

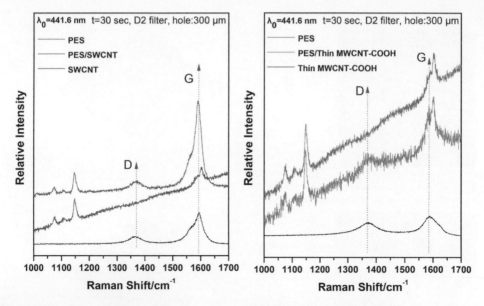

Figure 5.24.   Raman spectra showing the existence of CNTs within the membrane structure after CNT-infiltration with SW-CNTs (left) and Thin-MW-CNT-COOH (right).

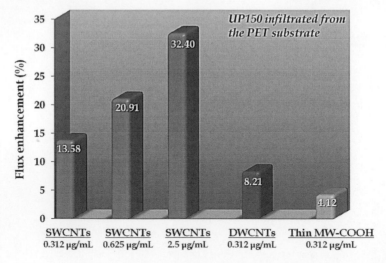

Figure 5.25.   Percentage of flux increase for membranes infiltrated with CNTs from the PET side compared to unmodified membranes.

unmodified and CNT-modified membranes in the dead-end filtration unit. The chemical composition of the model textile wastewater is presented in Table 5.4. For the collected permeates UV-vis spectra were recorded and from the absorption peaks of the dyes the concentration of the permeates were calculated. A representative diagram of the retention % is presented in Figure 5.27.

In the corresponding retention diagram, one can observe that the rejection of the dye molecules was very high. However, when additional volumes of the model foulant were left to permeate the CNT-membranes, the retention was the same as that of the pristine membrane and consequently it was concluded that the initial retention was due to adsorption effects.

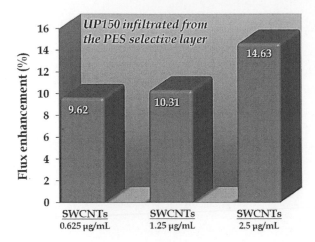

Figure 5.26. Percentage of flux increase for membranes infiltrated with CNTs from the PES side compared to unmodified membranes.

Table 5.4. Chemical composition of the textile model foulant.

No	Dyestuff	Concentration [$\mu g\,mL^{-1}$]
1	Remazol Brilliant Blue R	50
2	Acid Red 4	50
3	NaCl	2500
4	NaHCO$_3$	1000
5	Glucose	2000
6	Albatex DBC (detergent)	50

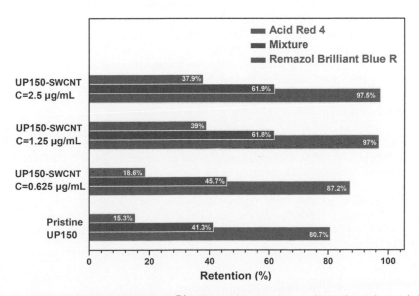

Figure 5.27. Retention of Remazol Brilliant Blue R – Acid Red 4 mixture of UP150 membranes infiltrated with SW-CNTs of various concentration from the PES side and of the pristine membrane.

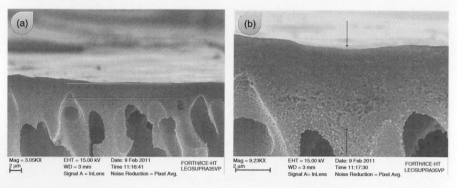

Figure 5.28.   SEM micrographs of the skin layer of the commercial PES membrane. (a) As indicated, the thickness of this region was ~2 μm resulting prohibited for the CNTs extrusion the external pores of the selective layer. (b) Micrograph as (a) of high magnitude allowing the observation of the dense pore structure at the top of membrane.

5.8   CONCLUSIONS AND FUTURE PROSPECTIVES

The efforts of transforming a UF commercial membrane to a NF one were described showing representative measurements. The structure of the UP150 membrane seemed to favor the intrusion of CNTs from the substrate aiming to reach the ~40-nm-pores of the thin selective layer.

However, as it is observed from the above SEM micrographs, due to the anisotropic, dense, worm/sponge like structure of the ~2 μm skin layer, efficient CNT-infiltration was impossible. Moreover, taking into consideration the length of the modified CNTs, that is <1 μm, it would also not be sufficient.

Hence, alternative relevant approaches were studied concerning the fabrication of CNT-membranes, which are highlighted below as a future prospective of the current study and will be discussed in detail in subsequent work/s.

The adjustment of the parameters of the immersion precipitation phase separation technique may allow the preparation of quasi-ultrafiltration membranes – the so-called CNT-infiltrated *tailor-made membranes* – with tuned parameters. Such polymer asymmetric membranes with tailored morphological properties may serve as hosts for the encapsulation of carbon nanotubes. This would be a step forward for the efficient fabrication of CNT-infiltrated membranes, whereas an extended characterization study *via* appropriate techniques, like contact-angle and ζ-potential measurements, porosimetry and AFM, is considered as a prerequisite.

ACKNOWLEDGMENTS

The research leading to these results has received funding from the European Union Seventh Framework Programme (*FP7/2007-2013*) under *grant agreement* n° 246039 (BioNexGen).

The AFM measurements were carried out by *Dr. Daniel Jonson*, Research Officer in the Centre for Water Advanced Technologies and Environmental Research (CWATER), part of the Systems and Process Engineering Centre in the College of Engineering at Swansea University.

The SEM images were carried out by *Dr. Vassileios Dracopoulos*, Senior Application Scientist at Institute of Chemical Engineering Sciences, Foundation of Research and Technology-Hellas (FORT/ICE-HT).

REFERENCES

Anastassiou, A., Karahaliou, E.K., Alexiadis, O. & Mavrantzas, V.G. (2013) Detailed atomistic simulation of the nano-sorption and nano-diffusivity of water, tyrosol, vanillic acid, and p-coumaric acid in single wall carbon nanotubes. *The Journal of Chemical Physics*, 139, 164711.

Andre de Sousa, J.M. (2009) *Mixed Matrix Membranes: A new Platform for enzymatic Reactions*. PhD Thesis, University of Twente, Twente, The Netherlands.

Ansorge, W., Schuster, O., Welchs, F. & Dombrowski, K. (2014) Microfiltration membrane with improved filtration properties. Membrana GmbH, Patent US 8727136 B2.

Babu, J.S. & Sathian, S. (2011) The role of activation energy and reduced viscosity on the enhancement of water flow through carbon nanotubes. *The Journal of Chemical Physics*, 134, 194509.

Baek, Y., Kim, C., Seo, D.K., Kim, T., Lee, J.S., Kim, Y.H., Ahn, K.H., Bae, S.S., Lee, S.C., Lim, J., Lee, K. & Yoon, J. (2014) High performance and antifouling vertically aligned carbon nanotube membrane for water purification. *Journal of Membrane Science*, 460, 171–177.

Calvert Investments (2013) Calvert global water fund. Available from: http://www.calvert.com [accessed December 2016].

Cheng, Q., Debnath, S., Gregan, E. & Byrne, H. (2010) Ultrasound-assisted SW-NTs dispersion: effects of sonication parameters and solvent properties. Dublin Institute of Technology, ARROW@DIT, Nanolab, Dublin, Ireland.

Choi, J.H., Jegal, J. & Kim, W.N. (2006) Fabrication and characterization of multi-walled carbon nanotubes/polymer blend membranes. *Journal of Membrane Science*, 284, 406–415.

Corry, B. (2008) Designing carbon nanotube membranes for efficient water desalination. *Journal of Physical Chemistry* B, 112, 1427–1434.

Dai, K., Shi, L., Zhang, D. & Fang, J. (2006) NaCl adsorption in multi-walled carbon nanotube/active carbon combination electrode. *Chemical Engineering Science*, 61, 428–433.

Das, R., Ali, Md.E., Hamid, S.B.A., Ramakrishna, S. & Chowdhury, Z.Z. (2014) Carbon nanotube membranes for water purification: a bright future in water desalination. *Desalination*, 336, 97–109.

Elimelech, M. & Phillip, W.A. (2011) The future of seawater desalination: energy, technology, and the environment. *Science*, 333, 712–717.

Goh, P.S., Ismail, A.F. & Ng, B.C. (2013) Carbon nanotubes for desalination: performance evaluation and current hurdles. *Desalination*, 308, 2–14.

Goh, P.S., Ismail, A.F. & Ng, B.C. (2014) Directional alignment of Carbon nanotubes in polymer matrices: contemporary approaches and future advances. *Composites* A, 56, 103–126.

Hillal, N., Al-Zoubi, H., Darwish, N.A., Mohammad, A.W. & Abu Arabi, M. (2004) A comprehensive review of nanofiltration membranes: treatment, pretreatment, modeling and atomic force microscope. *Desalination*, 170, 281–308.

Hinds, B.J., Chopra, N., Rantell, T., Andrews, R., Gavalas, V. & Bachas, L.G. (2004) Aligned multiwalled carbon nanotube membranes. *Science*, 303, 62–65.

Holt, J.K., Noy, A., Huser, T., Eaglesham, D. & Bakajin, O. (2004) Fabrication of a carbon nanotube embedded silicon nitride membrane for studies of nanometer-scale mass transport. *Nano Letters*, 4, 2245–2250.

Holt, J.K., Park, H.G., Wang, Y., Stadermann, M., Artyukhin, A.B., Grigoropoulos, C.P., Noy, A. & Bakajin, O. (2006) Fast mass transport through sub-2-nanometer carbon nanotubes. *Science*, 312, 1034–1037.

Hummer, G., Rasaiah, J.C. & Noworyta, J.P. (2001) Water conduction through the hydrophobic channel of a carbon nanotube. *Nature*, 414, 188–190.

Jiang, T., Kennedy, M.D., Van der Meer, W.G.J., Vanrolleghem, P.A. & Schippers, J.C. (2003) The role of blocking and cake filtration in MBR fouling. *Desalination*, 157, 335–343.

Joseph, S. & Aluru, N.R. (2008) Why are carbon nanotubes fast transporters of water? *Nano Letters*, 8, 452–458.

Judd, S. (2006) *The MBR Book: Principles and Applications of Membrane Bioreactors in Water and Wastewater Treatment*. 1st edition, to Typesetter Elsevier, Amsterdam. Chapter 2, 21–121.

Kannam, S.K., Todd, B.D., Hansen, J.S. & Daivis, P.J. (2013) How fast does water flow in carbon nanotubes? *The Journal of Chemical Physics*, 138, 094701.

Kar, S., Bindal, R.C. & Tewari, P.K. (2012) Carbon nanotube membranes for desalination and water purification: challenges and opportunities. *Nano Today*, 7, 385–389.

Khin, M.M., Nair, A.S., Babu, V.J., Murugan, R. & Ramakrishna, S. (2012) A review on nanomaterials for environmental remediation. *Energy & Environmental Science*, 5, 8075–8109.

Kim, S., Jinschek, J.R., Chen, H., Sholl, D.S. & Marand, E. (2007) Scalable fabrication of carbon nanotube/polymer nanocomposite membranes for high flux gas transport. *Nano Letters*, 7, 2806–2811.

Lee, J., Kim, M., Hong, C.K. & Shim, S.E. (2007) Measurement of the dispersion stability of prostine and surface-modified multiwalled carbon nanotubes in various nonpolar and polar solvents. *Measurement Science and Technology*, 18, 3707–3712.

Li, H. & Zou, L. (2011) Ion-exchange capacitive deonization: a new strategy for brackish water desalination. *Desalination*, 275, 62–66.

Li, J., Ng, H.T., Cassell, A., Fan, W., Chen, H., Ye, Q., Koehne, J., Han, J. & Meyyappan, M. (2003) Carbon nanotube nanoelectrode array for ultrasensitive DNA detection. *Nano Letters*, 3, 597–602.

Liu, Z. & Shen, Z. (2000) Organizing single-walled carbon nanotubes on gold using a wet chemical self-assembling technique. *Langmuir*, 16, 3569–3573.

Madaeni, S.S., Zinadini, S. & Vatanpour, V. (2013) Preparation of superhydrophobic nanofiltration membrane by embedding multiwalled carbon nanotube and polydimethylsiloxane in pores of microfiltration membrane. *Separation and Purification Technology*, 111, 98–107.

Majumder, M., Chopra, N., Andrews, R. & Hinds, B.J. (2005) Nanoscale hydrodynamics: Enhanced flow in carbon nanotubes. *Nature*, 438, p. 44.

Mauter, M.S. & Elimelech, M. (2008) Environmental applications of carbon-based nanomaterials. *Environmental Science and Technology*, 42, 5843–5859.

Nagasawa, S. & Yudasaka, M. (2000) Effect of oxidation on single-wall carbon nanotubes. *Chemical Physics Letters*, 328, 374–380.

Nasrabadi, A.T. & Foroutan, M. (2011) Ion-separation and water-purification using single walled carbon nanotube electrodes. *Desalination*, 277, 236–243.

Ohlrogge, K. & Ebert, K. (2006) *Membranen: Grundlagen, Verfahren und Industrielle Anwendungen.* Wiley-VCH, Weinheim, Germany.

Qu, X., Alvarez, P.J. & Li, Q. (2013) Applications of nanotechnology in water and wastewater treatment. *Water Research*, 47, 3931–3946.

Sharma, A., Tripathi, B. & Vijay, Y.K. (2010) Dramatic improvement in properties of magnetically aligned CNT/polymer nanocomposites. *Journal of Membrane Science*, 361, 89–95.

Sotirchos, E. & Mitri, S. (2007) Advanced catalysts for the large scale production of high purity carbon nanotubes with chemical vapor deposition. Patent GR 1005879, Published 2008.

Sparreboom, W., Van den Berg, A. & Eijkel, J.C.T. (2010) Transport in nanofluidic systems: A review of theory and applications. *New Journal of Physics*, 12, 015004–0150027.

Van Hooijdonk, E., Bittencourt, C., Snyders, R. & Colomer, J.F. (2013) Functionalization of vertically aligned carbon nanotubes', *Beilstein International Journal of Nanotechnology*, 4, 129–152.

Vatanpour, V., Madaeni, S.S., Moradian, R., Zinadini, S. & Astinchap, B. (2011) Fabrication and characterization of novel antifouling membrane prepared from oxidized multiwalled carbon nanotube/polyethersulfone. *Journal of Membrane Science*, 375, 284–294.

Walther, J.H., Ritos, K., Cruz-Chu, E.R., Megaridis, C.M. & Koumoutsakos, P. (2013) Barriers to superfast water transport in carbon nanotube membranes. *Nano Letters*, 13, 1910–1914.

Wiltshire, J.G. & Khlobystov, A.N. (2004) Comparative studies on acid and thermal based selective purification of HiPCO produced single-walled carbon nanotubes. *Chemical Physics Letters*, 386, 239–243.

Yang, H.Y., Han, Z.J., Yu, S.F., Pey, K.L., Ostrikov, K. & Karnik, R. (2013) Carbon nanotube membranes with ultrahigh specific adsorption capacity for water desalination and purification. *Nature Communications*, 4, Article 2220.

Zhao, W. & Song, C. (2002) Water soluble and optically pH-sensitive single-walled carbon nanotubes from surface modification. *Journal of the American Chemical Society*, 124, 12,418–12,419.

Zimmerman, C.M., Singh, A. & Koros, W.J. (1997) Tailoring mixed matrix composite membranes for gas separations. *Journal of Membrane Science*, 137, 145–154.

CHAPTER 6

Effects of the solvent ratio on carbon nanotube blended polymeric membranes

Evrim Celik-Madenli, Ozgur Cakmakcı, Ilkay Isguder, Nevzat O. Yigit,
Mehmet Kitis, Ismail Koyuncu & Heechul Choi

6.1   INTRODUCTION

Membrane operations are used in a vast range of industries. Hence, there is neither a single
type of membrane nor a single technology (Cardew and Le, 1998). Membrane operations use
a membrane to execute a particular separation. Due to physical and/or chemical differences
between the membrane and the permeating components, membrane can transport some elements
more readily than others (Mulder, 1997). Membrane operations can be classified according to
the driving force, the mechanism of separation, the structure of the membrane and the phases
in contact (Aptel and Buckley, 1996). Most of the membrane operations are governed by the
pressure, concentration, electrical potential or temperature gradients (Cardew and Le, 1998).
In pressure driven membrane operations the driving force is the pressure difference across the
membrane. Main pressure driven membrane operations are reverse osmosis, nanofiltration, ultrafiltration, and microfiltration. Typical properties of the pressure driven operations are given in
Table 6.1.

Osmosis may be defined as the spontaneous movement of solvent through a semipermeable
membrane from a dilute solution to a more concentrated solution. Membrane in osmosis allows
solvent to pass but retains the dissolved solids. Transport from the dilute solution to the concentrated solution through the membrane is spontaneous in osmosis. Because of the concentration
difference, head develops in the concentrated solution, which is called osmotic pressure. If
pressure is applied that is higher than osmotic pressure, the flow will be reversed through the
membrane. Pure water will flow from the concentrated solution to the dilute solution through
the membrane, which is called reverse osmosis (Dawson and Merces, 1986). Reverse osmosis
operations are mainly used for sea water and brackish water desalination, recovery of organic
and inorganic materials from chemical processes, treatment of hazardous wastes, food, beverage,
pulp and paper industries (Williams *et al.*, 1992).

In terms of membrane selectivity, nanofiltration lies between reverse osmosis and ultrafiltration
(Aptel and Buckley, 1996). The only difference between the nanofiltration and reverse osmosis is
that the network structure is denser on reverse osmosis. Even though monovalent ions like Na^+
and Cl^- are poorly rejected by nanofiltration, the rejection of bivalent ions such as Ca^{2+} and
CO_2^{2-} is very high (Mulder, 1997). The main applications of nanofiltration are softening, removal
of pesticides, lignin, disinfection by-products, sulfate, color and sugar concentration (Cardew
and Le, 1998).

Microfiltration is a separation process which uses porous membranes with pore sizes between
0.1 to 10 μm to retain suspensions and emulsions. Microfiltration is a membrane operation which
lies between the conventional filters and ultrafiltration (Baker, 2004; Mulder, 1997). The main
applications of the microfiltration operations are sterilization of beverages and pharmaceuticals, separation of oil-water emulsions, wastewater treatment, cell harvesting, metal recovery as
colloidal oxides or hydroxides, and continuous fermentation (Mulder, 1997).

Table 6.1.   Summary of the pressure driven membrane processes (Cardew and Le, 1998; Mulder, 1997).

Membrane operation	Material	Pressure [MPa]	Size range [μm]	Separation principle	Flux range [L m^{-2} h^{-1} MPa^{-1}]
Reverse osmosis	Polymers	1–10	<0.001	Size, charge, affinity	0.5–14
Nanofiltration	Polymers	0.5–2	~0.001	Size, charge, affinity	14–120
Ultrafiltration	Polymers, ceramics	0.1–0.5	0.001–0.1	Size, charge	100–500
Microfiltration	Polymers, ceramics, metals	<0.2	0.1–1	Size	>500

Ultrafiltration (UF) can be considered as a porous membrane. The size and shape of the solutes, which are relative to the pore size of the membrane, are rejected by the UF membrane. The transport of the solvent is directly proportional to the applied pressure (Mulder, 1997).

The structure of the UF membranes is generally anisotropic, with a finely porous layer supported on a more open microporous substrate. The separation is performed on the finely porous layer, and the microporous substrate provides mechanical strength to the membrane (Baker, 2004). In the UF operations, fouling is the most severe issue still to be solved (Rahimpour and Madaeni, 2007). Non-specific adsorption and deposition of the macromolecules on the membrane surface or in the pores increases the hydraulic resistance to the flow, and by that means severely reduces the permeation flux (Chen *et al.*, 2006; Shi *et al.*, 2007). Membrane surface chemistry is very important in the performance of UF operations (Reddy and Patel, 2008). It is well known that increasing membrane hydrophilicity can effectively minimize membrane fouling (Galiano *et al.*, 2015, Wang *et al.*, 2006).

Immersion precipitation is a well-known technique for preparing asymmetric membranes. In this process a homogenous mixture of polymer and solvent is cast on a support and immersed in a coagulation bath of non-solvent. The solvent in the cast film is exchanged with the non-solvent in the coagulation bath by immersing the cast film in a coagulation bath. Depending on the polymer and coagulation conditions, many phase changes occur, namely; liquid-liquid phase change, solid-liquid phase change or both. Afterwards the polymer film solidifies and forms the asymmetric membrane with a dense top layer and porous sub-layer. There are many variables which affect the sub-layer formation of the membrane, such as composition of the dope solution, coagulant temperature, evaporation time, nature and the temperature of the gelation media (Chakrabarty *et al.*, 2008; Mosqueda-Jimenez *et al.*, 2004; Rahimpour and Madaeni, 2007).

There are three components in this process: a polymer, a volatile solvent and a non-solvent for the polymer. The parameters that influence the ultimate structure of the membrane prepared by the phase inversion method are:

- *Choice of solvent*: Aprotic solvents like N-methyl pyrrolidinone and dimethyl formamide are generally the best solvents, which can dissolve a wide range of polymers. The casting solutions based on these solvents precipitate rapidly in water leaving porous membranes (Baker, 2004).
- *Choice of polymer*: The ideal polymer for the phase inversion method should be soluble in the suitable water miscible solvent. The polymer should also be tough, amorphous, not brittle thermoplastic, with a glass transition temperature of more than the expected use temperature. Polysulfone, cellulose acetate and polyetherimide are examples of polymers that meet these specifications (Baker, 2004).
- *Polymer concentration*: Higher polymer concentration at the thin film/non-solvent interface is formed by increasing the initial concentration of the polymer in the casting solution. This results in a lower porosity, less porous top layer and a lower flux (Mulder, 1997).

- *Precipitation medium*: The casting solution precipitation medium used is almost always water. Organic based solvent precipitation media like methanol generally cause a slow precipitation of the casting solution, which results in a denser membrane structure with lower flux (Baker, 2004).
- *Composition of the casting solution*: Membrane properties can be tailored by adding a small amount of modifiers like polymeric additives (e.g. polyvinyl pyrrolidinone), low solubility parameter solvents (e.g. acetone), and salts (e.g. zinc chloride) (Baker, 2004).

Mixed matrix membranes (MMMs), prepared by blending inorganic and organic materials, exhibit the properties of both ceramic and organic polymers (Huang *et al.*, 2008). There has been a great interest in polymer – inorganic nanocomposites over the last two decades. The addition of inorganic nanoparticles in a polymer matrix can provide high performance novel materials, which can be used in many industrial fields (Li *et al.*, 2010). Many studies in the preparation and characteristics of MMMs have reported the advantages of the inorganic fillers, such as suppression of macrovoids, enhanced mechanical strength and lifetime, and superior permeability with unchanged retention properties (Huang *et al.*, 2008).

Carbon nanotubes (CNTs) can be defined as tiny cylinders of single or multiple graphene layers that are closed at the end by half fullerene (C_{60}). The diameter of the CNTs range from 2.5 nm to 30 nm and with a length from a few tens of nanometers to several micrometers. There are three main structures of the CNTs, namely: (i) zigzag structure, (ii) armchair structure and (iii) helical structure (Harris, 1999). CNTs have unique mechanical properties (extremely high Young's modulus, stiffness and flexibility) and electronic properties (Santos *et al.*, 2006), high electrical and thermal conductivities, low coefficient of thermal expansion and high aspect ratio (Ash *et al.*, 2004). The unique properties of the CNTs make them attractive candidates for polymer composites.

CNTs are chemically inert and insoluble in water or organic solvents. Applications of the CNTs are hindered because of their stable structure and insolubility (Maser *et al.*, 2008). Functionalization of CNTs can increase their solubility and processability, which will allow the combination of the unique properties of the CNTs with other materials (Hirsch and Vostrowsky, 2005). There will be functional groups attached to the nanotube surface after functionalization. CNTs react readily with other chemical agents, with the help of functional groups, and form homogenous dispersions or well aligned materials (Zhang *et al.*, 2002).

CNTs can be functionalized covalently or non-covalently by adding functional groups to their surfaces (Najafi *et al.*, 2006). Both covalent and non-covalent functionalizations are exohedral derivatization. CNTs can be functionalized by filling the tubes with atoms or small molecules, which is called endohedral functionalization (Hirsch and Vostrowsky, 2005).

In the covalent functionalization method, functional groups are covalently linked on the CNT surface. Large numbers of functional groups can be introduced to the CNTs by covalent functionalization, such as amides, esters, thiol groups, fluorine, carbenes, nitrenes, etc.

CNTs can be oxidized by nitric acid, sulfuric acid, mixtures of both, piranha (mixture of sulfuric acid and hydrogen peroxide), gascous oxygen, and ozone at elevated temperatures. The oxidation of CNTs introduces carboxylic groups and other oxygen-bearing groups at the end of the tubes and at defect sites of the tubes. Oxidative treatment introduces defects on the nanotube surface, cuts and shortens the CNTs and opens the tips of the CNTs (Hirsch and Vostrowsky, 2005).

Non-covalent functionalization of the CNTs is based on the supramolecular complexation by the combined effects of hydrophobic force and van der Waals' attraction, electrostatic and π-stacking interactions (Hirsch and Vostrowsky, 2005; Xie and Soh, 2005). Functionalization of the CNT can be performed without destroying the CNT side walls. There are several types of non-covalent functionalization of the CNTs, which are briefly described in the following sub-sections (Bianco *et al.*, 2007).

In π-π stacking method, CNTs are functionalized by the π-π stacking interactions between the conjugated materials and the CNTs side walls. CNTs irreversibly adsorb the compounds with pyrene moiety, such as N-succinimidyl-1-pyrenebutanoate. Then various molecules with primary and secondary amines can be covalently attached by the succinimidyl ester group. A variety of molecules can be immobilized on CNTs by π-π stacking interactions, such as biomolecules (e.g. protein, DNA), gold nanoparticles, phthalocyanines, porphyrins, and polymers with conjugated structures (Bianco *et al.*, 2007).

Hydrophobic interactions are based on the association of the CNTs with amphiphilic molecules in aqueous media. Hydrophobic parts of the amphiphilic molecules interact non-covalently with the aromatic surface of the CNTs. CNTs have been wrapped with water soluble polymers (e.g. polyvinylpyrrolidone, polystyrenesulfonate) and surfactants (e.g. deoxycholic acid, taurodeoxycholic acid) by hydrophobic interactions (Bianco *et al.*, 2007).

Immersion precipitation is a well-known technique to prepare asymmetric membranes. In this process the homogenous mixture of polymer and solvent is cast on a support and immersed in a coagulation bath of non-solvent. The solvent in the cast film is exchanged with the non-solvent in the coagulation bath by immersing the cast film in a coagulation bath. Depending on the polymer and coagulation conditions many phase changes occur, namely: liquid-liquid phase change, solid-liquid phase change or both. Afterwards the polymer film solidifies and forms the asymmetric membrane with a dense top layer and porous sub-layer. There are many variables which affect the sub-layer formation of the membrane, such as composition of the dope solution, coagulant temperature, evaporation time, nature and the temperature of the gelation media (Chakrabarty *et al.*, 2008; Figoli *et al.*, 2015; Mosqueda-Jimenez *et al.*, 2004; Rahimpour and Madaeni, 2007). Aprotic solvents like N-methyl pyrrolidinone and dimethyl formamide are generally the best solvents and can dissolve a wide range of polymers. The casting solutions based on these solvents precipitate rapidly in water leaving porous membranes (Baker, 2004). However, recently most efforts are being made in the direction of employing "greener solvents" for replacing the classical aprotic solvents used in membrane preparation (Figoli *et al.*, 2014).

To date, several authors have shown the successful preparation of CNT blended polymeric membranes. For instance, Choi *et al.* (2006) prepared polysulfone/CNT composite membranes for filtration. Prior to the composite membrane synthesis, they functionalized CNTs with a mixture of sulfuric acid and nitric acid. They showed that hydrophilicity of the composite membranes increased with an increasing amount of CNT. Moreover, the permselective properties of the composite membranes depended on the amount of CNT.

Cong *et al.* (2007) prepared brominated poly(2,6-diphenyl-1,4-phenylene oxide)/CNT composite membranes for gas separation. They functionalized CNTs with carboxyl groups prior to composite membrane synthesis. They showed that the composite membranes had an increased tensile strength and CO_2 permeation but the selectivity of CO_2/N_2 did not change compared to the bare polymeric membrane.

Nechifor *et al.* (2009) prepared polysulfone/CNT composite membranes for hemodialysis. They used polysulfone and single walled carbon nanotubes (SW-CNTs) or amino double wall carbon nanotubes for composite membrane synthesis. Lead adsorption in membranes with double wall carbon nanotubes was higher than in the membranes with SW-CNTs. Composite membranes with 5% amino double wall carbon nanotubes presented superior retention for the removal of lead and mercury.

Brunet *et al.* (2008) prepared polysulfone/CNT composite membranes. In their work, CNT addition did not improve the hydrophilicity or permeability of the composite membranes. Also, composite membranes did not show any antibacterial activity. Moreover, composite membranes presented higher roughness and lower mechanical properties (lower elongation to failure) compared to the bare polymeric membranes.

Qiu *et al.* (2009) prepared polysulfone composite membranes with isocyanate and isophthaloyl chloride functionalized CNTs. They showed that the morphology and the permeation properties of the composite membranes were influenced by the amount of CNTs. Moreover, they showed that the composite membranes had a lower protein adsorption than the polysulfone membranes. However, the effects of the solvent ratio on these membranes has yet to be determined. Based on these considerations and the body of previous research, the objective of this work is to synthesize CNTs blended polyethersulfone (PES) membranes by using different ratios of N-methyl-2-pyrrolidone (NMP) to dimethylformamide (DMF) and then to determine protein fouling properties. To characterize the blend membranes, pure water flux tests were employed. Also, water contact angles of the membranes were determined. Then, to determine the fouling resistances of the blend membranes, permeation tests with bovine serum albumin (BSA) were conducted.

6.2   EXPERIMENTAL

6.2.1   *Materials*

PES (Veradel 3000P) was kindly supplied by Solvay Specialty Polymers (Germany). Multi-walled carbon nanotubes (MW-CNTs) were purchased from Nanostructured and Amorphous Materials Inc. (USA) and NMP and DMF were purchased from Sigma Aldrich (USA). In addition, BSA having an approximate molecular weight of 66 kDa, molecular size of 14 nm × 4 nm × 4 nm, and an isoelectric point (IEP) at pH 4.7–4.9 (Nakamura and Matsumoto, 2006) was purchased from Fluka (USA).

6.2.2   *CNT functionalization*

CNTs were functionalized to improve the dispersion in organic solvent and the interfacial bonding between the polymer matrix and CNTs (Zhu *et al.*, 2004) following a reported procedure (Liu *et al.*, 1998). In brief, CNTs were ultrasonicated in a mixture of 3:1 concentrated sulfuric acid and nitric acid for 9 h. Then, the CNTs were washed until the neutral pH was reached, followed by drying in a vacuum oven at 100°C overnight (Fig. 6.1).

The morphology of the bare and functionalized CNTs were then analyzed by transmission electron microscopy (TEM; Tecnai G2 F20, FEI, USA). The functional groups of the bare and functionalized CNTs were determined by Fourier transform infrared spectroscopy (FTIR, Spectrum Two, Perkin Elmer, USA).

6.2.3   *Membrane fabrication and characterization*

Membrane phenomena were first observed in the 18th century. Even though the elements of modern membrane science were developed in the 1960s, they were used only on a laboratory scale. Their cost, unreliability, low speed and unselectivity prohibited their widespread use as a separation process (Baker, 2004). In the last few decades membrane processes were developed within different industries (Blanco *et al.*, 2001). Membrane is the basic element of the membrane processes. The membrane is a permselective barrier or interface between two phases (Mulder, 1997).

Membrane has the ability of transporting some constituents more than others, so that separation is achieved (Mulder, 1997). The general benefits of the membrane technology are: (i) clean technology with easy operation; (ii) high value product recovery (Nath, 2008); (iii) easy application of hybrid systems; (iv) various and adjustable membrane properties; (v) continuous separation; (vi) no additive requirement (Mulder, 1997).

The general drawbacks of the membrane technology are (i) fouling (Mulder, 1997; Nath, 2008); (ii) short membrane lifetime (Mulder, 1997); (iii) cost (Nath, 2008).

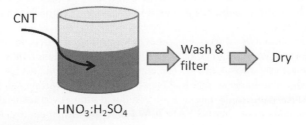

Figure 6.1.   CNT functionalization.

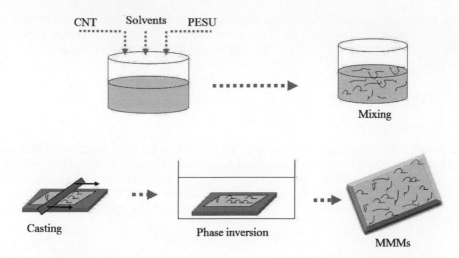

Figure 6.2.   CNT/PES MMMs preparation.

Table 6.2.   Compositions of the membranes.

Membrane	PES [%]	CNT [%]	NMP:DMF ratio
N:D-1:9	15	0.5	1:9
N:D-2:8	15	0.5	2:8
N:D-5:5	15	0.5	5:5
N:D-8:2	15	0.5	8:2
N:D-9:1	15	0.5	9:1

Producing organic-inorganic composite membranes with excellent separation performance and adaptable to rigorous environmental conditions has been attracting a great deal of interest (Yang *et al.*, 2006). CNTs have an exceptionally high aspect ratio and high strength and stiffness. Their unique properties make them attractive candidates for polymer composites (Gojny *et al.*, 2004).

CNTs blended PES membranes were synthesized *via* the phase inversion method as described in our previous publications (Celik and Choi, 2012; Celik *et al.*, 2011a, 2011b). In brief, casting solutions were prepared by ultrasonicating the 0.5% (with respect to PES by weight) of functionalized CNTs in a mixture of NMP and DMF and dissolving 15% of PES. The blend solutions were stirred at room temperature and ultrasonicated to remove the air bubbles. The blend solutions were then casted on a glass plate by using a casting knife and a film applicator (1133N, Sheen, USA) and immersed into a coagulation bath of DI water for phase inversion. After complete coagulation, the membranes were peeled off and washed with and stored in DI water until use (Fig. 6.2). The compositions of the membranes fabricated are given in Table 6.2. Note that the membranes marked as N:D-2:8 refer to membranes prepared in a casting solution in which the ratio of the NMP to DMF was 2 to 8.

The surface hydrophilicities of the membranes were evaluated based on the dynamic sessile drop method using a contact angle goniometer (T200, KSV, USA). Membranes were dried at 60°C overnight for contact angle measurements. The contact angles were measured by dropping 2 μL of DI water onto a dry membrane surface and the contact angle was measured. The average value and the corresponding standard deviation were calculated after seven measurements. One-way complete statistical analysis of variance (ANOVA) test at a confidence level of 95% applied to the results of contact angle measurements.

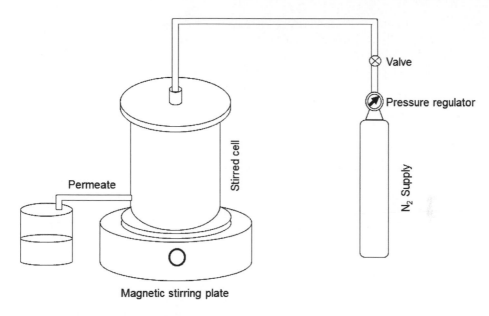

Figure 6.3.   Dead-end membrane test unit.

For determining the water contents of the membranes, the wet membranes were weighed first. These wet membranes were dried for 24 h at 60°C and weighed to determine the dry weights. Membrane water contents were determined as follows:

$$\text{Water content} = \frac{\text{weight of wet membrane} - \text{weight of dry membrane}}{\text{weight of wet membrane}} \times 100 \qquad (6.1)$$

In order to minimize the experimental errors, membrane water contents were measured at least three times and average values were reported.

6.2.4   *Water permeability experiments*

A dead-end membrane test unit (HP4750, Sterlitech, USA) was used for the permeation tests (Fig. 6.3). The test unit was activated by using compressed air as a pressure source. The effective membrane area used for permeation measurements was 14.6 cm^2. Membranes were initially operated at 0.28 MPa for 2 h for membrane compaction; after that, the pressure was reduced to 0.21 MPa for pure water flux determination. The pure water flux (J_{wv}) and hydraulic resistance of the membranes (R_{vm}, m^{-1}) were determined using:

$$J_{wv} = \frac{V}{A \Delta t} = \frac{\Delta P}{\mu R_{vm}} \qquad (6.2)$$

where V is the volume of the water permeated [L], A is the effective membrane area [m^2], and Δt is time [h], ΔP is the pressure [Pa], and μ water viscosity at 22°C [Pa s] (0.0096 Pa s).

The fouling behavior of the produced membranes were determined by protein aqueous filtration which was done by 1 g L^{-1} BSA filtration at 0.21 MPa for 1 h. The initial (J_{pi}) and final protein flux (J_{pf}) after 1 h of BSA filtration and BSA rejection ratio (R) was determined as follows:

$$R = \frac{C_f - C_p}{C_f} \times 100 \qquad (6.3)$$

$$J_{pf} = \frac{V}{A \Delta T} = \frac{\Delta P}{\mu R_{fm}} \qquad (6.4)$$

where, C_p and C_f [mg L^{-1}] are the BSA concentrations of the permeate and feed solutions measured with a UV-vis spectrophotometer at 280 nm, respectively, R_{fm} [m^{-1}] is the hydraulic resistance of the fouled membrane, which is the sum of the hydraulic resistance of the adsorbed protein and the hydraulic resistance of the concentration polarization layer.

6.3   RESULTS AND DISCUSSION

6.3.1   *CNT functionalization*

CNTs are chemically inert and insoluble in water or organic solvents. Applications of the CNTs are hindered because of their stable structure and insolubility (Maser *et al.*, 2008). Functionalization of CNTs can increase their solubility and processability, and will allow the combination of the unique properties of the CNTs with other materials (Hirsch and Vostrowsky, 2005). As shown in Figure 6.4, raw CNTs (a) were 10–30 μm in length, though treatment with a strong acid mixture shortened the lengths of the CNTs (b) to 250 nm to 2 μm.

The conversion of the terminal carbons of CNTs to carboxylic groups by functionalization is shown by the FTIR (Fig. 6.5). In the figure, even though there were no peaks observed on the raw CNTs (Fig. 6.5b), three new peaks at ~3440 cm^{-1} (−OH), ~1630 cm^{-1} (>C=O) (Kaniyoor *et al.*, 2009; Kim *et al.*, 2005), and ~1460 cm^{-1} (O−C=O) (Bolbukh *et al.*, 2010) were observed on functionalized CNTs (Fig. 6.5a). These observations indicate that the surfaces of the CNTs are functionalized by a strong acid mixture and the terminal carbons are converted to carboxylic groups.

6.3.2   *Membrane fabrication and characterization*

Polymeric membranes are widely used in membrane operations but they have low mechanical and chemical resistances. In addition, the hydrophobic nature of polymeric membranes causes fouling (Peng *et al.*, 2007). Hence, increased hydrophilicity might lower the membrane fouling. Blending CNTs in the polymeric membranes increases the membrane hydrophilicity (Celik *et al.*, 2011a). As shown in Figure 6.6, blending CNTs increased the hydrophilicities of the membranes. However, changing the solvent ratio did not result in changes in the contact angles in CNT blended membranes (by ANOVA tests) which shows similar hydrophilicities. As a result, even though CNT

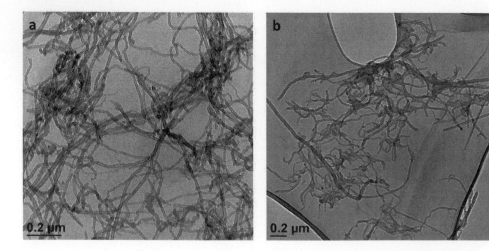

Figure 6.4.   Transmission electron micrographs of raw CNTs (a) and functionalized CNTs (b).

addition increases the membrane hydrophilicity, membrane hydrophilicity is not dependent on the solvent ratios.

One of the structural parameters of the membranes is membrane water content, which depends on the porosities of the membranes. Water content is also indicative for the membrane

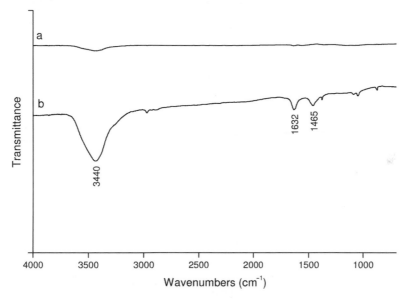

Figure 6.5. The FTIR spectra of raw CNTs (a) and functionalized CNTs (b). Source: Celik-Madenli and Cakmakci (2017).

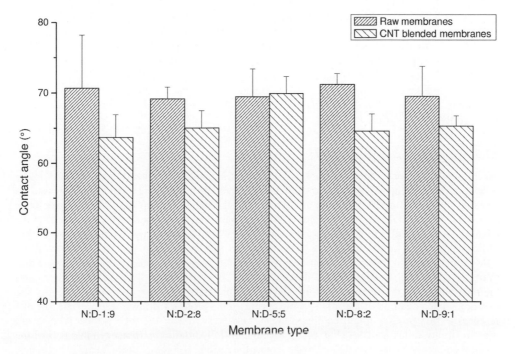

Figure 6.6. Contact angles of the blend membranes.

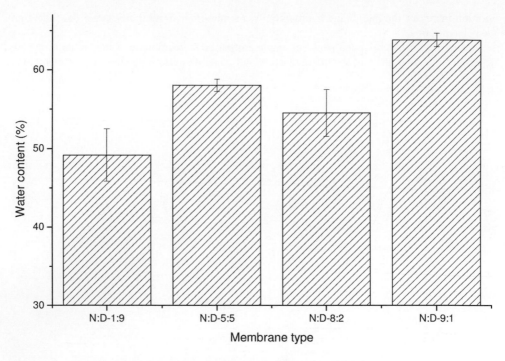

Figure 6.7.   Water contents of the blend membranes as a function of solvent ratio.

hydrophilicity and flux behaviors (Arthanareeswaran *et al.*, 2004). As shown in Figure 6.7, water contents of the membranes increased by increasing NMP ratio. Since the hydrophilicities of the membranes are not dependent on solvent ratio (Fig. 6.6), the increase in water content might be indicating the increase in porosities of the membranes by increasing NMP ratio.

The cross-section morphologies of the forward osmosis membranes, synthesized by using pure PES and CNT doped PES membranes as a support layer, are characterized by SEM micrographs and shown in Figure 6.8. The support layer of both the membranes showed a typical asymmetric membrane structure with a dense top layer, a porous sub-layer, and fully developed macropores at the bottom.

6.3.3   *Water permeation experiments*

An important parameter in the design and economic feasibility of filtration membranes is the permeate flux. The pure water fluxes of the membranes fabricated (Fig. 6.9) were increased by increasing the NMP ratio, which is consistent with the water content data of the membranes (Fig. 6.7). Moreover, Bruggen and Kim (2012) demonstrated the higher permeabilities of NMP membranes than DMF membranes, which is consistent with our findings, as the pure water flux of N:M-9:1 is three times higher than N:M-1:9.

Membrane fouling causes a decline in permeability and also reduces the membrane life (Yang *et al.*, 2007). Hydrodynamic conditions and chemical interactions of the foulants and the membranes are the main factors influencing the membrane fouling (Hua *et al.*, 2008). The hydrodynamic conditions were all the same during all the protein filtration tests. Hence, the differences in flux profiles might be due to the surface properties of the membranes. As a result, protein filtration flux profiles (Fig. 6.10), together with hydraulic resistances (Table 6.3) of the blend membranes, were investigated to determine the fouling behaviors of the blend membranes. The protein fluxes are given as relative fluxes because the pure water fluxes of all the blend

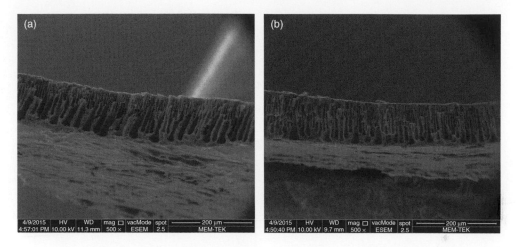

Figure 6.8.   SEM images of the pure PES support layers (a) and CNT doped PES support layer (b).

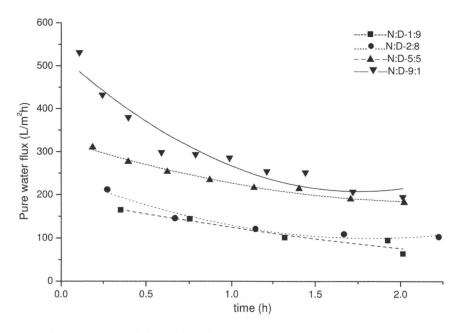

Figure 6.9.   Pure water fluxes of the blend membranes as a function of solvent ratio.

membranes are different. R_{vm} represents the hydraulic resistance of the virgin membranes due to the porosity, cross-sectional structure and pore size. R_{fm} represents the hydraulic resistance of the membranes due to fouling and concentration polarization (Huang *et al.*, 2008). As is shown in Figure 6.10 and Table 6.3, there are very slight differences between the relative fluxes or hydraulic resistances of the blend membranes with respect to solvent ratio. Even though the fouling resistances of the membranes increased by blending CNTs (Celik *et al.*, 2011b), changing solvent ratio does not affect the fouling resistances of the blend membranes.

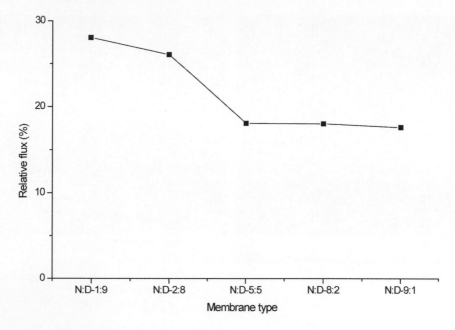

Figure 6.10.   Relative fluxes of the blend membranes.

Table 6.3.   Hydraulic resistances of the blend membranes.

Membrane type	R_{vm} $[\times 10^9$ m$^{-1}]$	R_{fm} $[\times 10^9$ m$^{-1}]$
N:D-1:9	1.19	3.45
N:D-2:8	0.76	3.69
N:D-5:5	0.43	3.03
N:D-8:2	0.79	2.53
N:D-9:1	0.40	2.46

6.4   CONCLUSIONS

CNTs blended membranes with different NMP:DMF ratios were prepared *via* the phase inversion method. From the findings of this study the following conclusions can be drawn:

- Even though the addition of CNTs improved the hydrophilicity of the blend membranes, changes in solvent ratio did not affect the hydrophilicity of the blend membranes.
- The water contents of the blend membranes with high NMP:DMF ratios were significantly higher than the blend membranes with lower NMP:DMF ratios.
- The pure water fluxes of the blend membranes increased significantly by increasing the NMP:DMF ratios.
- The fouling resistances of the blend membranes increased with CNT additions, while they are not dependent on the solvent ratio.

From this work, a fundamental understanding of the effects of the solvent ratio on CNT/PES MMMs fabrication can be achieved. As part of continuing efforts to understand the clear mechanism of the effects of the solvent ratio on CNT/PES MMMs, ongoing works are being made by applying different characterization methods.

ACKNOWLEDGMENTS

The funding was provided by the Scientific and Technological Research Council of Turkey (TUBITAK) through the Scientific and Technological Research Projects Funding Program (111R012).

REFERENCES

Aptel, P. & Buckley, C. (1996) Categories of membrane operations. In: American Water Works Association (ed) *Water Treatment Membrane Processes*. McGraw-Hill, New York City. 2.1–2.24.

Arthanareeswaran, G., Thanikaivelan, P., Srinivasn, K., Mohan, D. & Rajendran, M. (2004) Synthesis, characterization and thermal studies on cellulose acetate membranes with additive. *European Polymer Journal*, 40(9), 2153–2159.

Ash, B.J., Eitan, A. & Schadler, L.S. (2004) Polymer nanocomposites with particle and carbon nanotube fillers. In: Schwarz, J.A., Contescu, C.I. & Putyera, K. (eds) *Dekker Encyclopedia of Nanoscience and Nanotechnology*. CRC Press, Boca Raton, FL.

Baker, R.W. (2004) *Membrane Technology and Applications*. 2nd edition, Maidenhead, UK, McGraw-Hill.

Bianco, A., Wu, W., Pastorin, G. & Klumpp, C. (2007) Carbon nanotube based vectors for delivering immunotherapeutics and drugs. In: Kumar, C.S.S.R. (ed) *Nanomaterials for Medical Diagnosis and Therapy*. Wiley VCH, Weinham, Germany. pp. 85–142.

Blanco, J.F., Nguyen, Q.T. & Schaetzel, P. (2001) Novel hydrophilic membrane materials: sulfonated polyethersulfone Cardo. *Journal of Membrane Science*, 186(2), 267–279.

Bolbukh, Y., Gunko, G., Prikhod'ko, G., Tertykh, V., László, K., Tóth, A. & Koczka, B. (2010) Multi-walled carbon nanotubes modified with biocompatible compounds. *Chemistry, Physics and Technology of Surface*, 1(4), 389–400.

Bruggen, B.V. & Kim, J. (2012) Nanofiltration of aqueous solutions. In: Buonomenna, B.G. & Golemme, G. (eds) *Recent Developments and Progresses in advanced Materials for Membrane Preparation*. Sharjah, UAE, Bentham Books.

Brunet, L., Lyon, D.Y., Zodrow, K., Rouch, J.C., Caussat, B., Serp, P., Remigy, J.C., Wiesner, M.R. & Alvarez, P.J.J. (2008) Properties of membranes containing semi-dispersed carbon nanotubes. *Environmental Engineering Science*, 25(4), 565–575.

Cardew, P.T. & Le, M.S. (1998) Membrane processes: a technology guide. The Royal Society of Chemistry, London, UK.

Celik, E. & Choi, H. (2012) Carbon canotube/polyethersulfone composite membranes for water filtration. In: Escobar, I. & Van der Bruggen, B. (eds) *Modern Applications in Membrane Science and Technology*. Oxford University Press, Oxford, UK.

Celik, E., Liu, L. & Choi, H. (2011b) Protein fouling behavior of carbon nanotube/polyethersulfone composite membranes during water filtration. *Water Research*, 45, 5287–5294.

Celik, E., Park, H., Choi, H. & Choi, H. (2011a) Carbon nanotube blended polyethersulfone membranes for fouling control in water treatment. *Water Research*, 45, 274–282.

Celik Madenli, E. & Cakmakci, O. (2017) Preparation and characterization of PAN/CNT nanocomposite fiber supports for membrane filtration. *Desalination and Water Treatment*, 60, 137–143.

Chakrabarty, B., Ghoshal, A.K. & Purkait, A.K. (2008) Preparation, characterization and performance studies of polysulfone membranes using PVP as an additive. *Journal of Membrane Science*, 315(1–2), 36–47.

Chen, J.P., Mou, H., Wang, L.K. & Matsuura, T. (2006) Membrane filtration. In: Wang, L.K., Hung, Y.T. & Shammas, N.K. (eds) *Advanced Physicochemical Treatment Processes*. Humana Press Inc., Totowa, NJ. pp. 203–260.

Choi, J.H., Jegal, J. & Kim, W.N. (2006) Fabrication and characterization of multi-walled carbon nanotubes/polymer blend membranes. *Journal of Membrane Science*, 284(1–2), 406–415.

Cong, H.L., Zhang, J.M., Radosz, M. & Shen, Y.Q. (2007) Carbon nanotube composite membranes of brominated poly(2,6-diphenyl-1,4-phenylene oxide) for gas separation. *Journal of Membrane Science*, 294(1–2), 178–185.

Dawson, G.W. & Merccs, B.W. (1986) *Hazardous Waste Management*. Wiley, New York.

Figoli, A., Marino, T., Simone, S., Di Nicolo, E., Li, X.M., He, T., Tornaghi, S. & Drioli, E. (2014) Towards non-toxic solvents for membrane preparation: A review. *Green Chemistry*, 16, 4034–4059.

Figoli, A., Simone, S. & Drioli, E. (2015) Polymeric membranes. In: Hilal, N., Ismail, A.F. & Wright, C. (eds) *Membrane Fabrication*. CRC Press, Boca Raton, FL. pp. 3–44.

Galiano, F., Figoli, A., Deowan, S.A., Johnson, D., Alsoy Altinkaya, S., Veltri, L., De Luca, G., Mancuso, R., Hilal, N., Gabriele, B. & Hoinkis, J. (2015) A step forward to a more efficient wastewater treatment by membrane surface modification via polymerizable bicontinuous microemulsion. *Journal of Membrane Science*, 482, 103–114.

Gojny, F.H., Wichmann, M.H.G., Kopke, U., Fiedler, B. & Schulte, K. (2004) Carbon nanotube-reinforced epoxy-compo sites: enhanced stiffness and fracture toughness at low nanotube content. *Composites Science and Technology*, 64(15), 2363–2371.

Harris, J.F. (1999) *Carbon Nanotubes and Related Structures: New Materials for the Twenty-first Century*. Cambridge University Press, Cambridge, UK.

Hirsch, A. & Vostrowsky, O. (2005) Functionalization of carbon nanotubes. In: Schlüter, A.D. (ed) *Functional Molecular Nanostructures*. Springer, Amsterdam. pp. 193–237.

Hua, H., Li, N., Wu, L., Zhong, H., Wu, G., Yuan, Z., Lin, X. & Tang, L. (2008) Anti-fouling ultrafiltration membrane prepared from polysulfone-graft-methyl acrylate copolymers by UV-induced grafting method. *Journal of Environmental Science*, 20, 565–570.

Huang, Z.Q., Chen, K., Li, S.N., Yin, X.T., Zhang, Z. & Xu, H.T. (2008) Effect of ferrosoferric oxide content on the performances of polysulfone-ferrosoferric oxide ultrafiltration membranes. *Journal of Membrane Science*, 315(1–2), 164–171.

Kaniyoor, A., Jafri, R.I., Arockiadoss, T. & Ramaprabhu, S. (2009) Nanostructured Pt decorated graphene and multi walled carbon nanotube based room temperature hydrogen gas sensor. *Nanoscale*, 1(3), 382–386.

Kim, U.J., Furtado, C.A., Liu, X.M., Chen, G.G. & Eklund, P.C. (2005) Raman and IR spectroscopy of chemically processed single-walled carbon nanotubes. *Journal of the American Chemical Society*, 127(44), 15,437–15,445.

Li, S.N., Lin, M.M., Toprak, M.S., Kim, D.K. & Muhammed, M. (2010) Nanocomposites of polymer and inorganic nanoparticles for optical and magnetic applications. *Nano Reviews*, 1, 5214.

Liu, J., Rinzler, A.G., Dai, H., Hafner, J.H., Bradley, R.K., Boul, P.J., Lu, A., Iverson, T., Shelimov, K., Huffman, C.B., Rodriguez-Macias, F., Shon, Y.S., Lee, T.R., Colbert, D.T., Smalley, R.E. (1998) Fullerene pipes. *Science*, 280, 1253–1256.

Maser, W., Benito, A.M., Munoz, E. & Martinez, M.T. (2008) Carbon nanotubes: from fundamental nanoscale objects towards functional nanocomposites and applications. In: Vaseashta, A. & Mihailescu, I.N. (eds) *Functionalized nanoscale Materials, Devices and Systems. NATO Science for Peace and Security* Series-B: *Physics and Biophysics*. Springer, The Netherlands. pp. 101–119.

Mosqueda-Jimenez, D.B., Narbaitz, R.M., Matsuura, T., Chowdhury, G., Pleizier, G. & Santerre, J.P. (2004) Influence of processing conditions on the properties of ultrafiltration membranes. *Journal of Membrane Science*, 231(1–2), 209–224.

Mulder, M. (1997) *Basic Principles of Membrane Technology*. 2nd edition, Kluwer Academic Publishers, The Netherlands.

Najafi, E., Kim, J.Y., Han, S.H. & Shin, K. (2006) UV-ozone treatment of multi-walled carbon nanotubes for enhanced organic solvent dispersion. *Colloids and Surfaces* A: *Physicochemical and Engineering Aspects*, 284, 373–378.

Nakamura, K. & Matsumoto, K. (2006) Properties of protein adsorption onto pore surface during micro-filtration: effects of solution environment and membrane hydrophobicity. *Journal of Membrane Science*, 280(1–2), 363–374.

Nath, K. (2008) *Membrane Separation Processes*. Prentice-Hall of India, New Delhi, India.

Nechifor, G., Voicu, S.I., Nechifor, A.C. & Garea, S. (2009) Nanostructured hybrid membrane polysulfone-carbon nanotubes for hemodialysis. *Desalination*, 241(1–3), 342–348.

Peng, F.B., Hu, C.L. & Jiang, Z.Y. (2007) Novel poly(vinyl alcohol)/carbon nanotube hybrid membranes for pervaporation separation of benzene/cyclohexane mixtures. *Journal of Membrane Science*, 297(1–2), 236–242.

Qiu, S., Wu, L.G., Pan, X.J., Zhang, L., Chen, H.L. & Gao, C.J. (2009) Preparation and properties of functionalized carbon nanotube/PSF blend ultrafiltration membranes. *Journal of Membrane Science*, 342(1–2), 165–172.

Rahimpour, A. & Madaeni, S.S. (2007) Polyethersulfone (PES)/cellulose acetate phthalate (CAP) blend ultrafiltration membranes: preparation, morphology, performance and antifouling properties. *Journal of Membrane Science*, 305(1–2), 299–312.

Reddy, A.V.R. & Patel, H.R. (2008) Chemically treated polyethersulfone/polyacrylonitrile blend ultrafiltration membranes for better fouling resistance. *Desalination*, 221(1–3), 318–323.

Santos, C.V., Hernandez, A.L.M. & Castano, V.M. (2006) Carbon nanotube – polymer nanocomposites: principles and applications. In: Martin, D.A. (ed) *Focus on Nanotube Research*. Nova Science Publisher, New York, NY. pp. 97–126.

Shi, Q., Su, Y.L., Zhu, S.P., Li, C., Zhao, Y.Y. & Jiang, Z.Y. (2007) A facile method for synthesis of pegylated polyethersulfone and its application in fabrication of antifouling ultrafiltration membrane. *Journal of Membrane Science*, 303(1–2), 204–212.

Wang, Y.Q., Su, Y.L., Sun, Q., Ma, X.L. & Jiang, Z.Y. (2006) Generation of anti-biofouling ultrafiltration membrane surface by blending novel branched amphiphilic polymers with polyethersulfone. *Journal of Membrane Science*, 286(1–2), 228–236.

Williams, M.E., Bhattacharyya, D., Ray, R.J. & McCay, S.B. (1992) Selected applications. In: Ho, W.S.W. & Sirkar, K.K. (eds) *Membrane Handbook*. Van Nostrand Reinhold, New York, NY. pp. 312–354.

Xie, Y.H. & Soh, A.K. (2005) Investigation of non-covalent association of single-walled carbon nanotube with amylose by molecular dynamics simulation. *Materials Letters*, 59(8–9), 971–975.

Yang, Y.N., Wang, P. & Zheng, Q.Z. (2006) Preparation and properties of polysulfone/TiO_2 composite ultrafiltration membranes. *Journal of Polymer Science* B: *Polymer Physics,* 44(5), 879–887.

Yang, Y.N., Zhang, H.X., Wang, P., Zheng, Q.Z. & Li, J. (2007) The influence of nano-sized TiO_2 fillers on the morphologies and properties of PSFUF membrane. *Journal of Membrane Science*, 288(1–2), 231–238.

Zhang, N.Y., Me, J. & Varadan, V.K. (2002) Functionalization of carbon nanotubes by potassium permanganate assisted with phase transfer catalyst. *Smart Materials & Structures*, 11(6), 962–965.

Zhu, J., Peng, H.Q., Rodriguez-Macias, F., Margrave, J.L., Khabashesku, V.N., Imam, A.M., Lozano, K., Barrera, E.V. (2004) Reinforcing epoxy polymer composites through covalent integration of functionalized nanotubes. *Advanced Functional Materials*, 14(7), 643–648.

CHAPTER 7

Photocatalytic activity and synthesis procedures of TiO$_2$ nanoparticles for potential applications in membranes

Tiziana Marino, Marcel Boerrigter, Mirko Faccini, Christiane Chaumette, Lawrence Arockiasamy, Jochen Bundschuh & Alberto Figoli

7.1 INTRODUCTION

Heterogeneous photocatalysis based on oxide semiconductors is a promising technique for the prevention of microbial proliferation and to limit the growth of microorganisms. Among the studied photocatalysts, titanium dioxide (TiO$_2$) represents one of the most interesting materials, due to its low cost, biocompatibility, chemical and thermal stability, and notable optical and dielectric properties. It exists in three polymorphic forms: rutile, which is the most stable form, anatase and brookite, both of which are metastable and convert into rutile upon heating. TiO$_2$ nanoparticles can be efficiently obtained *via* different synthesis techniques, such as sol-gel, sol, hydrothermal, solvothermal, direct oxidation, chemical or physical vapor deposition, microwave, and reverse micelle methods, which offer the possibility to obtain well-controlled nanoparticle size and morphology. In the different TiO$_2$ applications, its antimicrobial action has attracted a lot of attention in the last few decades. Particularly promising are the hybrid or mixed matrix TiO$_2$-polymeric membranes, which allow separation and simultaneous photocatalytic reaction, without requiring any catalyst recovery operation. Several works have been published on the efficiency of the hybrid inorganic-organic membranes as antimicrobial systems, for bacteria, viruses, fungi and algae disruption. The TiO$_2$ incorporation in/on the polymeric membrane also allows the reduction of fouling and at the same time highly improves water permeability and self-cleaning ability.

In this chapter, a brief overview is given of the photocatalytic process, TiO$_2$ oxidative action, the environmental/human health risks associated with the use of this semiconductor, as well as an investigation of the TiO$_2$ nanoparticles synthesis methods. Examples evidencing the possibility to efficiently apply TiO$_2$-based polymeric membranes for disinfection applications are described. Finally, the difficulties in the optimization of the TiO$_2$-polymeric casting solution preparation, which still limits the use of TiO$_2$-based membranes on an industrial scale, are presented.

7.2 PHOTOCATALYSIS AND SEMICONDUCTORS

Since the first energy crisis in the early 1970s and the resulting necessity for a search for new alternative sources of energy, much research has been devoted to the development of efficient systems that would enable the use of renewable resources. In this context, heterogeneous photo-catalysis represents a promising technology to convert solar energy into chemical energy at an ambient temperature and pressure. Photocatalysis could be applied as an environmentally friendly, versatile and low-cost treatment alternative for water and air purification (Fujishima and Honda, 1972; Geng and Chen, 2011; Strini and Schiavi, 2011; Yu *et al.*, 2010), for the destruction of microorganisms such as bacteria (Gumy *et al.*, 2006) and viruses (Blake *et al.*, 1999), for nitrogen fixation (Petriconi and Papee, 1983), for the inactivation of cancer cells (Kabachkov *et al.*, 2011), for odor control (Liu T.-X. *et al.*, 2008), for the clean-up of oil spills (Hsu *et al.*, 2008) and for the water splitting to produce molecular hydrogen as high-energy and ecologically clean fuel

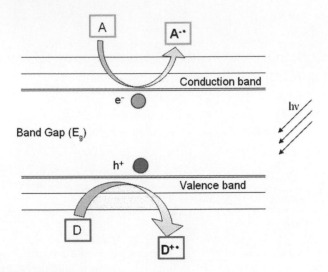

Figure 7.1.   Simplified reaction scheme of photocatalysis.

(Al-Rasheed, 2005; Primo *et al.*, 2011; Ravelli *et al.*, 2009; Silva *et al.*, 2011). Heterogeneous photocatalysis was defined by Palmisano and Sclafani (1997) as "a catalytic process during which one or more reaction steps occur by means of electron-hole pairs photogenerated on the surface of semiconducting materials illuminated by light of suitable energy". As a direct consequence of this definition, both catalyst and light are necessary to induce a chemical process. In fact, upon irradiation excited states of the photocatalyst are generated and can allow subsequent processes like reduction-oxidation reactions and substrate transformations. The basic mechanism of heterogeneous photocatalysis has been investigated by many research groups (Hoffmann *et al.*, 1995; Litter, 1999) and can be summarized as illustrated in Figure 7.1.

The irradiation of a semiconductor with light of energy equal or higher than its band gap energy (E_g) leads to the promotion of electrons (e^-) from the valence band to the conduction band, at the same time leaving positive holes (h^+) in the valence band. The photogenerated electron-hole pairs can promote oxidation-reduction reactions with electron donor (D in Fig. 7.1) and electron acceptor (A in Fig. 7.1) adsorbed on the catalysts surface or located within the electrical double layer surrounding the charged particles. The redox process competes with a possible electrons-holes recombination, which occurs within a few nanoseconds, with a consequent input energy dissipation as heat. The thermodynamic requirement necessary for the occurrence of such a process is a more negative potential for the electrons photogenerated in the conduction band in comparison with the reduction potential of the electron acceptor, and a more positive potential for the holes of the valence band compared with the oxidation potential of the donor.

An ideal photocatalyst should exhibit the following properties (Al-Rasheed, 2005):

- High photoactivity.
- Biological and chemical inertness.
- Stability towards corrosion.
- Mechanical robustness.
- Suitability towards visible or near UV light.
- Low cost.
- Lack of toxicity.

A wide range of inorganic semiconductors can be used for photocatalytic applications. Band gap energies of the most commonly used semiconductors are shown in Table 7.1.

Table 7.1.   Band gap energies of semiconductors used in photocatalytic processes.

Photocatalyst	Band gap energy [eV]
Si	1.1
WSe$_2$	1.2
WO$_3$	2.8
α-Fe$_2$O$_3$	2.2
V$_2$O$_5$	2.7
SiC	3.0
BaTiO$_3$	3.3
CdO	2.1
CdS	2.4
CdSe	1.7
Fe$_2$O$_3$	3.1
TiO$_2$ rutile	3.0
TiO$_2$ anatase	3.2
SrTiO$_3$	3.4
SnO$_2$	3.5
GaAs	1.4
SrTiO$_3$	3.4
ZnS	3.7
ZnO	3.2

Nevertheless, some of the photocatalysts do not ensure long-term activity during the photocatalytic process. Binary metal sulfides, such as CdS, CdSe or PbS, with narrow band gaps, which offer the advantage of absorbing the visible light, represent an example of unstable semiconductors for catalysis in aqueous media as they suffer photoanodic corrosion (Howe, 1998). To overcome this drawback, several works propose the addition of sulfide and sulfite to the contacting solution (Beydoun *et al.*, 1999; Fox and Dulay, 1993). These materials are also known to be toxic. Hematite (α-Fe$_2$O$_3$), having a band gap of about 2.2 eV, could be a potential photocatalyst in the visible range, but shows much lower activity than TiO$_2$ or ZnO, probably because of corrosion or the formation of short-lived metal-to-ligand or ligand-to-metal charge-transfer states (Trillas *et al.*, 1992). Although ZnO and TiO$_2$ anatase possess band gap energies of about 3.2 eV, zinc oxide is more unstable in illuminated aqueous solutions, with hydroxide species being formed on the particle surface leading to catalyst deactivation (Navaladian *et al.*, 2007). Wu (2004) also noticed a higher TiO$_2$ photocatalytic activity in comparison to that of SnO$_2$. WO$_3$ was investigated as a promising semiconductor, especially in view of its remarkable photostability in acidic media, which makes it a powerful material for many types of photocatalytic processes (Monllor-Satokca *et al.*, 2006). It has a band gap of about 2.8 eV, which is 0.4 eV narrower than that of TiO$_2$, so it can absorb more visible light from sunlight (Santato *et al.*, 2001). However, this oxide is also generally less active catalytically than TiO$_2$, and unsuitable for achieving an efficient oxidative reaction, probably due to the inability of the electrons generated in the conduction band to directly reduce molecular oxygen (Joshi *et al.*, 2011).

7.2.1   *TiO$_2$ properties*

Having high photocatalytic activity and chemical stability, non-toxicity and low cost, titanium dioxide has become the most widely used photocatalyst. It has been widely used as white pigment in paints, plastic, paper, cosmetics and foodstuffs (Gàzquez *et al.*, 2014). Three crystalline forms

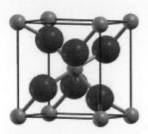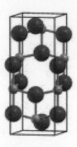

Figure 7.2.   The schematic conventional cells for rutile, brookite and anatase TiO₂ (from left to right). (adapted from Esch *et al.*, 2014).

Table 7.2.   Crystal structure parameters of anatase and rutile TiO₂ (Mathews and Antony, 2015).

	Anatase	Rutile
Crystal structure	tetragonal	tetragonal
Lattice constants [Å]	$a = 3.784$	$a = 4.5936$
	$c = 9.515$	$c = 2.9587$
Space group	$I4_1/amd$	$P4_2/mnm$
Molecule/cell	4	2
Volume/molecule [Å³]	34.061	31.216
Density [g/cm³]	3.79	4.13
Ti-O bond length [Å]	1.965(4)	1.949(4)
	1.965(2)	1.980(2)
O-Ti-O bond angle [°]	77.7	81.2
	92.6	90.0

exist in nature: anatase (tetragonal), rutile (tetragonal) and brookite (orthorhombic). The unit cells of the TiO₂ crystal structures are presented in Figure 7.2 (Esch *et al.*, 2014).

The big red spheres represent Ti atoms and the small grey spheres represent O atoms (Esch *et al.*, 2014).

Rutile is the most stable, chemically inert form, which can be excited by both visible and UV light (Austin and Lim, 2008; Hashimoto *et al.*, 2005). Anatase can be activated by UV light and can be transformed into rutile at high temperatures (400–700°C). Both of these two phases present a tetragonal ditetragonal dipyramidal crystal system but they differ from each other by the different space group lattices. Brookite undergoes a non-reversible phase transition into rutile when heated at temperatures between 700 and 900°C. TiO₂ can also exist in additional, less common, crystal structures, such as columbite, baddeleyite, hollandite, ramsdellite and monoclinic structures (Kavan *et al.*, 2011). The majority of these phases occur only at high pressure. Rutile is the thermodynamically most stable form, and, having a higher refractive index than anatase, is preferred for optical applications. On the other hand, anatase is the most suitable form for the photocatalytic processes, gas sensing, and solar cells, due to higher charge carrier mobility and its catalytic properties (Bagheri *et al.*, 2014; Banerjee, 2011; Yang *et al.*, 2014). The crystal structure data of the titania anatase and rutile is listed in Table 7.2 (Mathews and Antony, 2015).

This phase transition is affected by several factors, such as particle size, strain, pressure and texture. It has been reported that the temperature of anatase-rutile transition increases with increasing particle size and that a decrease in the nanoscale particle size, e.g. from 23 to 12 nm as reported by Li *et al.* (2004), leads to a higher anatase stability. TiO₂ has a band gap energy of ∼3.2 eV, hence its photoactivation requires radiation with light of wavelength less than or equal to ∼380 nm, with a maximum activation at approximately 340 nm (Mathews and Antony, 2015). The valence band of wide-gap rutile and anatase is made of O $2p$ states, while the conduction band is formed by Ti $3d$ states (Henrich and Cox, 1994). The electrical properties depend on the crystallographic direction. Being a wide-gap semiconductor, titania crystals present high resistivity ($\sim$1015 Ωcm) (Ardakani, 1994) and different types of ionic defects: bulk oxygen vacancies, titanium interstitials, titanium vacancies and reduced crystal surfaces, which are considered to generate shallow electron donor levels contributing to the electric conductivity of TiO$_2$ (Paxton and Thien-Nga, 1998). One of the most used commercial TiO$_2$ materials for photocatalytic oxidation applications is TiO$_2$ Degussa P25. It is composed of a mixture of 80% of anatase and 20% of rutile, with a surface area of 50 m^2 g^{-1} and an average particle size of 21 nm (Ohno *et al.*, 2001).

7.2.2 *Mechanisms of TiO$_2$ photocatalytic activity*

TiO$_2$ is a semiconductive oxide that, under irradiation, acts as a strong oxidizing material, lowering the activation energy for the decomposition of organic/inorganic molecules. Upon excitation with sufficient energy, an electron-hole pair on the TiO$_2$ surface is generated by the photon. For TiO$_2$, the recombination rate of electrons and hole is generally in the nanosecond timescale at room temperature.

The energy required to promote electrons depends on the semiconductor band gap, which represents the minimum energy necessary to convert the material into an electrically conductive substance.

The band gap energy for TiO$_2$ anatase is 3.2 eV and that for the rutile phase is 3.0 eV, which correspond to photons with a wavelength of 388 nm and 413 nm, respectively (De Lasa *et al.*, 2005). Upon excitation by light with wavelengths less than 388 and 413 nm, respectively, an electron-hole pair on the TiO$_2$ surface is generated by the photon energy. Consequently, the generated hole in the valence band can react with water/hydroxide ions, which are adsorbed on the semiconductor surface, producing hydroxyl radicals. The mechanism of the photocatalytic process using this semiconductor was proposed by several researchers (Kabra *et al.*, 2004; Litter, 1999; Teichner, 2008) according to the following equations:

$$\text{TiO}_2 \xrightarrow{h\nu} \text{TiO}_2(\text{e}_{\text{CB}}^-, \text{h}_{\text{VB}}^+) \tag{7.1}$$

$$\text{TiO}_2(\text{h}_{\text{VB}}^+) + \text{H}_2\text{O}_{\text{ads}} \rightarrow \text{TiO}_2 + \text{H}^+ + \text{HO}_{\text{ads}}^{\bullet} \tag{7.2}$$

$$\text{TiO}_2(\text{h}_{\text{VB}}^+) + \text{HO}_{\text{ads}}^- \rightarrow \text{TiO}_2 + \text{HO}_{\text{ads}}^{\bullet} \tag{7.3}$$

Donor (D) molecules will adsorb and react with a hole in the valence band and an acceptor (A) will also be adsorbed and react with the electron in the conduction band, following the reactions:

$$\text{TiO}_2(\text{h}_{\text{VB}}^+) + \text{D}_{\text{ads}} \rightarrow \text{TiO}_2 + \text{D}_{\text{ads}}^+ \tag{7.4}$$

$$\text{TiO}_2(\text{e}_{\text{CB}}^-) + \text{A}_{\text{ads}} \rightarrow \text{TiO}_2 + \text{A}_{\text{ads}}^- \tag{7.5}$$

Oxygen molecules can trap the electrons generated in the conduction band, forming superoxide ion ($\text{O}_2^{\bullet-}$), which can react with hydrogen ions deriving from water, thus leading to the formation of HO$_2^{\bullet}$ and H$_2$O$_2$. The decomposition of H$_2$O$_2$ may yield an OH radical:

$$\text{TiO}_2(\text{e}_{\text{CB}}^-) + \text{O}_{2\text{ads}} + \text{H}^+ \rightarrow \text{TiO}_2 + \text{HO}_2^{\bullet} \rightarrow \text{O}_2^{\bullet-} + \text{H}^+ \tag{7.6}$$

$$\text{HO}_2^{\bullet} + \text{TiO}_2(\text{e}_{\text{CB}}^-) + \text{H}^+ \rightarrow \text{H}_2\text{O}_2 \tag{7.7}$$

$$2\text{HO}_2^{\bullet} \rightarrow \text{H}_2\text{O}_2 + \text{O}_2 \tag{7.8}$$

$$\text{H}_2\text{O}_2 + h\nu \rightarrow 2\text{HO}^{\bullet} \tag{7.9}$$

$$\text{H}_2\text{O}_2 + \text{O}_2^{\bullet-} \rightarrow \text{HO}^{\bullet} + \text{O}_2 + \text{HO}^- \tag{7.10}$$

$$\text{H}_2\text{O}_2 + \text{TiO}_2(\text{e}_{\text{CB}}^-) \rightarrow \text{HO}^{\bullet} + \text{HO}^- + \text{TiO}_2 \tag{7.11}$$

A could also be a metal ion (*M*) with an appropriate redox potential to undergo a change of its oxidation state:

$$M^{n+} + TiO_2(ze_{CB}^-) \rightarrow M^{(n-z)+} \qquad (7.12)$$

The adsorption of organic species or water is favored by the superhydrophilicity on the TiO_2 surface, which leads to a contact angle less than 5° under UV light irradiation (Wang *et al.*, 1997). The superhydrophilicity derives from changes in the chemical conformation of the semiconductor surface. As reported above, the majority of the photogenerated holes give rise to hydroxyl radicals' generation by reacting with the adsorbed species, while the rest are captured by lattice oxygen vacancies and, reacting with the titania molecules, cause a weakening of the lattice titanium-oxygen ions bonds. Impeding these bonds, water can promote the formation of additional hydroxyl groups on the TiO_2 surface during the UV light irradiation. Takeuchi *et al.* (2005) reported that the remarkable wettability of the semiconductor surface decreases, until it disappears, if the UV light is turned off. The considerable influence of light changes on the titania photo-induced efficiency and its structure modifications has attracted more and more attention from researchers, who aim to optimize the semiconductor performance. By monitoring the hydrophilicity phenomenon, it could be possible to realize better water and air pollution control, self-cleaning glasses and antifogging. Modifications of the semiconductor matrix, especially with SiO_2 and In_2O_3 (Guan 2005; Guan *et al.*, 2003; Skorb *et al.*, 2008), increase the acidity and the hydroxyl content at the surface of the composite films and enhance the hydrophilic properties.

7.2.3   *Application of TiO₂ as antimicrobial agent*

TiO_2 has been used extensively to destroy microbial agents. Matsunaga *et al.* (1985) firstly proposed a photocatalytic mechanism for bacteria inactivation, which was subsequently reviewed by many researchers (Dunlop *et al.*, 2010; Robertson *et al.*, 2012). The TiO_2 photocatalytic activity is more effective than that of any other antimicrobial agent because it not only leads to the microbes being killed, but also to the decomposition of the cells of bacteria, fungi, algae and viruses. Almost all of the organic substances can be completely mineralized by TiO_2 under UV light irradiation. The cell membrane represents the main attack site for both the photogenerated hydroxyl radicals and oxygen reactive species, and undergoes firstly lipid peroxidation and subsequently complete destruction (Foster *et al.*, 2011; Nadtochenko *et al.*, 2006). However, the efficiency of titania is still reduced by some drawbacks, such as the quick recombination of the photogenerated electron-hole pairs (within 10–100 ns) with subsequent release of thermal energy or unproductive photons, the high reactivity which promotes fast backward or secondary reactions and the formation of undesirable by-products, and a low absorption in the visible region which determines its inability to use solar light (less than 5% is used in the case of TiO_2 anatase). Therefore, the development of semiconductor modifications, able to overcome these disadvantages, represents one of the main topics in photocatalytic research. The major practices involve catalyst modification by metal loading, ion doping into the semiconductor lattice, dye photosensitization and mixing with other semiconductors. Among the inorganic additives, metallic Ag nanoparticles show extraordinary biocidal properties when deposited/doped on/into titania, enhancing the photocatalytic activity by electron capture as well as promoting light absorption in the visible region (450–550 nm) (Page *et al.*, 2007; Reddy *et al.*, 2007; Yao *et al.*, 2008). Furthermore, the Ag-TiO_2 system shows antibacterial properties in the absence of light, due to the metal nanoparticles being directly in contact with the pathogen agents and/or the formation of toxic Ag species in the media, such as Ag^+, Ag, AgCl, and $AgCl_2$ (Kubacka *et al.*, 2013; Li *et al.*, 2011).

Necula *et al.* (2009) studied the antibacterial properties and the cytotoxicity of Ag nanoparticles deposited on the titania matrix *via* the electrochemical procedure. Gram-positive bacteria, *Staphylococcus aureus*, was chosen for their experiments, which demonstrated a good antimicrobial effect, leading to the complete killing of methicillin-resistant S. aureus within 24 hours in culture conditions. Zhao *et al.* (2011) incorporated Ag nanoparticles into titania nanotubes using a method based on the UV irradiation and $AgNO_3$ immersion. By controlling the $AgNO_3$

content and its immersion time in the support, it was possible to control the size and the amount of the metal nanoparticles. The Ag nanoparticles strongly adhered to the inner walls of the support along the entire length, showing exceptional ability to kill bacteria in the culture medium for several days. Furthermore, the bacterial adhesion was inhibited for at least 30 days with experiments, showing high potential in specific fields of application, such as biomedical implants, and more specifically for the prevention of initial and intermediate-stage infection after operations. Gupta *et al.* (2013) investigated the comparative photocatalytic activity of TiO_2 and Ag-doped TiO_2 nanoparticles prepared by the acid-catalyzed sol-gel technique (Akpan and Haamed, 2010). The semiconductor photocatalytic efficiency was verified by observing the gradual viable colony reduction of three different bacterial types (*Staphylococcus aureus*, *Pseudomonas aeruginosa*, *Escherichia coli*), under visible light irradiation. The viability of all the bacterial colonies fell to zero at 60 mg/30 mL culture when a 3 and 7% of Ag doping was used. On the contrary, unmodified titania showed only a poor photocatalytic activity. Liu Y. *et al.* (2008) prepared Ag/TiO_2 *via* the photoreduction procedure by preparing mesoporous anatase titania as support material and testing the antimicrobial activity using fluorescence labeled *Escherichia coli* by confocal laser scanning microscopy. The incorporation of Ag nanoparticles significantly improved the killing ability of the semiconductor, highlighting how the prepared Ag/TiO_2 composite films could contribute to the application of the photocatalysis technology in the environmental disinfection for the prevention of infectious diseases. Nithyadevi *et al.* (2015) described the chemical reduction synthesis of Ag nanoparticles and the subsequent hydrolysis of titanium precursor with ethanol as solvent. In order to reduce the metal particles agglomeration, cetyltrimethylbromide and sodium alginate were selected as surfactants. The bactericidal activity was performed on the gram-positive *Staphylococcus aureus* cells. The experimental tests evidenced a considerable germicidal effect when sodium alginate induced Ag-TiO_2 core shell nanoparticle treatment under laboratory conditions was used, while without any surfactant or using cetyltrimethylbromide the antimicrobial activity was less pronounced. Transition metals, producing additional energy levels within the semiconductor band gap and reducing the photon energy for the electron transfer, have also been investigated in order to improve titania antibacterial properties (Egerton *et al.*, 2006; Fu *et al.*, 2005; Karunakaran *et al.*, 2010; Sayilkan *et al.*, 2009; Yadav *et al.*, 2014).

In particular, copper (Karunakaran *et al.*, 2010; Yadav *et al.*, 2014), vanadium (Fu *et al.*, 2005), tin (Sayilkan *et al.*, 2009), iron (Egerton *et al.*, 2006) and nickel (Yadav *et al.*, 2014) have been proposed as effective dopant ions for photocatalytic bactericidal experiments. Yadav *et al.* (2014) reported the photocatalytic disinfection using Ni-doped TiO_2 under visible light irradiation. Ni-TiO_2 nanoparticles were synthesized by the sol-gel procedure and studied in photocatalytic antibacterial tests against gram-positive *Staphylococcus aureus*, *Bacillus subtilis* and gram-negative *Escherichia coli* and *Salmonella abony* under fluorescent visible light irradiation. The data obtained showed that the rate of inactivation of gram-positive bacteria was greater than that of gram-negative ones. In all four species, for gram-negative *Salmonella abony* a much higher irradiation time was required. The gram-negative species were less susceptible than the gram-positive species (Adams *et al.*, 2006; Caballero *et al.*, 2009; Foster *et al.*, 2011). This was probably related to the difference in cell wall structure. Gram-positive bacteria present a thick cell wall composed of layers of peptidoglycan and teichoic acids, while gram-negative bacteria are characterized by a relatively thin cell wall with an outer membrane containing lipopolysaccharides and lipoproteins bilayers. Consequently, gram-negative species are relatively more resistant to the hydroxyl radicals attack, opposing the absorption of different molecules to movements through the cell membrane (Tortora *et al.*, 2010).

7.2.4   *TiO₂-based membranes for disinfection applications*

Polymeric and ceramic materials have been employed as support for preparing TiO_2 based membranes (Kim *et al.*, 2003; Leong *et al.*, 2014; Li *et al.*, 2014; Ma *et al.*, 2009; Wang *et al.*, 2014). Semiconductor powders have been incorporated in the casting solution or, alternatively, TiO_2 nanofibers, nanowires or nanotubes have been used for preparing active membranes (Li *et al.*,

2014; Wang *et al.*, 2014). The catalyst particles can be efficiently deposited on the membrane surface or dispersed in the membrane matrix.

Many polymers, such as polyamide (Kwack and Kim, 2001), polyvinylidenefluoride (PVDF) (Damodar *et al.*, 2009), polyethersulfone (PES) (Wu *et al.*, 2008b), poly(vinylidene fluoride) (You *et al.*, 2012), polyurethane (PU) (Zhou *et al.*, 2008), polyamide and polyester (Ingole *et al.*, 2016), have been investigated for disinfection applications and water treatment. HO– groups present on the TiO_2 surface can interact with the CO– group forming hydrogen bonds (Souza and Quadri, 2013). Photoactive membranes with considerable disinfection efficiency against microorganisms have been prepared by several research groups, which used *E. coli* as an indicator (Kim *et al.*, 2003; Leong *et al.*, 2014; Liu *et al.*, 2012; Ma *et al.*, 2009). Liu *et al.* (2012) made Ag/TiO_2 nanofiber membranes for the simultaneous filtration and photocatalytic disinfection and degradation processes under visible irradiation. The obtained membranes showed remarkable antibacterial activity, improvement in fouling control and allowed both the oxidation reactions and the membrane filtration operation without needing any post-treatment process. $Ag-TiO_2$/hydroxyapatite (HAP, $Ca_{10}(PO_4)_6(OH)_2$)/Al_2O_3 bioceramic composite membranes were prepared by Ma *et al.* (2009) to integrate separation operation and photocatalytic microorganisms inactivation *via* a two-step approach based on the sol-gel method followed by calcination. Morphological analysis evidenced a microporous structure which included an $Ag-TiO_2$/HAP composite layer overlaid on $–Al_2O_3$ disk support. A superior ability in promoting interfacial charge-transfer reactions led to the reduction of electron-hole recombination in $Ag-TiO_2$ photocatalytic processes. *E. coli* inactivation tests confirmed the high bactericidal activity of the prepared membranes, which was attributed to the coupled performance of membrane separation, superior bacterial adsorption of HAP, enhanced photobiocide activity of $Ag-TiO_2$ nanocomposite, reduced fouling and at the same time the chemical activity and Ag bacterial adherence. Almost 100% of *E. coli* was eliminated using a TiO_2/PVDF membrane system under irradiation with UV light by Damodar *et al.* (2009). When a catalyst content of 4 wt% was used, complete bacteria inactivation was achieved. The presence of TiO_2 also improved the water permeability across the membrane and the self-cleaning ability. Furthermore, the fouling was reduced. Song *et al.* (2012) also proposed the preparation of PVDF-PEG-TiO_2 membrane for self-cleaning applications, which was highlighted by the color change – from yellowish to near white after UV light irradiation for 8 hours.

7.3 TiO_2 NANOMATERIALS

As already mentioned, titanium dioxide is commonly produced in two forms, either as anatase or as rutile. Most titanium dioxide in the anatase form is produced as a white powder; whereas various rutile grades are often off-white and can even exhibit a slight color, depending on the physical form, which affects the light reflectance. Titanium dioxide may be coated with thin films of inorganic oxides, such as alumina and silica, to improve photocatalytic properties in terms of stabilization against flocculation, electron-hole recombination rate, substrate adsorption, oxidation reaction efficiency and photostability (Schwarz *et al.*, 2004).

The titanium dioxide production is either from ilmenite (FeO/TiO_2), naturally occurring rutile or titanium slag. There are essentially two commercial process options for the production of titanium dioxide, namely the 'sulfate' or the 'chloride' process route. The sulfate process employs a simpler technology than that of the chloride route and can use lower grade, cheaper ores. However, in general a sulfate plant has higher production costs in comparison with a chloride plant due to the additional acid treatment process. However, the latter may require the construction of a chloralkali unit (Gàzquez *et al.*, 2014).

Both anatase and rutile forms of titanium dioxide can be produced by the sulfate process, whereas the chloride process yields only the rutile form. Titanium dioxide with high levels of purity can be prepared. Specifications for food use currently contain a minimum purity assay of 99.0% (Kuznesof and Rao, 2006).

Due to its brightness and high refractive index, titanium dioxide is most widely used as white pigment in products such as paints, coatings, plastics, paper, inks, fibers, food and cosmetics. In combination with other colors, soft pastel shades can be achieved. The high refractive index allows titanium dioxide to be used at relatively low levels to achieve its technical effect. Nowadays, the presence of nanotechnology in consumer products, such as cosmetics, plastics and electronics, is well established and continuously growing. Engineered nanoparticles often possess highly desirable properties at nanoscale, due to size confinement, dominance of interfacial phenomena, and quantum effects, which are desirable for commercial and medical applications. The properties of nanoparticles are often unique, since the origin of the physical behavior of the particles changes from classical physics to quantum physics with particle sizes below 100 nm. As a consequence, the behavior of particles at nanoscale may be completely different in comparison with identical bulk materials, and these nanoparticles may partly behave like new chemical substances (Vippola *et al.*, 2009). Engineered nanoparticles are defined as particles with at least one dimension smaller than 100 nm while exhibiting from zero to three dimensions (a dot, a wire, a tube, a particle) and potentially as small as atomic and molecular length scales ($\sim$0.2 nm). Nanoparticles can have amorphous or crystalline form and their surfaces can act as carriers for liquid droplets or gases (Zhang *et al.*, 2014). During the past decades, various synthesized TiO$_2$ nanoparticles with different shapes and new physical and chemical properties were manufactured, such as nanospheres, nanowires, nanotubes, nanorods, nanoflowers, nanorings, nanosprings, nanobowls, nanobelts, nanosheets, nanocages, nanorods, etc. Significant research into nanotechnology in the last decade has shown promising new applications for titanium dioxide. As an example, titanium dioxide nanoparticles are used in dye-sensitized solar cells, a relatively new photovoltaic technology which mimics the way in which plants convert sunlight into energy. There are a wide range of potential applications, which range from lightweight low-power markets to large-scale applications. Other areas of research for the application of titanium dioxide nanoparticles are: arsenic removal in water treatment facilities, in cancer treatments (ability to target and destroy cancer cells), in cement containing TiO$_2$ with self-cleaning properties and applications as photocatalysis to photo-/electrochromics and sensors (Gàzquez *et al.*, 2014). Independent from the type of TiO$_2$ nanomaterial, the same nanomaterial in bulk form shows little photocatalytic ability. The interesting properties of nano TiO$_2$ are obtained due to their low dimensionality and quantum size effect. The advantages of the nanocrystals obtained over their bulk counterparts in terms of potential applications are due to their high surface to volume ratio, increased number of delocalized carriers on the surface, improved charge transport and lifetime afforded by their dimensional anisotropy, and the efficient contribution in the separation of photogenerated holes and electrons (Bavykin *et al.*, 2006; Evtushenko *et al.*, 2011). Therefore, it is essential to have control over the particle size, shape, and distribution during the preparation of TiO$_2$ nanoparticles.

7.3.1 *Nanoparticles drawbacks*

Nanomaterials are often praised for their "new and unique" properties. However, because of these new properties, nanomaterials are also likely to differ from their conventional chemical equivalents with respect to their behavior in the environment and their kinetic and toxic properties. This raises concerns in connection to their widespread use, as this leads to an increased exposure to these nanomaterials for humans as well as the environment.

7.3.2 *Nanoparticles and health risks*

Since nanotechnology is an emerging field, there are many uncertainties about whether the properties of this new engineered nanomaterial also bring with them some potential health risks (Handy and Shaw, 2007). These questions arise due to the lack of knowledge and other factors that are essential for predicting the health risks of these new materials. Important factors are, for example, routes of exposure, translocation of materials once they enter the body, and interactions of the materials with the body's biological systems. The potential health risk following exposure to a substance is generally associated with the magnitude and duration of the exposure, the persistence of the material in the body, the inherent toxicity of the material, and the susceptibility or health status of the person exposed. More data is needed on the health risks associated with exposure to engineered nanomaterials (Yokel and MacPhail, 2011). Some nanomaterials may initiate catalytic reactions, depending on their composition and structure, which would not otherwise be anticipated based on their chemical composition.

7.3.3   *Nanoparticles and the environment*

The large-scale development and application of nanotechnologies will lead to gradual, as well as accidental, releases of engineered nanoparticles into the environment. The possible ways that they may be exposed to the environment, range over the whole lifecycle of products and applications that contain engineered nanoparticles, are:

- Discharge or leakage during production, transport and storage of intermediate and finished products.
- Discharge or leakage from waste streams.
- Release of nanoparticles during product use.
- Diffusion, transport and transformation in water, air and soil.

In applications like cosmetic products or food ingredients, diffuse sources of nanoparticles will be used. In this area, there will probably be the most significant quantitative release of nanoparticles in the coming years. Additionally, certain applications, such as environmental remediation with the help of nanoparticles, could lead to the deliberate release of nanoparticles into the environment.

The main criteria in the risk assessment of nanomaterials for the environment, and indirectly for human health, are toxicity, persistence and bioaccumulation. Some substances that degrade slowly and remain in the environment for a long time are called persistent contaminants. When contaminants are released into the environment, their persistence becomes an important concern. A substance that is relatively toxic may be a minor hazard if it breaks down quickly into non-hazardous substances before people can be exposed to it. Conversely, a mild toxic contaminant that remains for a long time in the environmental media to which humans are exposed can accumulate in human tissues and become a significant concern. Substances that can cause direct damage to organisms (high toxicity), that decay very slowly in the environment (high persistence) and that can concentrate in fatty tissues (high potential for bioaccumulation) are of particular concern.

For a specific risk assessment of engineered nanoparticles, the particular characteristics have to be taken into account. The available information about bulk material properties will not be sufficient to classify the environmental risk of the same material in the form of nanoparticles. The possible environmental effect therefore has to be assessed specifically for each type/class of nanomaterials.

7.3.4   *Synthesis of TiO$_2$ nanomaterials*

Nanostructured TiO$_2$ can be synthesized *via* a variety of preparation methods. The most common methods consist of the sol-gel method, sol method, hydrothermal method, solvothermal method, direct oxidation method, chemical or physical vapor deposition method, microwave method, and reverse micelle method.

7.3.4.1   *Sol-gel method*
The sol-gel method (Fig. 7.3) is a versatile process and nowadays is widely used for the preparation of nanostructured TiO$_2$. In a typical sol-gel process, a colloidal suspension, or a sol, is formed from the hydrolysis and polymerization reactions of the precursors, which are usually inorganic metal salts or metal-organic compounds such as metal alkoxides. Complete polymerization and loss of solvent leads to the transition from the liquid sol into a solid gel phase.

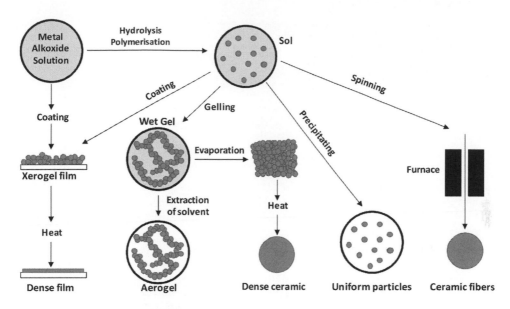

Figure 7.3.   Sol-gel method representation.

This simple, cost-effective and low-temperature synthesis procedure has also been favored and largely applied in catalyst preparation due to its potential to fabricate catalysts with high purity, homogeneity, fine-scale and controllable morphology. Various photocatalysts have been fabricated by the sol-gel process, including ZrO_2, $SrTiO_3$, ZnO, WO_3 and TiO_2 (Chen *et al.*, 2011; Djaoued *et al.*, 2013; Sun *et al.*, 2011; Yu *et al.*, 2011). Specifically for TiO_2, titanium alkoxides (such as titanium isopropoxide, titanium *n*-butoxide), alcohol, and acid/water are introduced into the reaction system. After stirring for several hours, densely cross-linked three-dimensional structures are built and terminated as TiO_2 gel (Bai *et al.*, 2005; Gupta and Tripathi, 2012; Macwan *et al.*, 2011).

7.3.4.2   *Hydrothermal method*
The hydrothermal technique is an important tool for advanced nanostructural material process-ing, covering the processing of electronics, catalysis, and ceramics (Cheng *et al.*, 2009). The hydrothermal technique is defined as any heterogeneous chemical reaction in the presence of an aqueous solvent where the temperature is elevated above room temperature and at a pressure greater than 1 atm ($\sim$101.3 kPa) in a closed system, reaching the pressure of vapor saturation. The hydrothermal synthesis covers the process of crystal growth, crystal transformation, phase equilibrium, and finally leads to fine to ultra-fine crystals (generally below 10 nm) (Byrappa and Adschiri, 2007).

7.3.4.3   *Solvothermal method*
The solvothermal method has earned lots of attention in preparing ceramic materials, such as ZrO_2, CeO_2, and Fe_2O_3, and is almost identical to the hydrothermal method except that the solvent used here is non-aqueous (Li *et al.*, 2012; Qin *et al.*, 2011; Yang *et al.*, 2013). However, the temperature and pressure can be elevated to much higher levels than those in a hydrothermal process. In addition, the solvothermal method normally allows better control than those of the hydrothermal method with respect to nanosize, crystal phase, narrow size distribution and minimal agglomeration (Chen and Mao, 2007; Nam and Han, 2003a, 2003b).

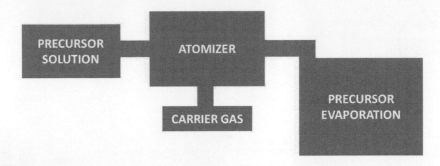

Figure 7.4.   CVD method representation.

7.3.4.4   *Direct oxidation method*

TiO_2 nanomaterials can be obtained by the oxidation of metallic Ti by oxidants such as O_2, H_2O_2 and acetone to form crystalline TiO_2, all of which can be classified as direct oxidation (Daothong *et al.*, 2007; Wu and Chen, 2008a, 2009). Since the vapor pressure of Ti is very low (0.133 Pa at 1577°C) and Ti has a high melting point (1668°C), direct oxidation of a Ti foil is difficult and may only result in scattered and sparse TiO_2 nanofibers with a low conversion. Because of this, direct oxidation of Ti is rarely used, and has only been reported by a few papers (Huo *et al.*, 2009; Jones and Hitchman, 2009).

7.3.4.5   *Chemical vapor deposition*

Chemical vapor deposition (CVD) (Fig. 7.4) is a widely applied material-processing technology, referring to a deposition process where chemical precursors are transported in a vapor state to decompose on a heated substrate to form a film. The films may be epitaxial, polycrystalline or amorphous depending on the materials and reactor conditions. This technology can deposit films with a conformal and elaborate coverage, which is a virtue not found by physical vapor deposition. As would be expected with the large variety of materials deposited, CVD can be conducted in various environments with enhanced equipment and different derivatives. CVD has become the major method of film deposition for the semiconductor industry due to its high throughput, high purity, and low cost of operation. Furthermore, CVD is commonly used in optoelectronics applications, optical coatings, and coatings of wear resistant parts.

Generally, CVD may be performed in hot-wall or cold-wall reactors, typically below 10^{-6} Pa to above atmospheric pressures, with or without carrier gases, and at temperatures typically ranging from 200–1600°C. The enhanced CVD processes involve the use of plasmas, ions, photons, lasers or combustion reactions to increase deposition rates and/or lower deposition temperatures. The derivatives of CVD entail metal-organic CVD, organo-metallic CVD and inorganic CVD (Jones and Hitchman, 2009; Sosnowchik *et al.*, 2010; Warwick *et al.*, 2011; Kodas and Hampden-Smith, 2007).

7.3.4.6   *Physical vapor deposition*

Physical vapor deposition (PVD) processes are atomistic deposition processes in which materials are vaporized from a solid or liquid source in the form of atoms or molecules and transported through a vacuum or low pressure gaseous/plasma environment to the substrate, where it condenses. The PVD processes are typically used for the deposition of films with thicknesses in the range of a few nanometers to thousands of nanometers; however, they can also be used to form multilayer coatings and very thick deposits. The main categories of PVD processing include thermal deposition, sputter deposition, arc vapor deposition, ion plating, etc. (Gonzàlez-Garcìa *et al.*, 2010; Helmersson *et al.*, 2006; Mattox, 2010).

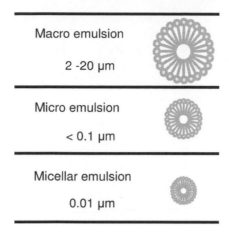

Figure 7.5.   Micelles and microemulsions dimensions.

7.3.4.7   *Microwave method*

The microwave-assisted method has been utilized for material synthesis since the late 1960s; since then it has received much attention due to the short reaction time and uniform heat distribution. Almost all reaction types have been tested by the microwave-assisted method. Moreover, the microwave-assisted heating technique offers cleaner, more cost-effective, more rapid heating, faster kinetics, higher yield, and better reproducibility of products, leading to enhanced structural and morphological properties of nanomaterials compared with conventional heating (Hoseinzadeh *et al.*, 2013; Oghbaei and Mirzaee, 2010).

7.3.4.8   *Microemulsions or micelles*

Micelles and microemulsions (Fig. 7.5) are liquid dispersions containing surfactant aggregates.

These are basically thermodynamically stable, isotropically dispersions of two immiscible liquids, such as oil and water, stabilized by a monolayer of surfactant, frequently in combination with a co-surfactant. The aqueous phase may contain salts and/or other ingredients, and the oil phase may actually be a complex mixture of different hydrocarbons and olefins. Surfactants are usually organic compounds that are amphiphilic, meaning that they contain both hydrophobic groups (their tails) and hydrophilic groups (their heads). Surfactants can lower the interfacial tension between the aqueous phase and the oil phase and stabilize the system. The co-surfactants are usually alcohols with carbon chains of moderate length (Chen and Mao, 2007; Wang *et al.*, 2008). There are two basic types of microemulsions: direct microemulsions (oil dispersed in water, o/w) and reverse microemulsions (water dispersed in oil, w/o).

7.3.5   *TiO₂ dispersion in solution*

For photocatalytic activity or anti-fouling properties in the systems containing engineered nanoparticles, it is crucial to obtain a well-dispersed titanium dioxide in order to benefit from the particular characteristics that are obtained at nanoscale.

The dispersion and stabilization of TiO₂ powders in liquid media are still a big problem, since numerous properties of the final solution depend strongly on the colloidal stability of particles and their distribution in certain volumes. The sedimentation behavior is often seen as the crucial stability criterion. According to the Derjaguin, Landau, Verwey and Overbeek theory (DLVO theory), there are two basic forces controlling the stability of colloidal suspensions: van der Waals and the electrostatic forces (Hunter *et al.*, 1981). Van der Waals forces are a weak attractive force between atoms or non-polar molecules caused by a temporary change in dipole moment arising from a brief shift of orbital electrons to one side of one atom or molecule, creating a similar shift in adjacent atoms or molecules. The electrostatic forces are related to the particle charge. The ζ potential is often used to measure these forces. For most suspensions, high values of this parameter indicate high stability, while low values imply coagulation (Fadda *et al.*, 2009; Veronovski *et al.*, 2010). Therefore, the stability of the suspension strongly depends on:

- Choice of solvents and additives (surfactants).
- Concentrations and particle size of TiO_2 nanoparticle.

As reported by Ng *et al.* (2013), the behavior control of both the agglomeration and dispersion of nanoparticles is also crucial for the organic-inorganic membrane preparation due to surface interaction, such as van der Waals, electrical and steric forces. Although many theories have been proposed to explain the surface phenomena that occur during the TiO_2-polymeric membrane preparation, the aggregation of nanoparticles still remains one of the major difficulties for obtaining a uniform defect-free membrane. For example, although the primary particle size of commercial TiO_2 nanoparticles, such as Evonik P25Aeroxide®, is about 25 nm, its particle size as powder or in dispersion is in the range of hundreds of nanometers due to agglomeration. This agglomeration leads not only to the uneven distribution but also to a potential reduction in its specific abilities obtained at nanoscale (Law *et al.*, 2009). In order to avoid agglomeration and enhance the colloidal stability of TiO_2 in solutions in general, physical and chemical methods are taken into account. The type and the TiO_2 content, as well as the type and the concentration of the polymer, solvent and additive(s) in the dope solution, can affect the final membrane structure and performance (Leong *et al.*, 2014; Ng *et al.*, 2013). Also, the preparation techniques used for making the membranes play a key role in the TiO_2 nanoparticles agglomeration prevention process: non-solvent/thermal/vapor induced phase separation (NIPS, TIPS and VIPS, respectively) offers the possibility to obtain welldispersed catalyst particles in the polymeric membrane matrix (Leong *et al.*, 2014; Wang *et al.*, 2014).

The surface modification of the TiO_2 nanoparticles can be achieved by using conventional methods like sonication and grinding (Razmjou *et al.*, 2011). Combining chemical and mechanical modifications seems to significantly affect hydrophilicity, pore size, roughness, surface free energy and protein absorption resistance. Chemical agents, such as aminopropyltriethoxysilane (Razmjou *et al.*, 2011a) and γ-amminopropyltriethoxysilane (Wu and Chen, 2008a) were also used to modify TiO_2 nanoparticles, and the experimental results confirmed that they improve the semiconductor dispersion in the casting solution. Tak *et al.* (2006) reported the fabrication of nanocomposite membranes *via* electrostatic self-assembly between the catalyst nanoparticles and sulfonic acid groups on the sulfonated-PES (SPES) membrane surface. The obtained membranes showed reduced fouling compared to the polymeric membrane without TiO_2. Benfer *et al.* (2001) used diethanolamine as a coupling agent, preparing TiO_2-based membranes *via* the sol-gel method. Similarly, Mansourpanah *et al.* (2009) investigated the effect of the PES/PI hybrid membranes immersion in an aqueous solution of diethanolamine followed by immersion in the TiO_2 aqueous suspension. This method allowed the increase of the hydroxyl groups on the membranes surface, at the same time improving the strength of interaction between the semiconductor nanoparticles and the membranes surface.

7.4   CONCLUSIONS AND FUTURE PERSPECTIVES

Photocatalysis is a challenging and low-cost technique for water and air treatment, for the destruction of microorganisms, for medical applications and for hydrogen production *via* water splitting. When irradiated with light, excited states of the photocatalyst are generated and can promote subsequent reaction, such as reduction-oxidation with substrate transformations. TiO_2 nanoparticles can be efficiently prepared by different synthesis methods, which include the sol-gel, hydrothermal and solvothermal techniques. Among these three methods, the sol-gel has been widely applied for the nanostructured TiO_2 synthesis, while the hydrothermal and solvothermal techniques are important tools for the TiO_2 nanoparticles synthesis for electronics and ceramics.

TiO$_2$ has been extensively studied and applied as white pigment in paper, paints, plastic, cosmetics and foodstuffs. This semiconductor is one of the most efficient catalysts for bacteria, viruses, fungi and algae elimination due to its action, which provides the death of the microorganisms and the decomposition of their cells. Almost all of the organic compounds are completely mineralized by TiO$_2$-UV light irradiation. TiO$_2$-polymeric membranes, mainly prepared by the phase inversion method, offer the possibility to overcome the catalyst recovery step, allowing the separation operation to occur concomitantly with the photocatalytic reaction. Chemical and physical techniques have been proposed to improve the catalyst dispersion in the dope solution. Chemical compounds, such as diethanolamine and sulfonic acid, as well as the TiO$_2$ nanoparticles, polymer and solvent content in the casting solution play a key role in the agglomeration prevention and, consequently, in the final membrane morphology and performance. Although the optimization of the membrane preparation procedure has not been reached, significant improvements in terms of antimicrobial, self-cleaning and anti-fouling properties, in addition to the photocatalytic efficiency have been registered. Thus, new perspectives exist in the TiO$_2$ photocatalysis area, which require systematic evaluation, as the research results should offer advantages in terms of economic costs and environmental impact for the commercial sector.

ACKNOWLEDGMENTS

The authors gratefully acknowledge financial support from the European Union's Seventh Framework Programme, within the NAWADES project, (grant agreement N° 308439) EU-FP7/project. Dr. Lawrence Arockiasamy and Dr. Alberto Figoli acknowledge the support of King Abdul Aziz City for Science and Technology (KACST), Kingdom of Saudi Arabia, under the National Plan for Science, Technology and Innovation (NPST); Contract award number: 12-ADV2611-02.

REFERENCES

Adams, L.K., Lyon, D.Y. & Alvarez, P.J.J. (2006) Comparative eco-toxicity of nanoscale TiO$_2$, SiO$_2$, and ZnO water suspensions. *Water Research*, 40, 3527–3532.

Akpan, U.G. & Hameed, B.H. (2010) Review: the advancements in sol-gel method of doped-TiO$_2$ photocatalysts. *Applied Catalysis* A: *General*, 375, 1–11.

Al-Rasheed, R.A. (2005) Water treatment by heterogeneous photocatalysis an overview. Presented at *4th SWCC Acquired Experience Symposium, 07 May 2005, Jeddah, Saudi Arabia*.

Ardakani, H.K. (1994) Electrical and optical properties of in situ "hydrogen-reduced" titanium dioxide thin films deposited. *Thin Solid Films*, 248, 234–239.

Austin, R. & Lim, S.F. (2008) The Sackler Colloquium on Pormoses and Perils in Nanotechnology for Medicine. *PNAS*, 105, 17,217–17,221.

Bae, T.H., Kim, I.C. & Tak, T.M. (2006), Preparation and characterization of fouling-resistant TiO$_2$ self-assembled nanocomposite membranes. *Journal of Membrane Science*, 275, 1–5.

Bagheri, S., Nurhidayatullaili, M.J. & Hamid, S.B.A. (2014) Titanium dioxide as a catalyst support in heterogeneous catalysis. *The Scientific World Journal*, Article ID 727496. Available from: http://dx.doi.org/10.1155/2014/727496 [accessed December 2016].

Bai, X., Song, H., Yu, L., Yang, L.M., Liu, Z.X., Pan, G., Lu, S., Ren, X., Lei, Y. & Fan, L. (2005) Luminescent properties of pure cubic phase Y$_2$O$_3$/Eu^{3+} nanotubes/nanowires prepared by a hydrothermal method. *Journal of Physical Chemistry* B, 109, 15,236–15,242.

Banerjee, A.N. (2011) The design, fabrication, and photocatalytic utility of nanostructured semiconductors: focus on TiO$_2$-based nanostructures. *Journal of Nanotechnology, Science and Applications*, 4, 35–65.

Bavykin, D.V., Friedrich, J.M. & Walsh, F.C. (2006) Protonated titanates and TiO$_2$ nanostructured materials: synthesis, properties, and applications. *Advanced Materials*, 18, 2807–2824.

Benfer, S., Popp, U., Richter, H., Siewert, C. & Tomandl, G. (2001) Development and characterization of ceramic nanofiltration membranes. *Separation and Purification Technology*, 22/23, 231–237.

Beydoun, D., Amal, R., Low, G. & Mcevoy, S. (1999) Role of nanoparticles in photocatalysis. *Journal of Nanoparticle Research*, 1, 439–458.

Blake, D.M., Maness, P., Huang, Z., Wolfrum, E.J. & Huang, J. (1999) Applications of photocatalytic disinfection. *Separation and Purification Technology*, 28, 1–50.

Byrappa, K. & Adschiri, T. (2007) Hydrothermal technology for nanotechnology. *Progress in Crystall Growth and Characterization of Materials*, 53, 117–166.

Caballero, L., Whitehead, K.A., Allen, N.S. & Verran, J. (2009) Inactivation of *Escherichia coli* on immobilized TiO_2 using fluorescent light. *Journal of Photochemistry and Photobiology* A: *Chemistry*, 202, 92–98.

Chen, X.B. & Mao, S.S. (2007) Titanium dioxide nanomaterials: Synthesis, properties, modifications, and applications. *Chemical Review*, 107, 2891–2959.

Chen, Y.J., Lunsford, S.K., Song, Y.C., Ju, H.X., Falaras, P., Likodimos, V., Kontos, A.G. & Dionysiou, D.D. (2011) Synthesis, characterization and electrochemical properties of mesoporous zirconia nanomaterials prepared by self-assembling sol-gel method with Tween 20 as a template. *Chemical Engineering Journal*, 170, 518–524.

Cheng, C.W., Liu, B., Yang, H.Y., Zhou, W.W., Sun, L., Chen, R., Fung, S.Y., Zhang, J., Gong, H., Sun, H. & Fan, H.J. (2009) Hierarchical assembly of ZnO nanostructures on SnO_2 backbone nanowires: Low-temperature hydrothermal preparation and optical properties. *ACS Nano*, 3, 3069–3076.

Damodar, R.A., You, S.-J. & Chou, H.-H. (2009) Study the self cleaning, antibacterial and photocatalytic properties of TiO_2 entrapped PVDF membranes. *Journal of Hazardous Materials*, 172, 1321–1328.

Daothong, S., Songmee, N., Thongtem, S. & Singjai, P. (2007) Size-controlled growth of TiO_2 nanowires by oxidation of titanium substrates in the presence of ethanol vapor. *Scripta Materialia*, 57, 567–570.

De Lasa, H., Serrano, B. & Solaices, M. (2005) *Photocatalytic Reaction Engineering*. Springer Science, Business Media, New York.

Djaoued, Y., Balaji, S. & Beaudoin, N. (2013) Sol-gel synthesis of mesoporous WO_3-TiO_2 composite thin films for photochromic devices. *Journal of Sol-Gel Science and Technology*, 65, 374–383.

Dunlop, P.S.M., Sheeran, C.P., Byrne, J.A., McMahon, M.A.S., Boyle, M.A. & McGuigan, K.G. (2010) Inactivation of clinically relevant pathogens by photocatalytic coatings. *Journal of Photochemistry and Photobiology* A: *Chemistry*, 216, 303–310.

Egerton, T.A., Kosa, S.A.M. & Christensen, P.A. (2006) Photoelectrocatalysis by titanium dioxide for water treatment. *Physical Chemistry Chemical Physics*, 8, 398–406.

Esch, T.R., Gadaczek, I. & Bredow, T. (2014), Surface structures and thermodynamics of low-index of rutile, brookite and anatase – a comparative DFT study. *Applied Surface Science*, 288, 275–287.

Evtushenko, Y.M., Romashkin, S. & Davydov, V. (2011) Synthesis and properties of TiO_2-based nanomaterials. *Theoretical Foundations of Chemical Engineering*, 45, 731–738.

Fadda, S., Cincotti, A., Concas, A., Pisu, M. & Cao, G. (2009) Modelling breakage and reagglomeration during fine dry grinding in ball milling devices. *Powder Technology*, 194, 207–216.

Foster, H., Ditta, I., Varghese, S. & Steele, A. (2011) Photocatalytic disinfection using titanium dioxide: spectrum and mechanism of antimicrobial activity. *Applied Microbiology and Biotechnology*, 90, 1847–1868.

Fox, M.A. & Dulay, M.T. (1993) Heterogeneous photocatalysis. *Chemical Reviews*, 93, 341–357.

Fu, G., Vary, P.S. & Lin, C.T. (2005) Anatase TiO_2 nanocomposites for antimicrobial coatings. *Journal of Physical Chemistry* B, 109, 8889–8898.

Fujishima, A. & Honda, K. (1972) Electrochemical photolysis of water at a semiconductor electrode. *Nature*, 238, 37–38.

Gàzquez, M.J., Bolívar, J.P., Garcia-Tenorio, R. & Vaca, F. (2014) A review of the production cycle of titanium dioxide pigment. *Materials, Sciences and Applications*, 5, 441–458.

Geng, Q. & Chen, N. (2011) Photocatalytic degradation of a gaseous benzene-toluene mixture in a circulated photocatalytic reactor. *Chemical Engineering and Technology*, 34, 400–408.

Gonzàlez-Garcìa, L., Lozano, G., Barranco, A., Míguez, H. & Gonzáez-Elipe, A.R. (2010) TiO_2-SiO_2 one-dimensional photonic crystals of controlled porosity by glancing angle physical vapor deposition. *Journal of Materials Chemistry*, 20, 6408–6412.

Guan, K. (2005) Relationship between photocatalytic activity, hydrophilicity and self-cleaning effect of TiO_2/SiO_2 films. *Surface and Coatings Technology*, 191, 155–160.

Guan, K., Lu, B. & Yin, Y. (2003) Enhanced effect and mechanism of SiO_2 addition in super-hydrophilic property of TiO_2 films. *Surface and Coatings Technology*, 173, 219–223.

Gumy, D., Morais, C., Bowen, P., Pulgarin, C., Giraldo, S., Hajdu, R. & Kiwi, J. (2006) Catalytic activity of commercial of TiO_2 powders for the abatement of the bacteria (*E. coli*) under solar simulated light: influence of the isoelectric point. *Applied Catalysis* B: *Environmental*, 63, 76–84.

Gupta, K., Singh, R.P., Pandey, A. & Beilstein, J. (2013) Photocatalytic antibacterial performance of TiO_2 and Ag-doped TiO_2 against *S. aureus*, *P. aeruginosa* and *E. coli*. *Nanotechnology*, 4, 345–351.

Gupta, S.M. & Tripathi, M. (2012) A review on the synthesis of TiO_2 nanoparticles by solution route. *Central European Journal of Chemistry*, 10, 279–294.

Handy, R.D. & Shaw, B.J. (2007) Toxic effects of nanoparticles and nanomaterials: implications for public health, risk assessment and the public perception of nanotechnology. *Health, Risk & Society*, 9, 125–144.

Hashimoto, K., Irie, H. & Fujishima, A. (2005) TiO_2, photocatalysis: a historical overview and future prospects. *Japanese Journal of Applied Physics*, 44, 8269–8285.

Helmersson, U., Lattemann, M., Bohlmark, J., Ehiasarian, A.P. & Gudmundsson, J.T. (2006) Ionized physical vapor deposition (IPVD): A review of technology and applications. *Thin Solid Films*, 513, 1–24.

Henrich, V.E. & Cox, P.A. (1994) *The Surface Science of Metal Oxides*. Cambridge University Press, Cambridge, UK.

Hoffmann, M.R., Martin, S.T., Choi, W. & Bahnemann, D.W. (1995) Environmental applications of semiconductor photocatalysis. *Chemical Reviews*, 95, 69–96.

Hoseinzadeh, H.R., Wan Daud, W.M.A., Sahu, J. & Arami-Niya, A. (2013) The effects of a microwave heating method on the production of activated carbon from agricultural waste: a review. *Journal of Analytical and Applied Pyrolysis*, 100, 1–11.

Howe, R.F. (1998) Recent developments in photocatalysis. *Developments in Chemical Engineering and Mineral Processing*, 6, 55–84.

Hsu, Y.Y., Hsiung, T.L., Wang, H.P., Fukushima, Y., Wei, Y.L. & Chang, J.E. (2008) Photocatalytic degradation of spill oils on TiO_2 nanotube thin films. *Marine Pollution Bulletin*, 57, 873–876.

Hunter, R.J., Ottewill, R.H. & Rowell, R.L. (1981) *Zeta Potential in Colloid Science*. Academic Press, London, UK.

Huo, K.F., Zhang, X.M., Fu, J.J., Qian, G.X., Xin, Y.C., Zhu, B.Q., Ni, H. & Chu, P.K. (2009) Synthesis and field emission properties of rutile TiO_2 nanowires arrays grown directly on a Ti metal self-source substrate. *Journal of Nanoscience and Nanotechnology*, 9, 3341–3346.

Ingole, P.G., Baig, M.I., Choi, W.K. & Lee, H.K. (2016) Synthesis and characterization of polyamide/polyester thin-film nanocomposite membranes achieved by functionalized TiO_2 nanoparticles for water vapor separation. *Journal of Materials Chemistry* A, 4, 5592–5604.

Jones, A.C. & Hitchman, M.L. (2009) Chemical vapor deposition: precursors, processes and applications. Royal Society of Chemistry, London, UK.

Joshi, U.A., Darwent, R.J., Yiu, H.H.P. & Rosseinsky, M.J. (2011) The effect of platinum on the performance of WO_3 nanocrystal photocatalysts for the oxidation of methyl orange and iso-propanol. *Journal of Chemical Technology and Biotechnology*, 86, 1018–1023.

Kabachkov, E., Kurkin, E., Nadtochenko, V. & Terentyev, A. (2011) Research of photocatalytic degradation of HeLa cells at the TiO_2 interface by ATR-FTIR and fluorescence microscopy. *Journal of Photochemistry and Photobiology* A: *Chemistry*, 217, 425–429.

Kabra, K., Chaudhary, R. & Sawhney, R. (2004) Treatment of hazardous organic and inorganic compounds through aqueous-phase photocatalysis: a review. *Industrial & Engineering Chemical Research*, 43, 7683–7696.

Karunakaran, C., Abiramasundari, G., Gomathisankar, P., Manikandan, G. & Anandi, V. (2010) Cu-doped TiO_2 nanoparticles for photocatalytic disinfection of bacteria under visible light. *Journal of Colloid and Interface Science*, 352, 68–74.

Kavan, L., Yum, J.H. & Grätzel, M. (2011) Optically transparent cathode for dye-sensitized solar cells based on graphene nanoplatelets. *ACS Nano*, 5, 165–172.

Kim, S.H., Kwak, S.-Y., Sohn, B.-H. & Park, T.H. (2003) Design of TiO_2 nanoparticle selfassembled aromatic polyamide thin-film-composite (TFC) membrane as an approach to solve biofouling problem. *Journal of Membrane Science*, 211, 157–165.

Kodas, T. & Hampden-Smith M.J. (2007) The chemistry of metal CVD. In: Kodas, T. & Hampden-Smith, M.J (eds) *Overview of Metal CVD*. John Wiley & Sons. Chapter 9, 429–498.

Kubacka, A., Muñoz-Batista, M.J., Ferrer, M. & Fernández-García, M. (2013) UV and visible light optimization of anatase TiO_2 antimicrobial properties: surface deposition of metal and oxide (Cu, Zn, Ag) species. *Applied Catalysis* B: *Environmental*, 140–141, 680–690.

Kuznesof, P. & Rao, M.V. (2006) Titanium dioxide chemical and technical assessment. Report prepared for the Food and Agricultural Organization of the United Nations (FAO), Rome, Italy. pp. 1–8.

Kwak, S.Y. & Kim, S.H. (2001) Hybrid organic/inorganic reverse osmosis (RO) membrane for bactericidal anti-fouling. 1. Preparation and characterization of TiO_2 nanoparticle self-assembled aromatic polyamide thin-film-composite (TFC) membrane. *Environmental Science and Technology*, 35, 2388–2394.

Law, W.S., Lam S.W., Gan, W.Y., Scott, J. & Amal, R. (2009) Effect of film thickness and agglomerate size on the superwetting and fog-free characteristics of TiO_2 films. *Thin Solid Films*, 517, 5425–5430.

Leong, S., Razmjou, A., Wang, K., Hapgood, K., Zhang, X. & Wang, H. (2014) TiO_2 based photocatalytic membranes: a review. *Journal of Membrane Science*, 472, 167–184.

Li, G.H., Hong, Z.L., Yang, H. & Li, D.N. (2012) Phase composition controllable preparation of zirconia nanocrystals via solvothermal method. *Journal of Alloys and Compounds*, 532, 98–101.

Li, M., Noriega-Trevino, M.E., Nino-Martinez, N., Marambio-Jones, C., Wang, J., Damoiseaux, R., Ruiz, F. & Hoek, E.M.V. (2011) Synergistic bactericidal activity of Ag-TiO₂ nanoparticles in both light and dark conditions. *Environmental Science and Technology*, 45, 8989–8995.

Li, W., Ni, C., Lin, H., Huang, C.P. & Shah, S.I. (2004) Size dependence of thermal stability of TiO₂ nanoparticles. *Journal of Applied Physics*, 96, 6663–6668.

Li, X., Fang, X., Pang, R., Li, J., Sun, X., Shen, J., Han, W. & Wang, L. (2014) Self-assembly of TiO₂ nanoparticles around the pores of PES ultrafiltration membrane for mitigating organic fouling. *Journal of Membrane Science*, 467, 226–235.

Litter, M. (1999) Heterogeneous photocatalysis: transition metal ions in photocatalytic systems. *Applied Catalysis* B: *Environmental*, 23, 89–114.

Liu, L., Liu, Z., Bai, H. & Sun, D.D. (2012) Concurrent filtration and solar photocatalytic disinfection/degradation using high-performance Ag/TiO₂ nanofiber membrane. *Water Research*, 46, 1101–1112.

Liu, T.-X., Li, X.Z. & Li, F.-B. (2008) AgNO₃-induced photocatalytic degradation of odorous methyl mercaptan in gaseous phase: mechanism of chemisorption and photocatalytic reaction. *Environmental Science and Technology*, 42, 4540–4545.

Liu, Y., Wang, X., Yang, F. & Yang, X. (2008) Excellent antimicrobial properties of mesoporous anatase TiO₂ and Ag/TiO₂ composite films. *Microporous and Mesoporous Materials*, 114, 431–439.

Ma, N., Fan, X., Quan, X. & Zhang, Y. (2009) Ag-TiO₂/HAP/Al₂O₃, bioceramic composite membrane: fabrication, characterization and bactericidal activity. *Journal of Membrane Science*, 336, 109–117.

Macwan, D.P., Dave, P.N. & Chaturvedi, S. (2011) A review on nano-TiO₂ sol-gel type syntheses and its applications. *Journal of Materials Science*, 46(11), 3669–3686.

Mansourpanah, Y., Madaeni, S.S., Adeli, M., Rahimpour, A. & Farhadian, A. (2009) Surface modification and preparation of nanofiltration membrane from polyethersulfone/polyimide blend-use of a new material (polyethyleneglycol-triazine). *Journal of Applied Polymer Science*, 112, 2888–2895.

Mathews, T. & Antony, R.P. (2015) Titania nano-architectures for energy. In: Babu Krishna Moorthy, S. (ed) *Thin Film Structures in Energy Applications*. Springer International Publishing Switzerland. pp. 129–165.

Matsunaga, T., Tomoda, R., Nakajima, T. & Wake, H. (1985) Photoelectrochemical sterilization of microbial cells by semiconductor powders. *FEMS Microbiology Letters*, 29, 211–214.

Mattox, D.M. (2010) *Handbook of Physical Vapor Deposition (PVD) Processing*. William Andrew, Elsevier Ltd, Oxford, UK.

Monllor-Satokca, D., Borja, L., Rods, A., Gomez, R. & Salvador, P. (2006) Photo-electrochemical behavior of nanostructured WO₃ thin-film electrodes: the oxidation of formic acid. *ChemPhysChem*, 7, 2540–2551.

Nadtochenko, V., Denisov, N., Sarkisov, O., Gumy, D., Pulgarin, C. & Kiwi, J. (2006) Laser kinetic spectroscopy of the interfacial charge transfer between membrane cell walls of *E. coli* and TiO₂. *Journal of Photochemistry and Photobiology* A: *Chemistry*, 181, 401–407.

Nam, W.S. & Han, G.Y. (2003a) A photocatalytic performance of TiO₂ photocatalyst prepared by the hydrothermal method. *Korean Journal of Chemical Engineering*, 20, 180–184.

Nam, W.S. & Han, G.Y. (2003b) Characterization and photocatalytic performance of nanosize TiO₂ powders prepared by the solvothermal method. *Korean Journal of Chemical Engineering*, 20, 1149–1153.

Navaladian, S., Janet, C.M. & Viswanath, R.P. (2007) On the possible treatment procedures for organic contaminants. In: Kaneco, S., Viswanathan, B. & Katsumata, H. (eds) *Photo/Electrochemistry & Photobiology in the Environment, Energy and Fuel*. Research Signpost, Kerala, India. pp. 1–51.

Necula, B.S., Fratila-Apachitei, L.E., Zaat, S.A.J., Apachitei, I. & Duszczyk, J. (2009) In vitro antibacterial activity of porous TiO₂-Ag composite layers against methicillin-resistant *Staphylococcus aureus*. *Acta Biomaterialia*, 5, 3573–3580.

Ng, L.Y., Mohammad, A.W., Leo, C.P. & Hilal, N. (2013) Polymeric membranes incorporated with metal/metal oxide nanoparticles: a comprehensive review. *Desalination*, 308, 15–33.

Nithyadevi, D., Suresh Kumar, P., Mangalaraja, D., Ponpandian, N., Viswanathan, C. & Meena, P. (2015) Improved microbial growth inhibition activity of bio-surfactant induced Ag-TiO₂ core shell nanoparticles. *Applied Surface Science*, 327, 504–516.

Oghbaei, M. & Mirzaee, O. (2010) Microwave versus conventional sintering: a review of fundamentals, advantages and applications. *Journal of Alloys and Compounds*, 494, 175–189.

Ohno, T., Sarukawa, K. & Matsumura, M. (2001) Photocatalytic activities of pure rutile particles isolated from TiO₂ powder by dissolving the anatase component in HF solution. *Journal of the Physical Society of Japan* B, 105, 2417–2420.

Page, K., Palgrave, R.G., Parkin, I.P., Wilson, M., Savin, S.L.P. & Chadwick, A.V. (2007) Titania and silver-titania composite films on glass-potent antimicrobial coatings. *Journal of Materials Chemistry*, 17, 95–104.

Palmisano, L. & Sclafani, A. (1997) *Heterogeneous Photocatalysis*. John Wiley & Sons, Ltd., Chichester, UK.

Paxton, A.T. & Thien-Nga, L. (1998) Electronic structure of reduced titanium dioxide. *Physical Review* B, 57, 1579–1584.

Petriconi, G. & Papee, H.M. (1983) Photocatalytic fixation of nitrogen by aqueous suspensions of silver iodide. *Water, Air & Soil Pollution*, 20, 273–276.

Primo, A., Marino, T., Corma, A., Molinari, R. & Garcìa, H. (2011) Efficient visible-light photocatalytic water splitting by minute amounts of gold supported on nanoparticulate CeO$_2$ obtained by a biopolymer templating method. *Journal of the American Chemical Society*, 133, 6930–6933.

Qin, D.D., Tao, C.L., In, S.I., Yang, Z.Y., Mallouk, T.E. & Bao, N.Z. (2011) Facile solvothermal method for fabricating arrays of vertically oriented α-Fe$_2$O$_3$ nanowires and their application in photoelectrochemical water oxidation. *Energy & Fuels*, 25, 5257–5263.

Ravelli, D., Dondi, D., Fagnoni, M. & Albini, A. (2009) Photocatalysis. A multi-faceted concept for green chemistry. *Chemical Society Reviews*, 38, 1999–2011.

Razmjou, A., Mansouri, J. & Chen, V. (2011) The effect of mechanical and chemical modification of TiO$_2$ nanoparticles on the surface chemistry, structure and fouling performance of PES ultrafiltration membranes. *Journal of Membrane Science*, 378, 73–84.

Reddy, M.P., Venugopal, A. & Subrahmanyam, M. (2007) hydroxyapatite-supported Ag-TiO$_2$ as *Escherichia coli* disinfection. *Water Research*, 41, 379–386.

Robertson, P.K.J., Robertson, J.M.C. & Bahnemann, D.W. (2012) Removal of microorganisms and their chemical metabolites from water using semiconductor photocatalysis. *Journal of Hazardous Materials*, 211/212, 161–171.

Santato, C., Odziemkowski, M., Ulmann, M. & Augustynski, J. (2001) Crystallographically oriented meso-porous WO$_3$ films: synthesis, characterization, and applications. *Journal of the American Chemical Society*, 123, 10,639–10,649.

Sayilkan, F., Asiltürk, M., Kiraz, N., Burunkaya, E., Arpac, E. & Sayilkan, H. (2009) Photocatalytic antibacterial performance of Sn^{4+}-doped TiO$_2$ thin films on glass substrate. *Journal of Hazardous Materials*, 162, 1309–1316.

Schwarz, S., Petzold, G. & Wienhold, U. (2004) Stabilizing titanium dioxide. *European Coatings Journal*, 34, 07–08/2004.

Silva, C.G., Juàrez, R., Marino, T., Molinari, R. & Garcìa, H. (2011) Influence of excitation wavelength (UV or visible light) on the photocatalytic activity of titania containing gold nanoparticles for the generation of hydrogen or oxygen from water. *Journal of the American Chemical Society*, 133, 595–692.

Skorb, E.V., Shchukin, D.G. & Sviridov, D.V. (2008) Hybrid silica-zirconia films loaded with titania nanoparticles and titania-based nanocontainers: novel materials for thin-film photocatalysts and photocontrollable coatings, molecular and nanoscale systems for energy conversion. In: Varfolomeev, S. (series ed) *Molecular and Nanoscale Systems for Energy Conversion*. Nova Science Publishers, New York, NY. pp. 75–87.

Song, H., Shao, J., He, Y., Liu, B. & Zhong, X. (2012) Natural organic matter removal and flux decline with PEG-TiO$_2$-doped PVDF membranes by integration of ultrafiltration with photocatalysis. *Journal of Membrane Science*, 405–406, 48–56.

Sosnowchik, B.D., Lin, L. & Englander, O. (2010) Localized heating induced chemical vapor deposition for one-dimensional nanostructure synthesis. *Journal of Applied Physics*, 107, 051101.

Souza, V.C. & Quadri, M.G.N. (2013) Organic-inorganic hybrid membranes in separation processes: a 10-year review. *Brazilian Journal of Chemical Engineering*, 30, 683–700.

Strini, A. & Schiavi, L. (2011) Low irradiance toluene degradation activity of a cementitious photocatalytic material measured at constant pollutant concentration by a successive approximation method. *Applied Catalysis* B: *Environmental*, 103, 226–231.

Sun, Y.M., Seo, J.H., Takacs, C.J., Seifter, J. & Heeger, A.J. (2011) Inverted polymer solar cells integrated with a low temperature annealed sol-gel derived ZnO film as an electron transport layer. *Advanced Materials*, 23(14), 1679–1683.

Takeuchi, M., Sakamoto, K., Martra, G., Coluccia, S. & Anpo, M.J. (2005) Mechanism of photoinduced superhydrophilicity on the TiO$_2$ photocatalyst surface. *Journal of Physical Chemistry* B, 109, 15,422–15,428.

Teichner, S.J. (2008) The origins of photocatalysis. *Journal of Porous Materials*, 15, 311–314.

Tortora, J.G., Funke, R.B. & Case, L.C. (2010) *Microbiology: An Introduction*. 10th edition, Pearson Education Inc., New York, NY.

Trillas, M., Pujol, M. & Domenech, X. (1992) Phenol photodegradation over titanium dioxide. *Journal of Chemical Technology & Biotechnology*, 55, 85–90.

Veronovski, N., Andreozzi, P., La Mesa, C. & Sfiligoj-Smole, M. (2010) Stable TiO_2 dispersions for nanocoating preparation. *Surface & Coatings Technology*, 204, 1445–1451.

Vippola, M., Bard, D., Sarlin, E., Tuomi, T. & Tossavainen, A. (2009) Nanoatlas of selected engineered nanoparticles. Finnish Institute of Occupational Health, Helsinki, Finland.

Wang, R., Hashimoto, K., Fujishima, A., Chikuni, M., Kojima, E., Kitamura, A., Shimohigoshi, M. & Watanabe, T. (1997) Light-induced amphiphilic surfaces. *Nature*, 388, 431–432.

Wang, Y.D., Zhou, A.N. & Yang, Z.Y. (2008) Preparation of hollow TiO_2 microspheres by the reverse microemulsions. *Materials Letters*, 62, 1930–1932.

Wang, Y., He, Y., Lai, Q. & Fan, M. (2014) Review of the progress in preparing nano TiO_2: an important environmental engineering material. *Journal of Environmental Sciences*, 26, 2139–2177.

Warwick, M.E., Dunnill, C.W., Goodall, J., Darr, J.A. & Binions, R. (2011) Hybrid chemical vapor and nanoceramic aerosol assisted deposition for multifunctional nanocomposite thin films. *Thin Solid Films*, 519, 5942–5948.

Wu, C.H. (2004) Comparison of azo dye degradation efficiency using UV/single semiconductor and UV/coupled semiconductor systems. *Chemosphere*, 57, 601–608.

Wu, G.S. & Chen, A.C. (2008a) Direct growth of F-doped TiO_2 particulate thin films with high photocatalytic activity for environmental applications. *Journal of Photochemistry and Photobiology* A, 195, 47–53.

Wu, G., Gan, S., Cui, L. & Xu, Y. (2008b) Preparation and characterization of PES/TiO_2composite membranes. *Applied Surface Science*, 254, 7080–7086.

Wu, Y.H., Long, M., Cai, W.M., Dai, S.D., Chen, C., Wu, D.Y. & Bai, J. (2009) Preparation of photocatalytic anatase nanowire films by in situ oxidation of titanium plate. *Nanotechnology*, 20, 185703.

Yadav, H.M., Otari, S.V., Koli, V.B., Mali, S.S., Hong, C.K., Pawar, S.H. & Delekar, S.D. (2014) Preparation and characterization of copper-doped anatase TiO_2 nanoparticles with visible light photocatalytic antibacterial activity. *Journal of Photochemistry and Photobiology* A: *Chemistry*, 280, 32–38.

Yang, H.K., Moon, B.K., Choi, B.C., Jeong, J.H. & Kim, K.H. (2013) Crystal growth and photoluminescence properties of Sm^{3+} doped CeO_2 nanophosphors by solvothermal method. *Journal of Nanoscience and Nanotechnology*, 13, 6060–6063.

Yang, S., Zheng, Y.C., Hou, Y., Yang, X.H. & Yang, H.G. (2014) Anatase TiO_2 with nanopores for dye-sensitized solar cells. *Physical Chemistry Chemical Physics*, 16, 23,038–23,043.

Yao, Y., Ohko, Y., Sekiguchi, Y., Fujishima, A. & Kubota, Y. (2008) Self-sterilization using silicone catheters coated with Ag and TiO_2 nanocomposite thin film. *Journal of Biomedical Materials Research* B, 85B, 453–460.

Yokel, R.A. & MacPhail, R.C. (2011) Engineered nanomaterials: exposures, hazards, and risk prevention. *Journal of Occupational Medicine and Toxicology*, 6, 1–27.

You, S.-J., Semblante, G.U., Lu, S.-C., Damodar, R.A. & Wei, T.-C. (2012) Evaluation of the antifouling and photocatalytic properties of poly(vinylidene fluoride) plasma-grafted poly(acrylicacid) membrane with self-assembled TiO_2. *Journal of Hazardous Materials*, 237–238, 10–19.

Yu, H., Ouyang, S.X., Yan, S.C., Li, Z.S., Yu, T. & Zou, Z.G. (2011) Sol-gel hydrothermal synthesis of visible light driven Cr-doped $SrTiO_3$ for efficient hydrogen production. *Journal of Materials Chemistry*, 21(30), 11,347–11,351.

Yu, Q.L., Ballari, M.M., Brouwers, H.J.H. & Yu, Q.L. (2010) Indoor air purification using heterogeneous photocatalytic oxidation. II. Kinetic study. *Applied Catalysis* B: *Environmental*, 99, 58–65.

Zhang, J., Zhou, P. & Liu, J. (2014) New understanding of the difference of photocatalytic activity among anatase, rutile and brookite TiO_2. *Physical Chemistry Chemical Physics*, 16, 20,382–20,386.

Zhao, L., Wang, H., Huo, K., Cui, L., Zhang, W., Ni, H., Zhang, Y., Wu, Z. & Chu, P.K. (2011) Antibacterial nano-structured titania coating incorporated with silver nanoparticles. *Biomaterials*, 32, 5706–5716.

Zhou, H., Chen, Y., Fan, H., Shi, H., Luo, Z. & Shi, B. (2008) Water vapor permeability of the polyurethane/TiO_2 nanohybrid membrane with temperature sensitivity. *Journal of Applied Polymer Science*, 109, 3002–3007.

CHAPTER 8

Application of nanosized TiO$_2$ in membrane technology

Alberto Figoli, Tiziana Marino, Silvia Simone, Marcel Boerrigter,
Mirko Faccini, Christiane Chaumette & Enrico Drioli

8.1 INTRODUCTION

The typical techniques for eliminating impurities from water sources include settling and granular media filtration, sand filters, coagulation and flocculation, while desalination is traditionally carried out using energy-consuming distillation. However, in recent decades, membrane-based techniques have attracted much attention. They are now widely accepted as viable alternatives to conventional processes. Membrane systems offer considerable advantages, such as high efficiency of impurities removal, the ability to select the class of contaminants to be removed (from algae and bacteria to monovalent and divalent salts), the obviation of the need for additional chemicals to achieve separation, a small footprint and ease of automation and scaling up, coupled with reduced energy consumption.

Membrane-based desalination represents one of the most attractive techniques for ensuring the availability of fresh water in many arid areas of the world. It is widely accepted that reverse osmosis (RO), combined with water pretreatment *via* microfiltration (MF) and ultrafiltration (UF), and eventually integrated with emerging contactor-based processes, is a viable, cost-effective and sustainable alternative to traditional desalination techniques. Furthermore, membrane techniques can be applied to avoid contamination of the environment and water sources with industrial and municipal effluents, as well as to ensure the supply of safe drinking water to large and small communities around the world. Membrane filtration can be applied to efficiently and cost-effectively remove from drinking water solid particulates, microorganisms, contaminants such as iron and manganese, and toxic contaminates like arsenic. Thus, the treatment of water and wastewater to obtain drinking water and reduce the environmental impact of residential and industrial waste represents one of the most interesting applications of membrane science and technology.

Membranes can be produced either from inorganic or polymeric materials. Although inorganic membranes offer some advantages, such as high chemical, thermal and mechanical resistance, their preparation is often very costly. On the other hand, polymeric membranes, both as flat sheets and hollow fibers, can be easily manufactured *via* the phase inversion technique, which is, indeed, the most versatile and widespread process for membrane preparation (Figoli *et al.*, 2015). Recently, the introduction of different nanomaterials into polymeric membranes, for producing the so-called mixed matrix membranes (MMMs), has been proposed as a viable strategy to combine the typical advantages of inorganic materials with the ease of preparation of polymeric membranes. Titanium dioxide (TiO$_2$), in particular, is among the most investigated materials, because it can endow membranes with excellent photocatalytic, antibacterial, anti-fouling and UV-cleaning properties.

In this chapter, the most recent advances regarding the preparation and application of TiO$_2$ membranes are presented. We show how the combination of the most commonly used polymeric materials, such as poly(vinylidene fluoride) (PVDF) and poly(ether sulfone) (PES), with TiO$_2$ nanoparticles (NPs) enables the preparation of membranes with superior performance coupled to outstanding fouling resistance. The application of TiO_2 membranes to water and wastewater treatment will be discussed as one of the most interesting applications of this technology.

8.2 TiO_2 NANOPARTICLES IN MEMBRANES

Nowadays, the use of membrane technologies is widely accepted as a more economic, environmentally friendly and sustainable alternative to several industrial processes. In the literature, a membrane is typically defined as an "interphase between two adjacent phases acting as a selective barrier, regulating the transport of substances between two compartments" (Ulbricht, 2006)

Although, as pointed out by several authors, these technologies offer several advantages, both in terms of costs and of reduced environmental impact, and are in perfect agreement with the philosophy of process intensification (Drioli *et al.*, 2004), fouling and, above all, biofouling still represent the greatest obstacle to more widespread adoption of membrane processes (Flemming, 1997; Wang *et al.*, 2000). Fouling negatively affects plant productivity, creating additional resistance to membrane transport. Furthermore, it often alters the membrane cutoff (meaning that some valuable compounds can be lost in the permeate). Consequently, operational costs increase while membrane lifetime reduces. Fouling makes necessary the use of chemical agents for cleaning, and thus one of the main advantages of any membrane process, that is, the obviation of the need for additional chemicals, is lost. There are different strategies to address this problem, such as physical and chemical cleaning (e.g. back-flushing or treatment with sodium hypochlorite, enzymes etc.), improvement of the system fluid-dynamic conditions, and membrane modification. In particular, smooth and hydrophilic surfaces are less susceptible to fouling (Field, 2010; Hashino *et al.*, 2011; Lee *et al.*, 2004).

According to the literature, different strategies have been explored to improve the hydrophilicity of the membrane surface and so increase its resistance to fouling, such as:

- Blending with other polymers and/or hydrophilic additives (Nunes and Peinemann, 1992; Wilhelm *et al.*, 2002).
- Surface treatments, such as coating, chemical grafting and plasma treatment (Steen *et al.*, 2002; Sui *et al.*, 2012; Zhang *et al.*, 2008).

Recently, the introduction of different filler types (several types of metal oxide particles, silica, montmorillonite and carbon-based materials such as nanotubes, graphene oxide etc.) has been proposed as a viable strategy for modifying/improving membrane properties (Anadao *et al.*, 2010; Ng *et al.*, 2013; Shen and Lua 2012; Wang *et al.*, 2012; Zhao *et al.*, 2013).

As already indicated above, titanium dioxide represents one of the most interesting inorganic materials, thanks to its photocatalytic properties coupled to non-toxicity, biocompatibility and low cost. The advantages of using nanomaterials and the possibility of producing different TiO_2-based nanomaterials have also been highlighted. The association between the possible advantages derivable from TiO_2 and those typical of membrane processes is, therefore, a very attractive concept.

Nevertheless, it should be pointed out that coupling membranes and TiO_2 is not a recent idea. In early works, TiO_2 was used as a suspended material in order to exploit its photocatalytic properties, and membrane filtration was typically applied to separate and recover the NPs after the reaction. It was shown that this method was effective in reducing membrane fouling (Lee *et al.*, 2001; Xi and Geissen, 2001). However, the approach was complex and the possibility of unrecycled particles being lost to the treated water was a major concern.

On the other hand, as previously described, several authors (Clarizia *et al.*, 2004; Molinari *et al.*, 2010) have pointed out that the combination of inorganic and polymeric materials to produce the so-called MMMs represents one of the most interesting solutions, which allows combination of the advantages of inorganic materials with the simplicity of preparation and reduced costs of polymer membranes. Immobilization of a TiO_2 photocatalyst in a polymeric membrane therefore represents an interesting and viable approach, making it possible to combine membrane separation and photocatalysis in one simple step and, at the same time, overcome the difficulty of recovering suspended TiO_2 NPs.

In the literature, several studies have examined the preparation of polymeric membranes containing TiO_2. The purposes of such a combination and the advantages arising from this coupling are extensively acknowledged. NPs may confer on the membranes (especially on inherently hydrophobic, but much used, materials such as PVDF) better hydrophilicity and, consequently, better resistance to fouling and biofouling, often united to an improvement in mechanical properties and performance (permeability and rejection) too. Furthermore, membranes can be endowed with self-cleaning properties, coupling the separation process to UV-photocatalytic degradation of contaminants.

Although these aspects and benefits are widely acclaimed by several studies, there is an energetic debate as to how to implement TiO_2/polymeric MMMs, that is, about which is the simplest, most advantageous and safest approach to producing such membranes, as well as about what the optimal TiO_2 loading is to achieve the best properties.

In this section, the strategies for producing such membranes, as well as the most significant and recent examples of TiO_2 membranes reported in the literature, will be described. The application of such membranes to water treatment will be examined in the subsequent section. In the authors' opinion, photocatalytic and anti-fouling TiO_2 membranes represent one of the most promising approaches for water remediation. The full exploitation of their advantages may achieve results competitive with both conventional technologies and classically applied polymer membranes.

8.2.1   *Strategies for TiO₂ immobilization*

One of the major drawbacks connected to the use of NPs is the difficulty of obtaining uniform dispersion of fillers in the polymeric matrix or, eventually, on the membrane surface. Different strategies have been proposed in the literature, and the debate on which is the most convenient and useful methodology for preparing stable and high-performing TiO_2 membranes is still ongoing.

Methods for producing hybrid TiO_2/polymer membranes reported in the literature include:

- Filtration of NP aqueous suspension and subsequent deposition of TiO_2 NPs on the membrane surface (Erdei *et al.*, 2008; Ho *et al.*, 2010; Xi and Geissen, 2001).
- Direct addition of TiO_2 NPs to the casting solution (Artale *et al.*, 2001; Kim *et al.*, 2003; Li *et al.*, 2009; Kwak *et al.*, 2001; Molinari *et al.*, 2004).
- Electrostatic self-assembly between TiO_2 NPs and anionic polymers (Bae *et al.*, 2006).

Teow *et al.* (2012) proposed an *in-situ* colloidal precipitation method to be carried out directly in the membrane coagulation bath. Recent works have suggested the use of coupling agents to modify NPs and, hence, improve TiO_2 dispersion, such as aminopropyltriethoxysilane (Razmjou *et al.*, 2011a) and γ-aminopropyltriethoxysilane (Wu *et al.*, 2008). Other researchers have suggested polymer modification using sulfonic acid (Bae *et al.*, 2006), diethanolamine (Mansourpanah *et al.*, 2009) and 4-vinylpyridine-co-acrylamide (Essawy *et al.*, 2008), followed by electrostatic self-assembly between TiO_2 NPs and the modified polymer functional groups.

Physical blending is one of the easiest and most commonly used ways of preparing TiO_2/polymer membranes. It is usually carried out during membrane preparation *via* temperature- or non-solvent-induced phase inversion (TIPS or NIPS) by introducing TiO_2 NPs directly into the casting solution. Both flat sheet and hollow fiber membranes have been prepared by using different polymeric materials. In addition, various strategies, involving different NP types and concentrations, have been proposed to improve the dispersion of the NPs, preventing their aggregation and, thus, achieving optimal performance.

Shi *et al.* (2012) prepared hybrid TiO_2/PVDF membranes *via* the TIPS method, using dimethylphthalate (DMP) as diluent, and examined the effect of the TiO_2 percentage on the membrane morphology, performance, and thermal and mechanical properties. It was found that there exists an optimal concentration of TiO_2 (0.45%) that creates membranes with improved pure water flux, higher porosity, higher contact angle (due to higher porosity and roughness of the membrane

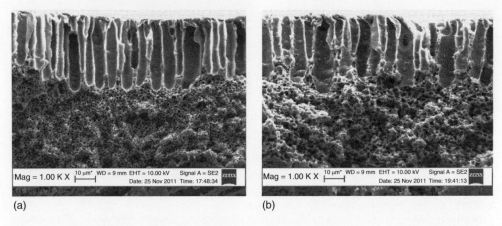

Figure 8.1.   Cross-section of ultrafiltration membranes: (a) neat; (b) mixed matrix (Ngang *et al.*, 2012).

surface), greater elongation at break and narrower pore size distribution. Higher TiO$_2$ dosages reduced these membrane properties and also resulted in particle agglomeration and, hence, non-homogeneous distribution of the filler in the polymeric matrix.

Poly(sulfone) (PS) hollow fibers were produced *via* NIPS, using dimethylacetamide (DMAc) as solvent and polyvinyl pyrrolidone (PVP; $MW = 30$ kDa) as additive (Hamid *et al.*, 2011). TiO$_2$ NPs were added directly to the dope solution in concentration of 2 wt%. Filtration of water and humic acid (HA) solution showed improved hydrophilicity, increased smoothness, higher water flux and better fouling resistance compared to the unmodified PS membrane.

Ngang *et al.* (2012) reported that self-cleaning PVDF membranes (Fig. 8.1), with 100% flux recovery ratio (*FRR*) after filtration of methylene blue solutions, can be obtained by loading TiO$_2$ NPs (1.5%) directly into the solvent (DMAc) prior to the polymer. Uniform filler dispersion was ensured by sonication for 15 min. It was found that membrane hydrophilicity and permeation flux were greatly improved.

Li *et al.* (2009) prepared microporous hybrid TiO$_2$/PES (Fig. 8.2) membranes by using a combination of vapor- and diffusion-induced phase separation (VIPS and DIPS). Dope solutions were prepared, with TiO$_2$ fillers dispersed in the solvent before the polymer. The cast PES films were exposed to air (relative humidity 50%) for 30 s before being immersed in the coagulation bath. Results showed that membrane hydrophilicity and water permeability were improved by the addition of TiO$_2$ NPs; in particular, a maximum flux of 3711 L m^{-2} h^{-1} was obtained at a TiO$_2$ concentration of 4%. However, better filler dispersion was observed at lower NP loadings, and the authors estimated that optimal TiO$_2$ concentration was around 1–2%.

Vatanpour *et al.* (2012) reported the preparation of mixed matrix PES nanofiltration membranes using three different types of TiO$_2$ NPs of different size (8, 20 and 15–25 nm). In general, the membranes obtained showed improved hydrophilicity. However, the best results in terms of water permeability were obtained with particles of 20 nm, which also showed better dispersion at high concentration (4%). Membrane susceptibility to biofouling was generally improved by the addition of TiO$_2$ particles, as demonstrated by experiments in whey filtration. Here, the best results were obtained with a membrane containing 20 nm particles, which showed the highest water flux recovery percentage (90.8%) after whey filtration.

Other authors have claimed that the best approach is to have NPs only on the membrane surface, because they then have a greater chance of exposure to substrate molecules. However, this approach does not ensure NP immobilization nor minimize leaching into the filtrated product.

Teow *et al.* (2012) prepared TiO$_2$/PVDF MMMs by adding the NPs to the coagulation bath (TiO$_2$ 0–0.1 g L^{-1}). Flat membranes were prepared using three different solvents: *N*-methyl-2-pyrrolidone (NMP), *N,N*-dimethylacetamide (DMAc) and *N,N*-dimethyl formamide (DMF).

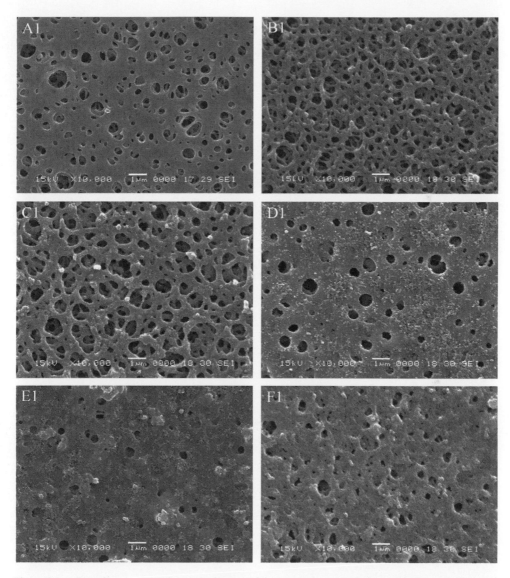

Figure 8.2. Scanning electron microscope images of the top-surface morphology of PES membranes with different TiO$_2$ content: (A1) 0 wt%; (B1) 1 wt%; (C1) 2 wt%; (D1) 3 wt%; (E1) 4 wt%; (F1) 5 wt% (Li *et al.*, 2009).

NMP resulted in smaller surface particles and narrower particle size distribution on the membrane surface. These membranes had a larger pore size and higher pure water flux (209.79 L m^{-2} h^{-1} at TiO$_2$ 0.01 g L^{-1}) but lower rejection of HA (94.70%). In contrast, optimal pore size and narrow pore size distribution, coupled with good pure water permeability (43.21 L m^{-2} h^{-1}) and higher rejection of HA (98.28%), were obtained using DMAc as the solvent with 0.01 g L^{-1} of TiO$_2$ in the coagulation bath.

Zhang *et al.* (2012) prepared hybrid TiO$_2$/PVDF membranes by impregnating PVDF film, pretreated with cetyltrimethyl ammonium bromide (CTAB) or sodium dodecyl sulfate (SDS), in a TiO$_2$ suspension.

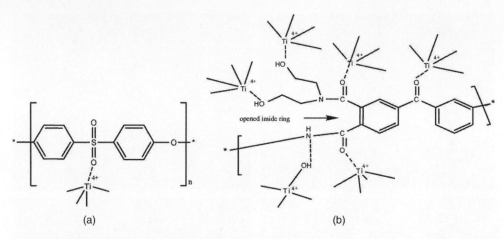

Figure 8.3.   Mechanism of self-assembly of TiO_2 nanoparticles: (a) on PES surface; (b) on DEA-modified surface of polyimide (Mansourpanah *et al.*, 2009).

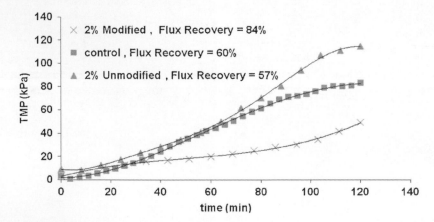

Figure 8.4.   Fouling behavior for modified, unmodified and control membranes (0.5 wt% BSA solution was filtered at constant flux at $5\,L\,m^{-2}\,h^{-1}$ for 2 h) (Razmjou *et al.*, 2011a).

In order to improve adhesion of TiO_2 NPs on the membrane surface, Mansourpanah *et al.* (2009) functionalized the surface of PES/polyimide-blend membranes (Fig. 8.3) with hydroxyl groups by treatment with diethanolamine (DEA). Results showed a very uniform settlement of TiO_2 nanoparticles on the membrane surface.

The modification of TiO_2 NPs is another viable strategy that was proposed in the literature to overcome the problem of particle agglomeration (Li *et al.*, 2014; Razmjou *et al.*, 2011a, 2011b, 2012a, 2012b; Safarpour *et al.*, 2015; Wu *et al.*, 2008; Yu *et al.*, 2009). Razmjou *et al.* (2011a) prepared PES UF membranes with modified (mechanically or mechanically and chemically) TiO_2 NPs (Fig. 8.4). Thus, TiO_2 NPs of 20 nm were mechanically ground to increase bulk density four-fold. For chemical modification, mechanically modified TiO_2 nanoparticles were functionalized by aminopropyltriethoxysilane (APTES) as a silane coupling agent. The hybrid membranes prepared showed better resistance to fouling; in particular, a flux recovery percentage of 84% was observed at the optimal concentration of TiO_2, equal to 2%.

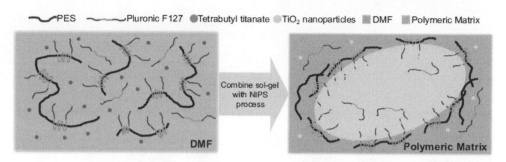

Figure 8.5.   A scheme illustrating the self-assembly of TiO₂ nanoparticles around pores of a PES membrane, accompanying the synchronous formation of a PES polymeric membrane during a NIPS process (Li *et al.*, 2014b).

In subsequent work, Razmjou *et al.* (2012b) used the same methods to modify TiO₂ particles, which were entrapped in hollow fiber PES UF membranes. The results showed that better NP dispersion was achieved after both chemical and mechanical treatment.

Wu *et al.* (2008) also modified TiO₂ NPs, in order to avoid aggregation, using APTES. They observed improved hydrophilicity and permeability upon addition of TiO₂ to PES membranes. However, in this case, the best results were obtained at lower loadings of TiO₂ (0.5%). Above this concentration, membrane performance was found to diminish.

Interesting research to elucidate the best strategy for preparing TiO₂/polymeric membranes was carried out by Rahimpour *et al.* (2008). Using PES membranes, they compared the performance of TiO₂-entrapped membranes, UV-irradiated TiO₂-entrapped membranes and UV-irradiated TiO₂-deposited membranes. UV irradiation improved the performance of TiO₂-entrapped membranes, which showed lower flux decline compared to a neat PES membrane. However, a comparison of UV-irradiated TiO₂-entrapped and UV-irradiated TiO₂-deposited membranes demonstrated that the optimal technique for the modification of PES membranes to minimize membrane fouling is deposition of TiO₂ NPs onto the membrane surface.

Similarly, Yu *et al.* (2009) compared the performance of TiO₂/PVDF hollow fibers obtained by the sol-gel method and blending. TiO₂ concentration was varied from 0–5 wt% in both cases. The addition of TiO₂ improved membrane properties such as hydrophilicity, flux and fouling resistance. The optimal TiO₂ concentration was 1 wt%. The sol-gel method resulted in better membranes due to improved dispersion and the reduction of particle aggregates, which negatively affect flux, hydrophilicity, pore size and mechanical properties.

Recently, Li *et al.* (2014b) proposed an innovative sol-gel method with self-assembly of TiO₂ particles around membrane pores during phase inversion (Fig. 8.5). PES flat sheet membranes, with Pluronic F-127 (generic name, poloxamer 407) as additive, were produced *via* NIPS. The membranes obtained showed improved properties, stable performance and reduced TiO₂ leaching. The surfactant Pluronic F-127 played a crucial role: its amphiphilic nature ensured better interaction between TiO₂ and the polymeric matrix.

Another viable strategy for improving TiO₂ dispersion has recently been proposed by Safarpour *et al.* (2015). PVDF flat sheet membranes, with PVP (29 kDa) as additive, were produced *via* NIPS, while novel reduced graphene oxide (rGO)/TiO₂ nanocomposites were synthesized *via* a hydrothermal method. The performance of bare PVDF, rGO/PVDF, TiO₂/PVDF and rGO/TiO₂/PVDF membranes was compared, and different rGO/TiO₂ ratios were tested as well. The best results were obtained with an rGO/TiO₂ ratio of 70/30. TiO₂ dispersion was improved by rGO and these membranes showed better hydrophilicity, smoothness and flux recovery compared to the others (Fig. 8.6).

The strategies proposed in the literature for preparing TiO₂/polymer MMMs are summarized in Figure 8.7.

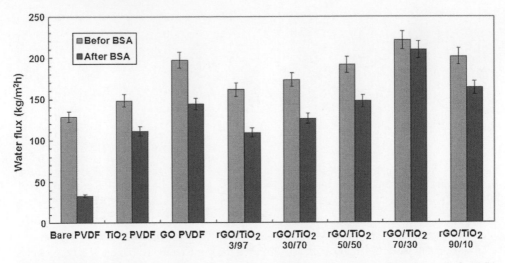

Figure 8.6. Water flux of the PVDF membranes prepared by Safarpour *et al.* (2015) before and after BSA filtration (90 min at 0.3 MPa; additive concentration = 0.05 wt%).

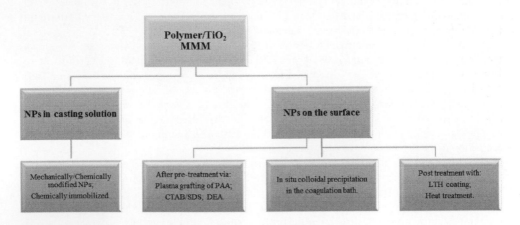

Figure 8.7. Strategies proposed for preparing TiO$_2$/polymer MMMs.

8.2.2 Examples of TiO$_2$ membranes

The use of TiO$_2$ NPs in combination with polymeric materials for preparing MMMs has been reported by several authors. In general, these studies aimed at preparing membranes with superior fouling resistance, the ability to degrade organic contaminants, self-cleaning properties, antibacterial properties, and improved mechanical resistance and/or rejection. Membranes were prepared from different polymeric materials, and authors proposed some viable approaches to improving dispersion of TiO$_2$ NPs. In addition, there have been several interesting applications of these membranes, particularly in the field of wastewater remediation. Literature analysis showed that the most widely used materials for membrane preparation are PVDF and PES, although other materials were also reported. Examples of PVDF and PES membranes containing TiO$_2$ particles, either entrapped or coated on the membrane surface, recently reported in the literature are summarized in Table 8.1.

Table 8.1. Summary of the properties and applications of the most recent TiO$_2$/polymer MMMs reported in the literature.

Polymer material	TiO$_2$ type	Membrane preparation technique/reagents	Main results/application	Reference
Nanoparticles in dope solution				
PVDF (FR904, M_w = 380,000; Shanghai 3F Ltd.)	Anatase TiO$_2$ (20 nm; Meidilin Nanometer Materials Development Co.)	TIPS using DMP as diluent	TiO$_2$ promotes heterogeneous nucleation resulting in uniform polymer spherulites (until 0.45%); higher TiO$_2$ percentage caused particle agglomeration. Optimal 0.45% TiO$_2$ results in membranes having improved pure water flux, contact angle, porosity and elongation at break and narrower pore size distribution.	Shi *et al.* (2012)
PVDF (Solef®; Solvay)	Anatase TiO$_2$ (20 nm; TitanPE Technologies)	NIPS using DMAc as solvent	Membrane hydrophilicity and permeation flux were greatly improved. NPs could provide extra adsorption sites for methylene blue. The MMM is photocatalytically active and self-cleaning.	Ngang *et al.* (2012)
PVDF (Solef® 6020; Solvay)	TiO$_2$ (20–30 nm; Degussa)	NIPS using DMAc as solvent and PEG-600 as pore former	Optimal concentrations of 12, 2 and 0.5% for PVDF, PEG and TiO$_2$, respectively, ensured good combination of membrane flux and rejection with no particle aggregation.	Song *et al.* (2012)
PVDF (Kynar® 741; Arkema Inc.)	98% anatase TiO$_2$ (20 nm; Degussa)	NIPS using acetone–butan-1-ol 81:19 or DMF as solvents	Optimal rejection of HA and diminished flux decline. Optimal UV irradiation time for self-cleaning of 30 min. The best membrane had a pore size of 0.96 μm, with a maximum porosity of 86% at 0.5 TiO$_2$/PVDF weight ratio. Pretreatment with ethanol improved the degradation rate of Brilliant Green and Indigo Carmine dyes.	Alaoui *et al.* (2009)
PES (Ultrason E 6020 P, BASF, Ludwigshafen, Germany)	TiO$_2$ (20 nm; Degussa)	NIPS using NMP as solvent	Membranes with improved flux and permeability and with antibacterial, photoactive and self-cleaning properties.	Susanto *et al.* (2009)

(continued)

Table 8.1. Continued.

Polymer material	TiO$_2$ type	Membrane preparation technique/reagents	Main results/application	Reference
PES (Jilin Jida High Performance Materials)	TiO$_2$ (21 nm; Degussa)	VIPS+NIPS using diethylene glycol:DMAc 1:1 as solvent	Membranes with improved hydrophilicity and permeability. Compared to the pure PES membrane, the hybrid membrane had higher breaking strength and low elongation ratio.	Li et al. (2009)
PES ($M_w = 58{,}000$ g mol^{-1}; BASF SE)	TiO$_2$ (20 nm; Degussa) mechanically or mechanically/ chemically modified	NIPS using DMAc as solvent and PVP ($M_w = 40{,}000$ g mol^{-1}) as pore former	Good dispersion of fillers was achieved after both chemical and mechanical modification of particles; however, best results were obtained at 2% concentration of TiO$_2$, which avoided agglomeration. The prepared UF membranes showed improved fouling resistance.	Razmjou et al. (2011a)
PES (Jilin University)	TiO$_2$ (30 nm; Hangzhou Dayang Chemical) chemically modified	NIPS using DMAc as solvent and H$_2$O and PVP as pore formers	The hydrophilicity, thermal stability, mechanical strength and anti-fouling property of PES membranes were enhanced by adding TiO$_2$ NPs. TiO$_2$ content above 0.5% resulted in impairment of membrane performance.	Wu et al. (2008)
PES ($M_w = 58{,}000$ g mol^{-1}; BASF SE)	TiO$_2$ (20 nm; Degussa) mechanically or mechanically/chemically modified	Dry-jet/wet spinning technique using NMP as solvent and PVP ($M_w = 40{,}000$ g mol^{-1}) as pore former	Mechanical and chemical modification of TiO$_2$ NPs ensured better dispersion of filler in the membrane. Some membrane properties, e.g. hydrophilicity, improved, but flux recovery was not observed as expected.	Razmjou et al. (2012b)
PES (Ultrason® E 6020 P, $M_w = 58{,}000$ g mol^{-1}; BASF SE)	TiO$_2$ (20 nm; Degussa) (8 or 15–25 nm; Millennium Inorganics)	NIPS using DMAc as solvent and PVP ($M_w = 25{,}000$ g mol^{-1}) as pore former	Particles of 20 nm showed better dispersion also at higher loading. Membranes with improved hydrophilicity and resistance to fouling.	Vatanpour et al. (2012)
PES (Radel A-300A; Solvay)	TiO$_2$ (rutile content = 61.2%, particle size = 70 nm)	Solvent evaporation (dense gas separation membranes) using DMF as solvent	MMM containing TiO$_2$ NPs showed improved CO$_2$/CH$_4$ selectivity compared to those containing Na-montmorillonite. Best results in terms of selectivity were obtained at TiO$_2$ concentration of 4%.	Liang et al. (2012)

Nanoparticles on membrane surface

Membrane	Nanoparticle	Method	Comments	Reference
PVDF (Solvay)	TiO$_2$ (20 nm; TitanPE Technologies) 85% anatase and 15% rutile	NIPS using NMP, DMAc or DMF as solvent (NPs in coagulation bath)	Membrane hydrophilicity and roughness improved with TiO$_2$ content in the bath until 0.01 g L^{-1}. Higher TiO$_2$ concentration could result in particle aggregation and membrane pore blocking. Pure water flux and rejection of HA can be modulated by changing the solvent type (due to different water miscibilities).	Teow *et al.* (2012)
PVDF membrane (2 μm; Mosu Science Equipment Corp.)	TiO$_2$ NPs with anatase structure were synthesized from tetrabutyl titanate	CTAB 2% or SDS 2% pretreated PVDF film was immersed in the TiO$_2$ suspension.	Membranes with improved surface hydrophilicity and permeability and with anti-fouling properties.	Zhang *et al.* (2012)
PVDF membrane (0.45 μm; Merck Millipore)	–	Plasma grafting of acrylic acid followed by dipping in aqueous TiO$_2$ suspension	Promoted adsorption and elution efficiency of Cu^{2+} compared to neat PVDF film. TiO$_2$ significantly enhanced membrane hydrophilicity; optimal concentration of 0.5% TiO$_2$ ensured the highest pure water flux and the best protein anti-fouling property.	You *et al.* (2012)
PVDF membrane (0.45 μm; Merck Millipore)	TiO$_2$ NPs were synthesized from titanium (IV) iso-propoxide	Coating of TiO$_2$ NPs onto the PVDF membrane	The MMM is photocatalytically active and able to degrade Reactive Black 5 dye under UV irradiation. Super-hydrophobic PVDF membrane with contact angle of 163°. Membranes with improved rejection of NaCl and anti-fouling properties compared to neat PVDF membrane.	Razmjou *et al.* (2012a)
PVDF (Alfa-Aesar) PES (Ultrason® E 6020 P; $M_w = 58,000$ g mol^{-1}; BASF SE) sulfonated by sulfuric acid	TiO$_2$ (20 nm; Degussa)	NIPS using DMAc as solvent and PVP ($M_w = 25,000$ g mol^{-1}) as pore former. Coating of TiO$_2$ NPs onto the PVDF/sulfonated PES membrane.	The produced membranes showed improved hydrophilicity, less tendency to fouling, improved BSA rejection and antibacterial properties.	Rahimpour *et al.* (2012)

(continued)

Table 8.1.   Continued.

Polymer material	TiO$_2$ type	Membrane preparation technique/reagents	Main results/application	Reference
PES (Ultrason® E 6020 P, $M_w = 58{,}000\,\mathrm{g\,mol^{-1}}$; BASF SE) Polyimide (PI)	TiO$_2$ (25 nm; Degussa)	NIPS using DMF/dioxane as solvent and PEG-600 as pore former. Virgin or DEA-treated PES/PI membranes were coated by dipping in aqueous TiO$_2$ suspension.	Treatment with DEA resulted in extra uniform settlement of TiO$_2$ NPs on the membrane surface. The presence of NPs and —OH groups improved both membrane hydrophilicity and flux, and an increment in flux recovery ratio of the membrane was also observed.	Mansourpanah et al. (2009)
PES (Ultrason® E 6020 P, $M_w = 58{,}000\,\mathrm{g\,mol^{-1}}$; BASF SE)	TiO$_2$ (25 nm; Degussa)	NIPS using DMAc as solvent and PVP ($M_w = 25{,}000\,\mathrm{g\,mol^{-1}}$) as pore former. TiO$_2$ NPs were both added to the casting solution or coated onto membrane surface by dipping neat membranes in aqueous TiO$_2$ suspension.	The anti-fouling property and long-term flux stability of PES membranes were significantly enhanced by addition of TiO$_2$, and further improved by UV irradiation. Better results obtained by coating with TiO$_2$ NPs. The optimum conditions for preparation of TiO$_2$-coated membrane were determined as: 0.03 wt% for concentration of TiO$_2$ colloidal suspension, 15 min immersion and 15 min UV irradiation with 160 W lamp.	Rahimpour et al. (2008)
PES ($M_w = 58{,}000\,\mathrm{g\,mol^{-1}}$; BASF SE)	TiO$_2$ nanoparticles were synthesized from titanium (IV) iso-propoxide	NIPS using DMAc as solvent and PVP ($M_w = 40{,}000\,\mathrm{g\,mol^{-1}}$) as pore former. PES membranes were coated with TiO$_2$ NPs by low-temperature hydrothermal process.	The NP layer was uniform and stable. Coated membranes exhibited lower protein adsorption, photocatalytic activity, long-term hydrophilicity, improvement in fouling performance and increase in flux recovery after filtration of HA.	Razmjou et al. (2011b)
PES	TiO$_2$ nanoparticles were synthesized from titanium (IV) iso-propoxide (particles 40 nm)	NIPS followed by dipping in aqueous TiO$_2$ suspension	Surface self-assembled TiO$_2$ NPs improved membrane hydrophilicity and anti-fouling ability. The prepared UF hybrid membrane also showed good separation performance (tests on PEG-5000).	Lui et al. (2005)
PES (Ultrason® E 6020 P, $M_w = 58{,}000\,\mathrm{g\,mol^{-1}}$; BASF SE) PVA (98% hydrolysis, $M_w = 72{,}000\,\mathrm{g\,mol^{-1}}$; Merck)	TiO$_2$ (20 nm; Aldrich)	PES membrane: NIPS using DMAc as solvent and PVP ($M_w = 25{,}000\,\mathrm{g\,mol^{-1}}$) as pore former. Coating with PVA cross-linked by glutaraldehyde. Immersion in TiO$_2$ suspension + heat treatment at 110°C for 5 min.	At optimal concentration of 0.1% TiO$_2$ the modified PES/PVA/TiO$_2$ membranes showed superior performance in terms of flux and NaCl rejection.	Pourjafar et al. (2012)

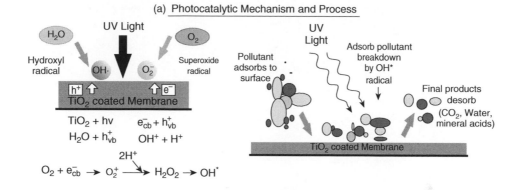

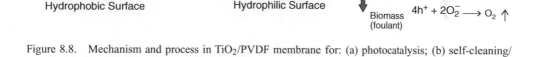

Figure 8.8.   Mechanism and process in TiO₂/PVDF membrane for: (a) photocatalysis; (b) self-cleaning/anti-fouling (modified from Damodar *et al.*, 2009).

PVDF is among the most frequently used polymeric materials, thanks to its outstanding properties, such as chemical and physical stability, but also because of the feasibility of preparing both flat sheet and hollow fiber membranes *via* phase inversion, both temperature-induced (Cui *et al.*, 2008; Ji *et al.*, 2008; Lin *et al.*, 2009; Ma *et al.*, 2009; Rajabzadeh *et al.*, 2008) and diffusion-induced (Buonomenna *et al.*, 2007; Figoli *et al.*, 2014; Simone *et al.*, 2010, 2014; Sukitpaneenit and Chung, 2009). PVDF is a hydrophobic material.

Therefore, as explained in the introduction to this chapter, PVDF membranes containing TiO₂ NPs can be prepared for their improved resistance to fouling and biofouling as well as their enhanced hydrophilicity, mechanical strength and performance. For instance, Damodar *et al.* (2009) prepared TiO₂/PVDF MMMs (Fig. 8.8) by adding different amounts of TiO₂ particles (0–4%) into the casting solution.

The addition of TiO₂ affected the membranes' pore size and hydrophilicity, improved membrane flux and permeability and, moreover, the modified membranes showed bactericidal (tests on *E. coli*), photoactive (tests on Reactive Black 5) and self-cleaning anti-fouling abilities (tests on BSA solution 1%) under UV light exposure.

PES is another material widely used for preparation of both flat sheet and hollow fiber membranes, due to its good chemical and thermal resistance, environmental endurance, potential for application across a wide pH range, easy processing and the possibility of obtaining a broad range of pore sizes. In fact, the preparation of PES membranes for MF, UF and nanofiltration (NF) has so far been reported by several studies (see, for example, Alsalhy *et al.*, 2014; Bolong *et al.*, 2010; Idris *et al.*, 2007; Li *et al.*, 2008; Susanto and Ulbricht, 2009; Susanto *et al.*, 2009).

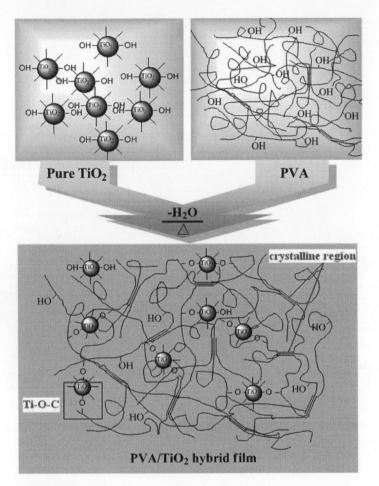

Figure 8.9.   Schematic illustration of the immobilization of TiO$_2$ in a PVA matrix by Ti-O-C chemical bonding (Lei *et al.*, 2012).

Alongside PVDF, PES is one of the polymeric materials most used for preparing MMMs loaded with TiO$_2$. Several studies reported the preparation of such membranes, especially with the aim of improving PES membranes' resistance to fouling through increased hydrophilicity. For instance, Razmjou *et al.* (2011b) prepared PES MMMs by depositing TiO$_2$ nanoparticles onto the surface of lab-prepared and commercial PES UF membranes. NPs were synthesized by sol-gel technology using titanium (IV) iso-propoxide as a precursor, with the addition of Pluronic F127 as a templating agent. Membrane coating was stabilized by a low-temperature heat treatment process. Results showed the formation of a uniform coating layer with good stability and durability. Coated membranes exhibited lower protein adsorption with respect to the control membrane, sustained photocatalytic activity and long-term hydrophilicity, which was coupled to a significant improvement in anti-fouling performance. An increase in flux recovery was observed after filtration of humic acid as a model foulant.

According to the literature, there are other polymeric materials that have been used to prepare MMMs loaded with TiO$_2$ (either entrapped or self-assembled onto the membrane surface). For instance, Lei *et al.* (2012) prepared a polyvinyl alcohol (PVA)/TiO$_2$ MMM with NPs immobilized in the polymeric matrix *via* a traditional solution-casting method followed by heat treatment (Fig. 8.9). This allowed the formation of Ti-O-C chemical bonds, chemically immobilizing TiO$_2$

in the polymeric matrix. The MMMs produced were tested for photocatalytic degradation of methyl orange dye. Results showed that hybrid TiO$_2$/PVA films loaded with 10 wt% NPs and treated at 140°C for 2 h exhibited the highest photocatalytic activity. These membranes offer two advantages: (i) the Ti-O-C chemical bonds made the TiO$_2$ nanoparticles difficult to leach from the hybrid films; (ii) the good swelling ability of the PVA matrix improved the contact between the TiO$_2$ and the dye molecules. Therefore, the immobilization studied in this work overcomes two typical problems often connected with the development of MMMs: instability of particles in the matrix and inaccessibility of the particles as a result of the membrane bulk.

8.3   APPLICATIONS OF TiO$_2$ MEMBRANES IN WATER TREATMENT

An analysis of the literature reveals that TiO$_2$/polymer MMMs have great potential for application in various fields, since they can degrade organic contaminants, kill bacteria and show self-cleaning properties due to their photocatalytical properties. These properties could help to achieve removal of pollutants in industrial effluents, extend membrane life and reduce maintenance costs in membrane bioreactor (MBR) systems. Water and wastewater treatment by processes such as MF, UF, NF, pervaporation or MBR represents one of the most interesting and promising applications for TiO$_2$/polymer membranes. Several interesting examples have already been reported in the literature; model foulants, in the form of dyes, humic acid and proteins, as well as simulated wastewater, were used to test membrane performance and demonstrate TiO$_2$'s effectiveness in preventing fouling and/or degrading contaminants.

Song *et al.* (2012) studied the preparation of photocatalytically active hybrid TiO$_2$/PVDF membranes, which were tested for natural organic matter removal in both dead-end and cross-flow UF experiments (Fig. 8.10). Poly(ethylene glycol), $M_n = 600$, (PEG-600) was used as a pore-forming additive, while DMAc was chosen as solvent. It was found that, as expected, membrane flux decreased while pepsin rejection gradually increased with increasing PVDF content, which was ranged from 10 to 16 wt%. An optimal PEG-600 concentration of 2% ensured a good combination of membrane flux and rejection. When the content of TiO$_2$ was increased from 0.25 wt% to 0.5 wt%, the membrane flux decreased, while rejection increased. Further increase in TiO$_2$ concentration resulted in higher flux and lower rejection; moreover, at TiO$_2$ concentrations above 2%, particle aggregates were visible on the membrane's surface. The PVDF-PEG-TiO$_2$ hybrid showed improved rejection of HA and less flux decline in comparison to a PVDF-PEG membrane. This was due to its smaller pore size and more hydrophilic and smoother surface. More-over, the PVDF-PEG-TiO$_2$ membrane was photocatalytically active, with optimal self-cleaning ability after 30 min of continuous UV irradiation.

Alaoui *et al.* (2009) reported the preparation of entrapped-TiO$_2$ PVDF membranes that showed appreciable degradation of Brilliant Green and Indigo Carmine dyes after pretreatment with ethanol, which increased the membrane wetting and, hence, the dye degradation rate.

You *et al.* (2012) prepared photocatalytically active TiO$_2$/PVDF membranes showing self-cleaning ability, with NPs self-assembled on the membrane surface (Figs. 8.11 and 8.12). In order to ensure immobilization of NPs on the membrane surface, poly(acrylic acid) (PAA) was plasma-grafted onto a commercial PVDF membrane, which provided functional groups that could support the NPs. It was found that the self-assembly of TiO$_2$ significantly enhanced membrane hydrophilicity, which increased with TiO$_2$ loading. An optimal concentration of 0.5% TiO$_2$ ensured the highest pure water flux and the best protein anti-fouling property. However, it was also found that at least 1.5% TiO$_2$ and 30 min of UV irradiation were necessary to recover the original performance of fouled membranes through the photodegradation of strongly bound foulants. The TiO$_2$-modified membranes were immersed in an aqueous solution of Reactive Black 5 dye and positioned directly under a UV lamp. It was found that the membrane with 3.0% TiO$_2$ loading displayed the best photocatalytic performance by removing 42% of the dye within 120 minutes of operation.

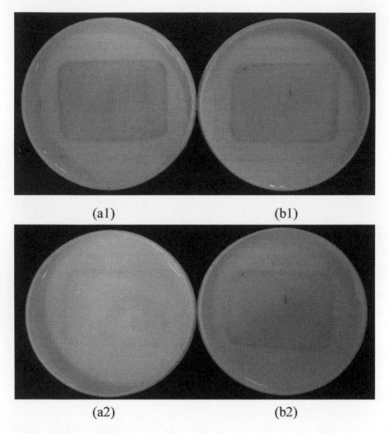

Figure 8.10.   The effect of photocatalysis on fouled TiO_2-doped membranes: (a1) PVDF-PEG-TiO_2 membrane; (b1) PVDF-PEG membrane; (a2) PVDF-PEG-TiO_2 membrane after 8 h UV irradiation; (b2) PVDF-PEG membrane after 8 h UV irradiation (Song *et al.*, 2012).

Liu *et al.* (2012) prepared a TiO_2/PVA/polyester composite membrane using a polyester filter cloth (22 μm). The prepared membranes were tested for the treatment, in an anoxic/oxic MBR (A/O-MBR), of synthetic wastewater containing terephthalic acid, glucose, NH_4Cl and $NaNO_3$, to simulate effluent from a polyester fiber production plant. The TiO_2/PVA/polyester composite membrane showed improved hydrophilicity and resistance to fouling by extracellular polymeric substances. This allowed a highly stabilized pure water flux and a high effluent flux to be obtained during the long-term filtration tests in the A/O-MBR system. Moreover, effluent qualities were similar to those obtained using commercial 0.1 μm PVDF membranes. The prepared TiO_2/PVA/polyester composite membranes showed great potential for reducing membrane costs, improving flux and reducing operation and maintenance costs in MBR systems for wastewater treatment.

Zhao *et al.* (2012) prepared TiO_2/poly(phthalazinone ether sulfone ketone) (PPESK) UF membranes for application in the purification of high-temperature condensed water. The results showed that the membranes produced had a sponge-like structure (suppression of macrovoids), improved mechanical strength and thermal stability, and higher hydrophilicity and porosity. By comparison with the neat PPESK membrane, the hybrid TiO_2/PPESK membrane showed improved antifouling properties, lower filtration resistances and better flux recovery during treatment of high-temperature condensed water containing excess oil and iron. The quality of the permeate

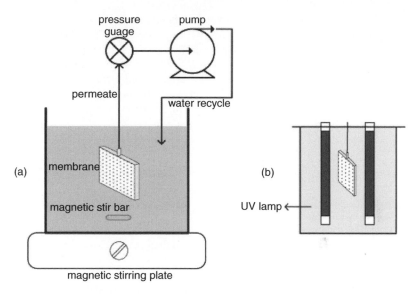

Figure 8.11.   Scheme of the setups used by You *et al.* (2012) for: (a) submerged flat sheet membrane reactor; (b) UV irradiation.

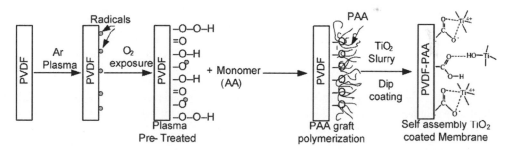

Figure 8.12.   The mechanism of plasma-induced graft polymerization of PAA on PVDF and the self-assembly of TiO₂ on the modified membrane (You *et al.*, 2012).

water met China's national Quality Criterion of Water and Steam for Steam Power Equipment standard, and the best results were obtained at a TiO₂ concentration of 2% in the casting solution.

Li *et al.* (2014a) prepared PVA/PVDF composite hollow fibers, by coating commercial PVDF hollow fibers with a PVA-TiO₂ layer. The prepared fibers, showing no rejection of inorganic salts, were applied to the desalination of three organic dyes: methyl orange, Congo red and methyl blue. Compared to the PVA/PVDF composite fibers, those containing TiO₂ showed improved properties in terms of higher separation efficiency, resistance to fouling and thermal stability.

The advantages of preparing hybrid TiO₂/polymer membranes become particularly evident when a UV light source is used to trigger TiO₂'s photocatalytic properties. Méricq *et al.* (2015) prepared flat sheet membranes *via* NIPS, by blending PVDF with different TiO₂ concentrations (0–7 wt%). Water and bovine serum albumin (BSA) solution filtration tests were carried out, and membrane cleaning was done using water and UV irradiation.

Jyothi *et al.* (2014) modified polysulfone flat sheet membranes, produced *via* NIPS. TiO₂ NPs were prepared by the sol-gel method and blended with the polymer. Membranes were surface-modified by UV treatment. The properties of the membranes obtained were improved by the addition of NPs after UV treatment. Exposure to UV light modified the membrane surface, roughness and hydrophobicity by breaking chemical bonds in the polymeric chains. Optimal results were obtained working with 2 wt% of TiO_2 and after 30 min of UV irradiation. The prepared membranes were successfully tested for removal of Cr(VI) from water.

Finally, Song *et al.* (2014) prepared PVDF hollow fibers *via* NIPS, blending 0–0.5 wt% TiO_2 with polymer 12 wt% and LiCl 0.5 wt%. The MMMs produced were tested for UF of pepsin and HA, coupled to photocatalysis. TiO_2/PVDF fibers showed improved hydrophilicity, rejection and fouling resistance. UV irradiation decreased flux decline by triggering the membranes' self-cleaning ability.

8.4   CONCLUSIONS AND FUTURE PERSPECTIVES

Although a number of very interesting studies have already been published, much work remains to be done. In fact, while each author identifies his own method as the best one, there is still no agreement on which is the most successful and convenient strategy for TiO_2 membrane preparation, and each method seems to have its advantages and drawbacks. Only a thorough evaluation that considers all of the factors involved, including the simplicity of the method itself, its cost, the actual performance of the resultant membrane, as well as the possible risk of particle release from the polymer matrix, is likely to lead to a resolution of this dispute. Careful selection of the polymer material is a crucial point. Although different polymer materials have already been tested successfully, the long-term resistance of the selected polymer upon UV irradiation should be carefully investigated. It is worth mentioning that the addition of titanium dioxide to the membranes becomes really interesting and commercially competitive when it could replace, at least in part, costly cleaning procedures. In fact, these latter are often expensive and time-consuming, and represent the only step of the membrane separation process that requires the use of additional chemicals. Indeed, the obviation of the need for additional chemicals is one of the major advantages of any membrane process, one which, for this reason, is being partially lost at the moment.

Finally, it is necessary to investigate carefully the effects of such materials on human health and the environment. Since these effects are often revealed only by long-term analysis, the need to establish such an assessment is an urgent one.

ACKNOWLEDGMENT

The authors gratefully acknowledge financial support from the NAWADES project (grant agreement no. 308439), within the European Union's Seventh Framework Programme (EU-FP7).

REFERENCES

Alaoui, O.T., Nguyen, Q.T., Mbareck, C. & Rhlalou, T. (2009) Elaboration and study of poly(vinylidene fluoride)-anatase TiO_2 composite membranes in photocatalytic degradation of dyes. *Applied Catalysis A: General*, 358, 13–20.

Alsalhy, Q.F., Salih, H.A., Simone, S., Zablouk, M., Drioli, E. & Figoli, A. (2014) Poly (ether sulfone) (PES) hollow-fiber membranes prepared from various spinning parameters. *Desalination*, 345, 21–35.

Anadao, P., Sato, L.F., Wiebeck, H. & Valenzuela-Diaz, F.R. (2010) Montmorillonite as a component of polysulfone nanocomposite membranes. *Applied Clay Science*, 48, 127–132.

Artale, M.A., Augugliaro, V., Drioli, E., Golemme, G., Grande, C., Loddo, V., Molinari, R., Palmisano, L. & Schiavello, M. (2001) Preparation and characterization of membranes with entrapped TiO_2 and preliminary photocatalytic tests. *Annali di Chimica* 91, 127–136.

Bae, T.Y., Kim, I.C. & Tak, T.M. (2006) Preparation and characterization of fouling-resistant TiO_2 self-assembled nanocomposite membranes. *Journal of Membrane Science*, 275, 1–5.

Bolong, N., Ismail, A.F., Salim, M.R., Rana, D., Matsuura, T. & Tabe-Mohammadi, A. (2010) Negatively charged polyethersulfone hollow fiber nanofiltration membrane for the removal of bisphenol A from wastewater. *Separation and Purification Technology*, 3, 92–99.

Buonomenna, M.G., Macchi, P., Davoli, M. & Drioli, E. (2007) Poly(vinylidene fluoride) membranes by phase inversion: the role the casting and coagulation conditions play in their morphology, crystalline structure and properties. *European Polymer Journal*, 43, 1557–1572.

Clarizia, G., Algieri, C. & Drioli, E. (2004) Filler-polymer combination: a route to modify gas transport properties of a polymeric membrane. *Polymer*, 45, 5671–5681.

Cui, Z., Du, C., Xu, Y., Ji, G. & Zhu, B. (2008) Preparation of porous PVDF membrane via thermally induced phase separation using sulfolane. *Journal of Applied Polymer Science*, 108, 272–280.

Damodar, R.A., You, S.J. & Chou, H.H. (2009) Study the self cleaning, antibacterial and photocatalytic properties of TiO$_2$ entrapped PVDF membranes. *Journal of Hazardous Materials*, 172, 1321–1328.

Drioli, E., Di Profio, G. & Fontananova, E. (2004) Membrane separations for process intensification and sustainable growth. *Fluid – Particle Separation Journal*, 16, 1–18.

Erdei, L., Arecrachakul, N. & Vigneswaran, S. (2008) A combined photocatalytic slurry reactor immersed membrane module system for advanced wastewater treatment. *Separation and Purification Technology*, 62, 382–388.

Essawy, A.A, Ali, E.A. & Abdel-Mottaleb, M.S.A. (2008) Application of novel copolymer-TiO$_2$ membranes for some textile dyes adsorptive removal from aqueous solution. *Journal of Hazardous Materials*, 157, 547–552.

Field, R. (2010) Fundamentals of fouling. In: Peinemann, K.V. & Nunes, S.P. (eds) *Membranes for Water Treatment*. Wiley-VCH, Weinheim, Germany. pp. 1–24.

Figoli, A., Simone, S., Criscuoli, A., Al-Jlil, S.A., Al Shabouna, F.S., Al-Romaih, H.S., Di Nicolò E., Al-Harbi, O.A. & Drioli, E. (2014) Hollow fibers for seawater desalination from blends of PVDF with different molecular weights: Morphology, properties and VMD performance. *Polymer*, 55, 1296–1306.

Figoli, A., Simone, S. & Drioli, E. (2015) Polymeric membranes. In: Hilal, N., Ismail, A.F. & Wright, C. (eds) *Membrane Fabrication*. CRC Press, Boca Raton, FL. pp. 1-44.

Flemming, H.-C. (1997) Reverse osmosis membrane biofouling. *Experimental Thermal and Fluid Science*, 14, 382–391.

Hamid, N.A.A., Ismail, A.F., Matsuura, T., Zularisam, A.W., Lau, W.J., Yuliwati, E. & Abdullah M.S. (2011) Morphological and separation performance study of polysulfone/titanium dioxide (PSF/TiO$_2$) ultrafiltration membranes for humic acid removal. *Desalination*, 273, 85–92.

Hashino, M., Hirami, K., Ishigami, T., Ohmukai, Y., Maruyama, T., Kubota, N. & Matsuyama, H. (2011) Effect of kinds of membrane materials on membrane fouling with BSA. *Journal of Membrane Science*, 384, 157–165.

Ho, D.P., Vigneswaran, S. & Ngo, H.H. (2010) Integration of photocatalysis and microfiltration in removing effluent organic matter from treated sewage effluent. *Separation Science and Technology*, 45, 155–162.

Idris, A., Zain, N.M. & Noordin, M.Y. (2007) Synthesis, characterization and performance of asymmetric polyethersulfone (PES) ultrafiltration membranes with polyethylene glycol of different molecular weights as additives. *Desalination*, 207, 324–339.

Ji, G., Zhu, L., Zhu, B., Zhang, C. & Xu, Y. (2008) Structure formation and characterization of PVDF hollow fiber membrane prepared via TIPS with diluent mixture. *Journal of Membrane Science*, 319, 264–270.

Jyothi, M.S., Nayak, V., Padaki, M., Balakrishna, R.G. & Ismail, A.F. (2014) The effect of UV irradiation on PSf/TiO$_2$ mixed matrix membrane for chromium rejection. *Desalination*, 354, 189–199.

Kim, S.H., Kwak, S.Y., Sohn, B.H. & Park, T.H. (2003) Design of TiO$_2$ nanoparticle self-assembled aromatic polyamide thin-film-composite (TFC) membrane as an approach to solve biofouling problem. *Journal of Membrane Science*, 211, 157–165.

Kwak, S.Y., Kim, S.H. & Kim, S.S. (2001) Hybrid organic/inorganic reverse osmosis (RO) membrane for bactericidal anti-fouling. 1. Preparation and characterization of TiO$_2$ nanoparticle self-assembled aromatic polyamide thin-film-composite (TFC) membrane. *Environmental Science and Technology*, 35, 2388–2394.

Lee, N., Amy G., Crouse, J. & Buison, H. (2004) Identification and understanding of fouling in low-pressure membrane (MF/UF) filtration by natural organic matter (NOM). *Water Research*, 38, 4511–4523.

Lee, S.A., Choo, K.H., Lee, C.H., Lee, H.I., Hyeon T., Choi, W. & Kwon, H.H. (2001) Use of ultrafiltration membranes for the separation of TiO$_2$ photocatalysts in drinking water treatment. *Industrial & Engineering Chemistry Research*, 40, 1712–1719.

Lei, P., Wang, F., Gao, X., Ding, Y., Zhang, S., Zhao, J., Liu, S. & Yang, M. (2012) Immobilization of TiO$_2$ nanoparticles in polymeric substrates by chemical bonding for multi-cycle photodegradation of organic pollutants. *Journal of Hazardous Materials*, 227/228, 185–194.

Li, J.F., Xu, Z.L. & Yang, H. (2008) Microporous polyethersulfone membranes prepared under the combined precipitation conditions with non-solvent additives. *Polymers for Advanced Technologies*, 19, 251–257.

Li, J.F., Xu, Z.L., Yang, H., Yu, L.Y. & Liu, M. (2009) Effect of TiO$_2$ nanoparticles on the surface morphology and performance of microporous PES membrane. *Applied Surface Science*, 255, 4725–4732.

Li, X., Chen, Y., Hu, X., Zhang, Y. & Hu, L. (2014a) Desalination of dye solution utilizing PVA/PVDF hollow fiber composite membrane modified with TiO$_2$ nanoparticles. *Journal of Membrane Science*, 471, 118–129.

Li, X., Fang, X., Pang, R., Li, J., Sun, X., Shen, J., Han, W. & Wang, L. (2014b) Self-assembly of TiO$_2$ nanoparticles around the pores of PES ultrafiltration membrane for mitigating organic fouling. *Journal of Membrane Science*, 467, 226–235.

Liang, C.-Y., Uchytil, P., Petrychkovych, R., Lai, Y.-C., Friess, K., Sipek, M., Reddy, M.M. & Suen, S.-Y. (2012) A comparison on gas separation between PES (polyethersulfone)/MMT (Na-montmorillonite) and PES/TiO$_2$ mixed matrix membranes. *Separation and Purification Technology*, 92, 57–63.

Lin, Y., Tang, Y., Ma, H., Yang, J., Tian, Y., Ma, W. & Wang, X. (2009) Formation of a bicontinuous structure membrane of polyvinylidene fluoride in diphenyl carbonate diluent via thermally induced phase separation. *Journal of Applied Polymer Science*, 114, 1523–1528.

Liu, L., Zhao, C. & Yang, F. (2012) TiO$_2$ and polyvinyl alcohol (PVA) coated polyester filter in bioreactor for wastewater treatment. *Water Research*, 46, 1969–1978.

Luo, M.-L., Zhao, J.-Q., Tang, W. & Pu, C.-S. (2005) Hydrophilic modification of poly(ether sulfone) ultrafiltration membrane surface by self-assembly of TiO$_2$ nanoparticles. *Applied Surface Science*, 249, 76–84.

Ma, W., Chen, S., Zhang, J., Wang, X. & Miao, W. (2009) Morphology and crystallization behavior of poly(vinylidene fluoride)/poly(methyl methacrylate)/methyl salicylate and benzophenone systems via thermally induced phase separation. *Journal of Polymer Science* B: *Polymer Physics*, 47, 248–260.

Mansourpanah, Y., Madaeni, S.S. & Rahimpour, A. (2009) Formation of appropriate sites on nanofiltration membrane surface for binding TiO$_2$ photo-catalyst: performance, characterization and fouling resistant capability. *Journal of Membrane Science*, 330, 297–306.

Méricq, J.-P., Mendret, J., Brosillon, S. & Faur, C. (2015) High performance PVDF-TiO$_2$ membranes for water treatment. *Chemical Engineering Science*, 123, 283–291.

Molinari, R., Pirillo, F., Loddo, V. & Palmisano, L. (2004) Photocatalytic degradation of dyes by using a membrane reactor. *Chemical Engineering and Processing*, 43, 1103–1114.

Molinari, R., Poerio, T., Granato, T. & Katovic, A. (2010) Fe-zeolites filled in PVDF membranes in the selective oxidation of benzene to phenol. *Microporous and Mesoporous Materials*, 129, 136–143.

Ng, L.Y., Mohammad, A.W., Leo, C.P. & Hilal, N. (2013) Polymeric membranes incorporated with metal/metal oxide nanoparticles: a comprehensive review. *Desalination*, 308, 15–33.

Ngang, H.P., Ooi, B.S., Ahmad, A.L. & Lai, S.O. (2012) Preparation of PVDF–TiO$_2$ mixed matrix membrane and its evaluation on dye adsorption and UV-cleaning properties. *Chemical Engineering Journal* 197, 359–367.

Nunes, S.P. & Peinemann, K.V. (1992) Ultrafiltration membranes from PVDF/PMMA blends. *Journal of Membrane Science*, 73, 25–35.

Pourjafar, S., Rahimpour, A. & Jahanshahi, M. (2012) Synthesis and characterization of PVA/PES thin film composite nanofiltration membrane modified with TiO$_2$ nanoparticles for better performance and surface properties. *Journal of Industrial and Engineering Chemistry*, 18, 1398–1405.

Rahimpour, A., Madaeni, S.S., Taheri, A.H. & Mansourpanah, Y. (2008) Coupling TiO$_2$ nanoparticles with UV irradiation for modification of polyethersulfone ultrafiltration membranes. *Journal of Membrane Science*, 313, 158–169.

Rahimpour, A., Jahanshahi, M., Mollahosseini, A. & Rajaeian, B. (2012) Structural and performance properties of UV-assisted TiO$_2$ deposited nano-composite PVDF/SPES membranes. *Desalination*, 285, 31–38.

Rajabzadeh, S., Maruyama, T., Sotani, T. & Matsuyama, H. (2008) Preparation of PVDF hollow fiber membrane from a ternary polymer/solvent/nonsolvent system via thermally induced phase separation (TIPS) method. *Separation and Purification Technology*, 63, 415–423.

Rajabzadeh, S., Maruyama, T., Ohmukai, Y., Sotani, T. & Matsuyama, H. (2009) Preparation of PVDF/PMMA blend hollow fiber membrane via thermally induced phase separation (TIPS) method. *Separation and Purification Technology*, 66, 76–83.

Razmjou, A., Mansouri, J. & Chen, V. (2011a) The effect of mechanical and chemical modification of TiO$_2$ nanoparticles on the surface chemistry, structure and fouling performance of PES ultrafiltration membranes. *Journal of Membrane Science*, 378, 73–84.

Razmjou, A., Mansouri, J., Chen, V., Lim, M. & Amal, R. (2011b) Titania nanocomposite polyethersulfone ultrafiltration membranes fabricated using a low temperature hydrothermal coating process. *Journal of Membrane Science*, 380, 98–113.

Razmjou, A., Arifin, E., Dong, G., Mansouri, J. & Chen, V. (2012a) Superhydrophobic modification of TiO$_2$ nanocomposite PVDF membranes for applications in membrane distillation. *Journal of Membrane Science*, 415–416, 850–863.

Razmjou, A., Resosudarmo, A., Holmes, R.L., Li, H., Mansouri, J. & Chen, V. (2012b) The effect of modified TiO$_2$ nanoparticles on the polyethersulfone ultrafiltration hollow fiber membranes. *Desalination*, 287, 271–280.

Safarpour, M., Khataee, A. & Vatanpour, V. (2015) Effect of reduced graphene oxide/TiO$_2$ nanocomposite with different molar ratios on the performance of PVDF ultrafiltration membranes. *Separation and Purification Technology*, 140, 32–42.

Shen, Y. & Lua, A.C. (2012) Preparation and characterization of mixed matrix membranes based on PVDF and three inorganic fillers (fumed nonporous silica, zeolite 4A and mesoporous MCM-41) for gas separation. *Chemical Engineering Journal*, 192, 201–210.

Shi, F., Ma, Y., Ma, J., Wang, P. & Sun, W. (2012) Preparation and characterization of PVDF/TiO$_2$ hybrid membranes with different dosage of nano-TiO$_2$. *Journal of Membrane Science*, 389, 522–531.

Simone, S., Figoli, A., Criscuoli, A., Carneval, M.C., Rosselli, A. & Drioli, E. (2010) Preparation of hollow fibre membranes from PVDF/PVP blends and their application in VMD. *Journal of Membrane Science*, 364, 219–232.

Simone, S., Figoli, A., Criscuoli, A., Carnevale, M.C., Alfadul, S., Al-Romaih, H., Al Shabouna, F., Al-Harbi, O.A. & Drioli, E. (2014) Effect of selected spinning parameters on PVDF hollow fibers morphology for potential application in desalination by VMD. *Desalination*, 344, 28–35.

Song, H., Shao, J., He, Y., Liu, B. & Zhong, X. (2012) Natural organic matter removal and flux decline with PEG-TiO$_2$-doped PVDF membranes by integration of ultrafiltration with photocatalysis. *Journal of Membrane Science*, 405–406, 48–56.

Song, H., Shao, J., Wang, J. & Zhong, X. (2014) The removal of natural organic matter with LiCl-TiO$_2$ doped PVDF membranes by integration of ultrafiltration with photocatalysis. *Desalination*, 344, 412–421.

Steen, M.L., Jordan, A.C. & Fisher, E.R. (2002) Hydrophilic modification of polymeric membranes by low temperature H$_2$O plasma treatment. *Journal of Membrane Science*, 204, 341–357.

Sui, Y., Wang, Z., Gao, X. & Gao, C. (2012) Antifouling PVDF ultrafiltration membranes incorporating PVDF-g-PHEMA additive via atom transfer radical graft polymerizations. *Journal of Membrane Science*, 413/414, 38–47.

Sukitpaneenit, P. & Chung, T.S. (2009) Molecular elucidation of morphology and mechanical properties of PVDF hollow fiber membranes from aspects of phase inversion, crystallization and rheology. *Journal of Membrane Science*, 340, 192–205.

Susanto, H. & Ulbricht, M. (2009) Characteristics, performance and stability of polyethersulfone ultrafiltration membranes prepared by phase separation method using different macromolecular additives. *Journal of Membrane Science*, 327, 125–135.

Susanto, H., Stahr, N. & Ulbricht, M. (2009) High performance polyethersulfone microfiltration membranes having high flux and stable hydrophilic property. *Journal of Membrane Science*, 342, 153–164.

Teow, Y.H., Ahmad, A.L., Lim, J.K. & Ooi, B.S. (2012) Preparation and characterization of PVDF/TiO$_2$ mixed matrix membrane via in situ colloidal precipitation method. *Desalination*, 295, 61–69.

Ulbricht, M. (2006) Advanced functional polymer membranes. *Polymer*, 47, 2217–2262.

Vatanpour, V., Madaeni, S.S., Khataee, A.R., Salehi, E., Zinadini, S. & Monfared, H.A. (2012) TiO$_2$ embedded mixed matrix PES nanocomposite membranes: Influence of different sizes and types of nanoparticles on antifouling and performance. *Desalination*, 292, 19–29.

Wang, Y., Kim, J.-H., Choo, K.-H., Lee, Y.-S. & Lee, C.-H. (2000) Hydrophilic modification of polypropylene microfiltration membranes by ozone-induced graft polymerization. *Journal of Membrane Science*, 169, 269–276.

Wang, Z., Yu, H., Xia, J., Zhang, F., Li, F., Xia, Y. & Li, Y. (2012) Novel GO-blended PVDF ultrafiltration membranes. *Desalination*, 299, 50–54.

Wilhelm, F.G., Punt, I.G.M. & Van Der Vegt, N.F.A (2002) Cation permeable membranes from blends of sulfonated poly(ether ketone) and poly(ether sulfone). *Journal of Membrane Science*, 199, 167–176.

Wu, G.P., Gan, S.Y., Cui, L.Z. & Xu, Y.Y. (2008) Preparation and characterization of PES/TiO₂ composite membranes. *Applied Surface Science*, 254, 7080–7086.

Xi, W. & Geissen, S.U. (2001) Separation of titanium dioxide from photocatalytically treated water by cross-flow microfiltration. *Water Research*, 35, 1256–1262.

You, S.-J., Semblante, G.U., Lu, S.-C., Damodar, R.A. & Wei, T.-C. (2012) Evaluation of the antifouling and photocatalytic properties of poly(vinylidene fluoride) plasma-grafted poly(acrylic acid) membrane with self-assembled TiO₂ *Journal of Hazardous Materials*, 237-238, 10–19.

Yu, L.-Y., Shen, H.-M. & Xu, Z.-L. (2009) PVDF–TiO₂ composite hollow fiber ultrafiltration membranes prepared by TiO₂ sol-gel method and blending method. *Journal of Applied Polymer Science*, 113, 1763–1772.

Zhang, C., Yang, F., Wang, W. & Chen, B. (2008) Preparation and characterization of hydrophilic modification of polypropylene non-woven fabric by dip-coating PVA (polyvinyl alcohol). *Separation and Purification Technology*, 61, 276–286.

Zhang, X., Wang, Y., You, Y., Meng, H., Zhang, J. & Xu, X. (2012) Preparation, performance and adsorption activity of TiO₂ nanoparticles entrapped PVDF hybrid membranes. *Applied Surface Science*, 263, 660–665.

Zhao, S., Wang, P., Wang, C., Sun, X. & Zhang, L. (2012) Thermostable PPESK/TiO₂ nanocomposite ultrafiltration membrane for high temperature condensed water treatment. *Desalination*, 299, 35–43.

Zhao, Y., Xu, Z., Shan, M., Min, C., Zhou, B., Li, Y., Li, B., Liu, L. & Qian, X. (2013) Effect of graphite oxide and multi-walled carbon nanotubes on the microstructure and performance of PVDF membranes. *Separation and Purification Technology*, 103, 78–83.

CHAPTER 9

Nanosized metal oxides (NMOs) and polyoxometalates (POMs) for antibacterial water treatment

Giulia Fiorani, Gloria Modugno, Marcella Bonchio & Mauro Carraro

9.1 INTRODUCTION

9.1.1 *Nanotechnology for antibacterial water treatment*

Securing and extending potable water provision is a formidable societal challenge, as the available supplies of freshwater are facing the problem of extended droughts, population growth and competing demands from different users (Lee and Schwab, 2005; Moe and Rheingans, 2006; Savage and Diallo, 2005; Vörösmarty *et al.*, 2000). Moreover, water sources can be contaminated by microorganisms, heavy metals, inorganic compounds, organic pollutants and many other emerging contaminants that can be harmful to human beings and the environment (Fatta-Kassinos *et al.*, 2011; Li *et al.*, 2011; O'Connor, 1996).

In order to guarantee a wide distribution of clean water, several purification techniques have been developed. Most of them are based on chemical and physical mechanisms, such as adsorption, catalysis, membrane separation and ionization (Ambashta and Sillanpää, 2010). However, well-established water treatment and distribution systems are not sufficient and are often too expensive to satisfy the demand for water with increasingly stringent quality standards (Qu *et al.*, 2013a). In this respect, recent advances in nanotechnology offer unprecedented opportunities to overcome some of the major constraints of the existing treatment technologies, and could enable the economic utilization of unconventional water sources (Qu *et al.*, 2013a).

Nanotechnologies are based on nanomaterials (e.g. carbon or metallic nanostructures, dendrimers, zeolites, etc., with at least one dimension smaller than 100 nm), which possess size-dependent properties (e.g. high specific surface area, fast dissolution, high reactivity, strong sorption) and/or discontinuous properties (e.g. superparamagnetism, localized surface plasmon resonance, quantum confinement effects). Owing to such behavior, they can be exploited to develop novel nanosorbents, nanocatalysts and photocatalysts, bioactive nanoparticles, nanostructured catalytic membranes and nanoparticle-enhanced filters, as well as to devise highly sensitive optical/electronic sensors to detect environmental pollutants (Qu *et al.*, 2013b).

Among nanomaterials under evaluation for water purification, metal-based nanoparticles (NPs) have emerged as a new kind of antimicrobial material (He *et al.*, 2013). Thanks to their large specific surface area, low cost and persistent bioactivity, metallic NPs could be integrated in large-scale disinfection processes, as well as in point-of-use/domestic treatment units. In addition, their biological activity may encompass a key role in controlling biofouling and biofilm formation, in order to extend the lifetime of existing treatment plants.

Several metallic nanosized materials exhibit size- and shape-dependent antimicrobial activity, arising from their specific interactions with the functional groups present on the outer membrane and on internal organelles. In addition, metal ions released from the surfaces of NPs can interfere with a number of biological molecules and metabolic processes, or catalyze the formation of reactive oxygen species (ROS) and radicals (Adams *et al.*, 2006; Aruoja *et al.*, 2009; Kasemets *et al.*, 2009; Wang *et al.*, 2003). As a result, metallic NPs can disrupt cellular membranes and

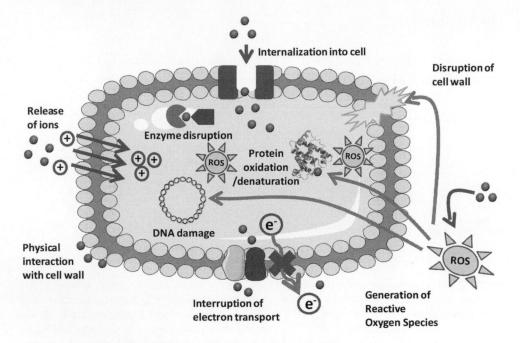

Figure 9.1.   Postulated mechanisms of action of antimicrobial metallic NPs.

membrane potentials, oxidize proteins and lipids, interfere with electron transport/respiration and deactivate cellular enzymes and DNA (Fig. 9.1).

In general, metallic nanomaterials show superior durability to organic antimicrobials, often retaining high biocompatibility and selectivity. In addition, they display a lower rate of resistance mutation, so they are active against antibiotic-resistant bacteria (Rai *et al.*, 2014). Unlike conventional disinfectants, such as chlorine and ozone, they give rise to minimal disinfection byproducts (e.g. halogenated disinfection byproducts, carcinogenic nitrosamines, bromates, etc.).

Metal oxide-based nanostructures represent a versatile and cost-effective family of compounds. In the following sections, nanosized metal oxides and polyoxometalates will be described, with the aim of offering new insights into the design of efficient, modular and multifunctional water treatment processes, characterized by low overall cost, high capacity and enhanced durability.

9.2   NANOSIZED METAL OXIDES (NMOs)

9.2.1   *Metal oxide nanoparticles with antimicrobial activity*

Nanosized metal oxides (NMOs) are characterized by high specific surface areas, accessible sorption sites and short intra-particle diffusion distances. These properties are useful within various environmental remediation processes, since they can (i) adsorb organic and metallic contaminants dissolved in water, (ii) implement separation technologies, and (iii) enable heterogeneous catalytic transformations (Hua *et al.*, 2012; Savage and Diallo, 2005).

Simple metal oxide NPs such as silver oxide (Ag_2O), titanium dioxide (TiO_2), zinc oxide (ZnO), copper oxide (CuO), iron oxide (Fe_2O_3), calcium oxide (CaO) and magnesium oxide (MgO), exhibit antimicrobial activity (Stoimenov *et al.*, 2002). Particle size and shape are pivotal parameters in promoting their inhibitory effect and can be tuned depending on the type of precursors, solvent, pH and temperature of the reaction mixture. The bactericidal effect of metal

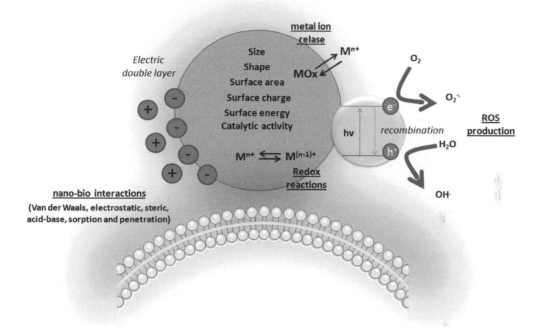

Figure 9.2. Properties of NMOs involved in antibacterial activity.

oxide NPs has, indeed, been attributed to their small size and high surface-to-volume ratio, which allows them to interact closely with microbial membranes, and to the simultaneous release of metal ions in solution (Morones *et al.*, 2005). Furthermore, oxidative stress can be generated upon production of ROS. Several metal oxides, indeed, are semiconducting materials and, when irradiated with energy higher than their band gap (3.2 eV), their frontier electrons jump from the valence band to the conduction band, forming electron (e^-) and electric hole (h^+) pairs. While dioxygen acts as an electron scavenger, forming the superoxide anion radical $O_2^{\bullet-}$, the positive electric holes react with water, generating hydroxyl radicals $OH^\bullet$ (Fig. 9.2). Although $O_2^{\bullet-}$ is not a strong oxidant, it is a precursor of other ROS, and it also becomes involved in the propagation of oxidative chain reactions. The hydroxyl radical, on the other hand, is a strong, non-selective oxidant ($E = 1.90$ V *vs.* NHE) that can damage all kinds of biomolecules. Hydrogen peroxide (H_2O_2) can form from both $O_2^{\bullet-}$ and $OH^\bullet$, and it can better diffuse across cell membranes. It is thus evident that ROS generation is a powerful feature of metal oxide NPs and can be exploited for disinfection applications.

Ag NPs are the most well-known and widely employed inorganic nanosized antimicrobial agents (Rizzello and Pompa, 2014). Their high biological activity is mainly due to the presence of Ag^+ ions on the NP surface, which favors electrostatic interactions with the negative charges of the bacterial cell wall and leads to cell death after membrane rupture (Sondi and Salopek-Sondi, 2004). Ag^+ can also induce a proton leakage through the bacterial membrane (Lok *et al.*, 2006), interact with functional biomolecules (e.g. membrane proteins, sulfur-containing intracellular proteins, DNA) (Allahverdiyev *et al.*, 2011) and affect metabolic processes (Egger *et al.*, 2009). Although antimicrobial Ag NPs can be incorporated within various support and/or permeable materials, such as superabsorbent poly(sodium/acrylate) cryogels, their practical use in water disinfection applications has been limited, due to aqueous solubility and dispersion issues (Loo *et al.*, 2015). On the other hand, it is also possible to use (slightly) soluble silver salts such as AgCl and Ag_2O. Ag_2O NPs, in particular, induce oxidative stress and apoptosis of *Escherichia coli*

(*E. coli*), which also loses its replication ability as a result of DNA damage. Ag_2O can also inhibit different bacterial strains (with minimum inhibitory concentration $= 9$–$18\,\mu g\,mL^{-1}$), being more effective against Gram-negative bacteria (Negi *et al.*, 2013).

TiO$_2$ NPs have unique properties, such as high stability, long lifetime, safety and negligible release of metal ions. The antimicrobial properties of TiO_2 are related to its crystal structure, shape and size (Fu *et al.*, 2005). The anatase form, in particular, is characterized by a high surface area and is more efficient for photocatalytic ROS generation. Under irradiation at $\lambda < 385\,nm$, TiO_2 acts as a self-cleaning material, and is able to oxidize most organic compounds, including otherwise recalcitrant ones, viruses and bacteria (Cermenati *et al.*, 1997; Mahmoodi *et al.*, 2006). For example, TiO_2 NPs have been used against *E. coli, Staphylococcus aureus* (*S. aureus*), *Pseudomonas putida* and *Listeria innocua* (Bonetta *et al.*, 2013), and also used to inhibit fungal biofilms (Haghighi *et al.*, 2013). The antibacterial activity observed is a result of lipid peroxidation, leading to enhanced membrane fluidity and disruption of cell integrity (Carré *et al.*, 2014), site-specific DNA damage (Roy *et al.*, 2010), and decrease or loss of respiratory activities due to oxidation of coenzyme A (Matsunaga *et al.*, 1985). The antibacterial and photocatalytic properties of TiO_2 NPs can be significantly enhanced upon doping with noble metal ions: doping with Au, Ag or Pd ions extends TiO_2 light absorption towards the visible/solar light spectrum (Chang *et al.*, 2006). In particular, Ag NPs of 5 nm diameter, grown on the surface of TiO_2, lead to longer absorption wavelengths ($\lambda > 400\,nm$) and enhanced antibacterial activity under irradiation (Kim *et al.*, 2006).

ZnO NPs show concentration-dependent antimicrobial effects against different bacterial strains and spores, being particularly efficient against Gram-positive bacteria (Azam *et al.*, 2012). The observed activity is likely to be related to NP internalization and release of zinc ions, leading to membrane leakage and interruption of electron transfer (Applerot *et al.*, 2009). Moreover, the abrasive surface texture of ZnO contributes to the mechanical damage of the cell membrane (Stoimenov *et al.*, 2002). However, zinc being an essential cofactor in a variety of cellular processes means that it requires relatively high concentrations ($>5\,mM$) to inhibit bacterial growth ($10^5\,CFU\,mL^{-1}$) (Nagarajan and Rajagopalan, 2008). On the other hand, ZnO is a n-type semiconductor characterized by the stability of its photo-excited state (Behnajady *et al.*, 2006) and, under UV irradiation, it exhibits photocatalytic activity which improves antimicrobial efficiency, thanks to ROS production (Cioffi and Rai, 2012).

CaO and MgO are low-cost, biocompatible and available materials which display antibacterial activity against both Gram-positive (e.g. *Bacillus subtilis, Bacillus megaterium*) and Gram-negative bacteria (e.g. *E. coli*) (Koper *et al.*, 2002). Their surface activity is related to an alkalinity increase upon hydration and to the production of ROS (superoxide anion radical) (Yamamoto *et al.*, 2010). In addition, MgO NPs can damage the cell membrane and cause the leakage of intracellular content (Leung *et al.*, 2014).

CuO NPs are characterized by high stability. They can cross the bacterial membrane of different bacterial strains, damaging intracellular enzymes (Mahapatra *et al.*, 2008). In particular, a size-dependent antibacterial activity was observed against two Gram-positive bacteria (*S. aureus* and *B. subtilis*) and two Gram-negative bacteria (*Pseudomonas aeruginosa* and *E. coli*), in which the restricted bacterial growth was attributed to internalization of NPs (Azam *et al.*, 2012).

Iron oxides show antibacterial activity, which has been ascribed to oxidative stress generated by ROS, causing damage to proteins and DNA (Touati, 2000). Indeed, Fe^{2+} reacts with endogenous H_2O_2 to produce hydroxyl radicals (Fenton reaction):

$$Fe^{2+} + H_2O_2 \rightarrow Fe^{3+} + OH^{\bullet} + OH^{-}$$
$$Fe^{3+} + H_2O_2 \rightarrow Fe^{2+} + OOH^{\bullet} + H^{+}$$

Fe_3O_4 NPs are also of interest for their paramagnetic properties, which allow easy separation and recovery upon application of a magnetic field (Yavuz *et al.*, 2006).

All these NMOs can be dispersed and used in slurry-type modules, in the dark or under irradiation, or retained by a suitable permeable support (Fig. 9.3). With regard to heterogeneous

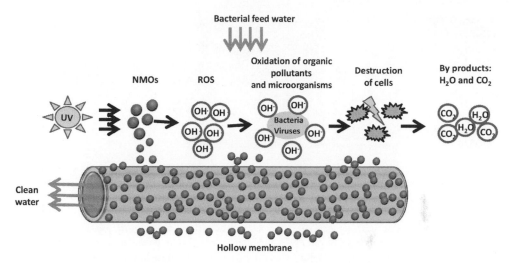

Figure 9.3. Schematic representation of a slurry-type photochemical reactor, involving heterogeneous photocatalysis by semiconducting NMOs (TiO$_2$, ZnO) and a hollow membrane for their retention.

photocatalytic oxidations with TiO$_2$, for example, an ultrafiltration membrane is usually required to separate the NPs. In this case, a micrometric NP cake deposited on the membrane can be useful in protecting the polymer from UV degradation.

However, NMOs are characterized by an increased surface energy, which leads to poor chemical stability. Consequently, NMOs are prone to agglomeration, due to van der Waals forces or other interactions, with a consequent reduction in surface-to-volume ratio. Moreover, their use in fixed beds or any other throughflow system may result in excessive pressure drops and/or head loss.

Within this scenario, a promising strategy to improve NMO applicability to real-world wastewater treatment consists of the dispersion of NMOs with antibacterial properties into porous supports or synthetic and naturally occurring polymers. Selected examples of such composite materials embedding NMOs are presented below.

9.2.2 *Metal oxide-based composite materials with potential application to water disinfection*

The synthesis of NMO-based materials involves three major pathways, consisting of (i) dispersion of the NPs within a preformed matrix, (ii) cross-linking of a matrix in the presence of NPs, and (iii) preparation *in-situ* of the NPs within a porous medium (Sarkar *et al.*, 2012). The resulting hybrid materials generally display improved permeability and robustness, owing to the hydrophilicity and mechanical and thermal stability of the NMOs. In some cases, contaminant degradation and self-cleaning capability may lead to an improved fouling resistance. Obviously, the immobilization should not affect the intrinsic characteristics of NMOs, nor lead to matrix decomposition (Chin *et al.*, 2006; Molinari *et al.*, 2002).

Among the naturally available matrixes, a suitable material for NMO immobilization is chitosan (CS), a cationic and hydrophilic polysaccharide, characterized by interesting biocompatibility, biodegradability and adsorption properties. Moreover, its chelating properties have been exploited for the removal of heavy metal ions and dyes from wastewater (Jayakumar *et al.*, 2010). CS can also establish electrostatic interactions with several macromolecules, including the lipopolysaccharides of Gram-negative bacterial surfaces, thus affecting the integrity of their outer membranes (Rabea *et al.*, 2003; Tang *et al.*, 2010).

CS derivatives show great potential as economical, environmentally friendly and sustainable materials. Their physico-chemical properties can benefit from chemical modification as well as from inclusion of inorganic NPs (Wan Ngah *et al.*, 2011), while the incorporation of NMOs leads

Figure 9.4.   Images of Fe_3O_4-CS hollow fibers obtained by scanning electron microscopy (cross-section at different magnification) and schematic representation of Fe_3O_4 reactivity.

to innovative materials with synergistic antibacterial activity. Sanuja *et al.* (2015) prepared a bio-nanocomposite film using CS, ZnO NPs and neem essential oil, as an additional hydrophobic antibacterial agent, by means of a solution casting method. The transparent film was characterized by improved tensile strength and elongation, and by decreased water solubility and swelling. The antibacterial activity of a film containing 0.5% (w/w) of ZnO was demonstrated against *E. coli* (Sanuja *et al.*, 2015). A very similar approach was previously used to incorporate Ag_2O NPs in CS films, which were used against *E. coli, S. aureus, B. subtilis* and *P. aeruginosa* (Tripathi *et al.*, 2011).

Several other CS-based composites are expected to be useful in water disinfection. For example, wet-spun CS hollow fibers were loaded with Fe_3O_4 NPs (Fig. 9.4) by a dipping-drying method. In this case, glutaraldehyde was used as cross-linker between CS chains, in order to stabilize the matrix in the form of hollow fibers, while Fe_2O_3 NPs were grown *in-situ* upon impregnation with Fe^{2+}/Fe^{3+} precursors, followed by reaction with NaOH. Despite the reduced adsorption efficiency compared to unfunctionalized CS, the resulting assembly retained Fenton-like reactivity in the presence of H_2O_2, as demonstrated by the oxidation of a model pollutant (reactive blue 19) (Seyed Dorraji *et al.*, 2015).

Liu *et al.* (2011) described the preparation of composite $CS/TiO_2/Fe_3O_4$ microspheres (Fig. 9.5). The CS microspheres, enclosing TiO_2 and Fe_3O_4 NPs, were fabricated by a simple electrospray technology. These materials showed the following properties: (i) a high capacity for contaminant adsorption, due to the strong binding ability of CS; (ii) *in-situ* self-cleaning under UV light; (iii) magnetic properties for cost-effective recovery and reuse (Liu *et al.*, 2011).

Besides CS, other biopolymers have been evaluated for support of NMOs. For example, Daoud *et al.* (2005) prepared TiO_2-coated cellulose fibers, wherein strong interactions between the cellulosic hydroxyl groups and the TiO_2 surface led to a well-adherent and resistant bactericidal surface. Scanning electron microscopy (SEM) images revealed that the film, characterized by low porosity, consisted of near-spherical grains of about 10 nm diameter. Despite the hospitable nature of cellulose, TiO_2 prevented the formation of biofilms, especially when irradiated (Daoud *et al.*, 2005).

Nanoporous graphene (NPG) and graphene oxide (GO) membranes are a novel class of mechanically robust, ultrathin, and fouling-resistant separation membranes that provide interesting opportunities for ionic/molecular sieving and water transport. In addition, the GO-based hybrid photocatalysts display promising performance in the degradation of pollutants. Liu *et al.* (2013) reported on a multifunctional nanocomposite (GO-TiO_2-Ag) integrating 2D GO sheets, 1D TiO_2 nanorods and 0D Ag NPs, which exhibited high photocatalytic degradation activity towards organic pollutants (e.g. Acid Orange 7 and phenol) and waterborne pathogens (e.g. *E. coli*). In particular, the bacterial inactivation efficiency (10^8 CFU mL^{-1}) of GO-TiO_2-Ag (100 µg mL^{-1} of nanocomposite containing 9.72% (w/w) of Ag), was 67% in dark conditions and 100% under solar irradiation, in 120 min (Liu *et al.*, 2013). This enhanced photocatalytic activity in respect of GO-TiO_2 and GO-Ag is a result of a suppressed recombination of photogenerated electron-hole pairs, owing to an effective electron transfer from TiO_2 to GO to Ag.

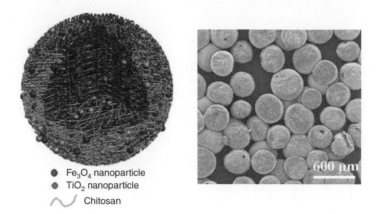

Fe$_3$O$_4$ nanoparticle
TiO$_2$ nanoparticle
Chitosan

Figure 9.5. Schematic representation (*left*) and optical image (*right*) of composite chitosan/TiO$_2$/Fe$_3$O$_4$ microspheres obtained *via* electrospray (adapted from Liu *et al.* (2011) with permission from the Centre National de la Recherche Scientifique (CNRS) and The Royal Society of Chemistry).

Fe$_3$O$_4$ was also anchored to GO via a one-step solvothermal route. Fe$_3$O$_4$ spheres, with sizes ranging from 200 to 250 nm, were distributed and firmly anchored onto the wrinkled graphene layers. The GO-Fe$_3$O$_4$ composites exhibited high adsorption capacity and fast adsorption rates for the removal of heavy metal ions (e.g. Pb^{2+}) and organic dyes (e.g. methylene blue) from aqueous solution. Due to oxidative stress generation, the composite showed antibacterial properties against *E. coli* (23% cell survival) in the presence of 100 µg mL^{-1} of GO-Fe$_3$O$_4$ (Santhosh *et al.*, 2014).

Several examples deal with the incorporation of metal and metal oxide NPs into polymeric membranes (e.g. poly(vinylidene fluoride), polyamide, polysulfone). These composite membranes combine polymer flexibility and processability with the thermal and mechanical stability of NMOs. The hybrid materials are also characterized by increased hydrophilicity, water permeability and fouling resistance (Homayoonfal *et al.*, 2013; Lee *et al.*, 2008; Ng *et al.*, 2013; Oh *et al.*, 2009). In this context, NMOs can be useful in improving the performance of membrane bioreactors (MBRs). MBR systems, indeed, have a pronounced fouling problem, which makes them less competitive for practical applications (Mei *et al.*, 2014). Because hydrophobic microbial products, such as extracellular polymeric substances, play a key role in membrane fouling, the modification of a polymeric surface by incorporation of a hydrophilic NMO represents a simple fouling-mitigation method. Indeed, a TiO$_2$-entrapped ultrafiltration membrane showed a lower flux decline compared to that of the neat polymeric membrane. In this case, however, a TiO$_2$-deposited membrane showed an even greater fouling-mitigation effect than that of the TiO$_2$-entrapped membrane (Bae and Tak, 2005; Lee *et al.*, 2008).

As an example of an antimicrobial polymeric system, NPs of silver and copper oxides were grown into a non-woven polyester fabric (1.27 mg cm^{-3}) and investigated as disinfectants. In particular, the nano-Ag$_2$O fabric removed more than 99% of bacteria (e.g. 5 × 10^3–5 × 10^4 CFU mL^{-1} of *E. coli*, *Salmonella typhi*, *S. aureus* and *Enterococcus faecalis*) in 24 h. In addition, the fabric removed more than 90% of coliforms, in 10 minutes, from an effluent of secondary treatment (Abou-Elela *et al.*, 2014).

9.3 POLYOXOMETALATES (POMs) AS ANTIMICROBIAL AGENTS

9.3.1 *Polyoxometalates*

The term polyoxometalates (POMs) refers to polyanionic and molecular oxides of early series transition metals. Their chemical composition is essentially represented by two types of general formula: $[M_mO_y]^{p-}$ and $[X_xM_mO_y]^{q-}$, where M is the main transition metal constituent (the *addenda atom*: typically W, Mo or V in their highest oxidation state), O is the oxygen atom and

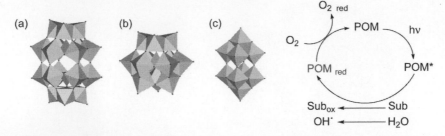

Figure 9.6. *Left:* Polyhedral representation of some polyoxotungstates: (a) Wells-Dawson type, $[\alpha\text{-}P_2W_{18}O_{62}]^{6-}$; (b) Keggin-type, $[\alpha\text{-}PW_{12}O_{40}]^{3-}$; (c) decatungstate, $[W_{10}O_{32}]^{4-}$. Color code: WO_6, turquoise octahedra; P, orange tetrahedra; Si, gray tetrahedron. *Right:* scheme of polyoxometalate photocatalytic activity.

X (the *heteroatom*) can be a non-metal – another element of the p-block – or a different transition metal. In the case of the first formula type, POMs are called isopolyanions, while in the second case they are known as heteropolyanions.

In order to form a POM, the dimensions of the metal ion M must be compatible with the formation of octahedral units, MO_6. In addition, empty and available d orbitals must be present to allow the formation of one or two terminal metal-oxygen double bonds within the octahedra, so as to limit the aggregation between MO_6 units, which requires, instead, the formation of μ-oxo bridges between two metal ions. In this way, terminal oxygens provide a barrier to linear polymerization and favor the formation of discrete molecular units, with dimensions usually ranging from one to a few nanometers (Pope, 1983).

POMs are characterized by an unmatched versatility in terms of structural features, resulting in a broad range of potential applications in catalysis, material science and medicine (Carraro and Gross, 2014; Long *et al.*, 2007, 2010; Pope and Müller, 1991). In particular, thanks to their redox behavior, POMs have been extensively employed as catalysts for the oxidation of organic substrates (Carraro *et al.*, 2014a), including the oxidative degradation of pollutants (Colmenares and Luque, 2014; Tzirakis *et al.*, 2009) and as synthetic enzymes (Bonchio *et al.*, 2007; Carraro *et al.*, 2011).

Similarly to semiconducting metal oxides, some POMs become strong oxidants under photoirradiation (Hiskia *et al.*, 2001). Indeed, POMs exhibit absorption bands in the region below 400 nm due to the occurrence of ligand-to-metal charge transfers, and can form an excited state characterized by a stronger reduction potential (up to 3 V increase) than the fundamental state. Depending on POM structure, such an excited state can react with organic substrates, *via* electron/hydrogen atom transfer, to give organic radicals prone to autooxidation in an oxygen atmosphere. In addition, hydroxyl radicals can be obtained by a similar reaction mechanism involving water as a substrate. Reduced POMs (heteropoly blues, HPBs) can be reoxidized by dioxygen, yielding superoxide anion radical and restoring its initial state (Fig. 9.6) (Bonchio *et al.*, 2005; Molinari, *et al.*, 2013a).

Several POMs have promising biomedical applications (antiviral, antitumoral and antibacterial), mainly stemming from their ability to oxidize cellular components (Rhule *et al.*, 1998) and to establish electrostatic interactions with proteins and enzymes, resulting in denaturation, hydrolysis and deactivation (Stephan *et al.*, 2013; Yamase, 2005). Within this scenario, the development of efficient POM-based antimicrobial systems suitable for water treatment, wound dressing and medical applications is an active multidisciplinary challenge (Fan *et al.*, 2012).

9.3.2 *POMs with antibacterial activity*

The most attractive feature of POMs in terms of their development as biologically active compounds is that they can be obtained from cheap precursors. In addition, molecular properties (redox potentials, polarity, shape, surface charge distribution and acidity) mediating the recognition by, and reactivity of, POMs with target biological macromolecules can be easily tuned. Hence, they promise significant economic and technical advantages over the organic drugs currently used.

Several **heteropolytungstates** were tested against methicillin-resistant *S. aureus* (MRSA) and vancomycin-resistant *S. aureus* (VRSA), which were sensitized towards β-lactam antibiotics. Yamase *et al.* (1996) have shown that POMs selectively accumulate in the membranes of MRSA cells. These same authors have shown that POMs inhibit the action of anion-sensitive enzymes, thus reducing the expression of penicillin-binding proteins with a low affinity for β-lactams, and suppressing the production of the β-lactamase enzyme. A synergy of action was thus observed in the presence of oxacillin (Fukuda *et al.*, 1999; Inoue *et al.*, 2006; Sami *et al.*, 2010).

Heteropolytungstates $K_{27}[KAs_4W_{40}O_{140}]$ and $K_{18}[KSb_9W_{21}O_{86}]$ exhibit antibacterial activity against Gram-negative *Helicobacter pylori*, with minimum inhibitory concentration (MIC) values $<256\ \mu g\,mL^{-1}$. *H. pylori* takes up the hydrophilic POM into the periplasmic space, thanks to pore-forming membrane proteins. A comparison with sodium tungstate (Na_2WO_4) has demonstrated the superior activity of the polyanionic metal oxides, while the internalization of photoactive POMs was highlighted by the formation of the corresponding HPB reduced complex (Inoue *et al.*, 2005).

Recently, the use of silicotungstic acid, $H_4SiW_{10}O_{40}$, was revisited, demonstrating a broad antibacterial activity against *S. aureus* (and MRSA), *P. aeruginosa*, *E. faecalis*, *E. coli*, multi-drug-resistant *P. aeruginosa* and *Klebsiella pneumoniae*, although with a relatively high MIC ($1.88\ mg\ mL^{-1}$) (Grama *et al.*, 2014).

Polyoxomolybdates are characterized by stronger ground-state reduction potentials than the corresponding tungstates. Their redox activity in a physiological environment was exploited against cancer cells (Hasenknopf, 2005). In one particular example, rhombic dodecahedral nanocrystals of $M_3PMo_{12}O_{40} \cdot nH_2O$ ($M = NH_4^+$, K^+ and Cs^+ salts) were prepared by hydrothermal treatment of phosphomolybdic acid ($M = H^+$) and used against *E. coli*. Compared to the commercial molybdate, the improved antiproliferative effect (at $25\ \mu g\ mL^{-1}$) suggested that they also exhibit morphology-dependent activity (He *et al.*, 2013). $[PMo_{12-x}V_xO_{40}]^n$, with $x = 0–3$, were employed in the formulation of different nanocomposites and were successfully used against both Gram-negative and Gram-positive bacteria (Kong *et al.*, 2007; Wu *et al.*, 2009; Yang *et al.*, 2009).

Polyoxovanadates, such as $(^tBuNH_3)_4[V_4O_{12}]$ and $(^tBuNH_3)_6[V_{10}O_{28}]$ have a potent antibacterial activity against *Streptococcus pneumoniae* ($MIC = 4–32\ \mu g\ mL^{-1}$) attributed to the altered transport of potassium and organic substrates (thymidine, uridine, leucine and glucose) through the membrane (Fukuda and Yamase, 1997).

Hybrid organic-inorganic POMs can be obtained upon association with organic counterions or *via* covalent modification (Dolbecq *et al.*, 2010; Berardi *et al.*, 2011). Suitable cations may contribute to the biological activity, as in the case of $[4-picH]_4[H_2V_{10}O_{28}]$, where the decavanadate was associated to 4-picoline. This decavanadate showed antibacterial activity against Gram-negative bacteria (*P. aeruginosa*) and fungi (*Aspergillus niger* and *Penicillium chrysogenum*), and the picolinium salts gave even better results than the corresponding sodium salt (Shahid *et al.*, 2014).

Organic pendants can be linked to Si, P, Ge and Sb atoms, which can be incorporated in the POM framework as heteroatoms. The resulting hybrid derivatives display improved biomedical proper ties, being more hydrolytically stable and less toxic than their inorganic precursors (Hasenknopf, 2005; Wang *et al.*, 1999). Organoantimony(III)-containing POMs, isolated as hydrated ammonium and alkali-metal salts, $(NH_4)_{12}[(PhSb^{III})_4(A-\alpha-Ge^{IV}W_9O_{34})_2] \cdot 20H_2O$, $Rb_9Na[(PhSb^{III})_4(A-\alpha-P^VW_9O_{34})_2] \cdot 20H_2O$ and $Rb_3[\{2-(Me_2NCH_2C_6H_4)Sb^{III}\}_3(B-\beta-As^{III}W_9O_{33})] \cdot 7H_2O$ (Fig. 9.7), were indeed found to be hydrolytically stable at physiological pH, and their antimicrobial activity against *E. coli* and *B. subtilis* were subsequently tested, being slightly higher ($MIC = 40–80\ \mu g\ mL^{-1}$) towards the latter, Gram-positive bacterium (Barsukova-Stuckart *et al.*, 2012; Meißner *et al.*, 2006).

The results obtained with hybrid POMs are encouraging, since they support the potential for decorating the antibacterial POMs with functional organic pendants, such as targeting ligands

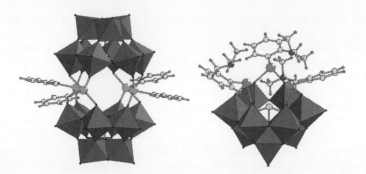

Figure 9.7.   Combined polyhedral/ball-and-stick representation of (*left*) $[(PhSb^{III})_4(A\text{-}\alpha\text{-}XW_9O_{34})_2]^{n-}$ (X = Ge or P) and (*right*) $[\{2\text{-}(Me_2NCH_2C_6H_4)Sb^{III}\}_3\ (B\text{-}\beta\text{-}As^{III}W_9O_{33})]^{3-}$. Color code: WO_6, red octahedra; X and As, yellow; Sb, green; C, light gray; N, blue; H, dark gray. Reprinted with permission from Barsukova-Stuckart *et al.* (2012). Copyright (2012) American Chemical Society.

and fluorescent labels. In this way, it would be possible to obtain insights into POM trafficking and mechanisms of action, while improving POM selectivity (Carraro *et al.*, 2014b; Geisberger *et al.*, 2013).

Due to the generally high solubility of POMs, a supporting matrix is required to enable their recovery and reuse. In addition, their negative charge is not convenient for their interaction with lipopolysaccharides on the outer membrane of Gram-negative bacteria. In the following paragraphs, selected strategies for POM immobilization/encapsulation are presented, including a few examples of insoluble composites with enhanced/synergistic antimicrobial activity.

9.3.3   *Polyoxometalate-based nanomaterials with potential application in water disinfection*

Thanks to their polyanionic and multimetallic surface, POMs can establish different and cooperative interactions with several supports, cations, macromolecules and nanostructures (Song and Tsunashima, 2012).

For example, POMs can interact with proteins and amino acids (AAs) (Stephan *et al.*, 2013). Kong *et al.* (2007) reported on the preparation of water-insoluble and biocompatible antimicrobials formed by combination of phosphomolybdic acid (PMA) with three different AAs (glycin (Gly), lysine (Lys) and histidine (His)). Under solvent-thermal conditions, AA-PMA nanorods with an average diameter of 50–80 nm were obtained. The nanostructures were tested on filter paper slices (Sherwood and De Beer, 1947) and showed a significant antibacterial activity against *E. coli*, with improvements by comparison with the individual components. To explore the convenient utilization of the nanorods, immobilization of AA-PMA was achieved through the preparation of multilayer films, composed of AA-PMA and polyethylenimine (PEI), by a layer-by-layer (LbL) self-assembly technique. The antiproliferative activity of {PEI/nanorod-Gly-PMA}$_3$ films, in particular, has indicated the potential of these nanostructured materials in wound dressing/membrane water disinfection applications (Kong *et al.*, 2007).

POMs can be used to stabilize noble metal NPs, since they act as polyelectrolytes, reducing the tendency to aggregate into bulk metallic precipitates (Keita *et al.*, 2009; Mitchell and de la Fuente, 2012). The antibacterial potential of POM/NP-based composites was demonstrated by preparing tyrosine-reduced gold NPs (AuNPs[Tyr]), functionalized with two different POMs – PMA and phosphotungstic acid (PTA) – and cationic amino acid lysine. With this approach, the core AuNPs[Tyr] act as a carrier for the antimicrobial POMs, while the presence of the cationic amino acid in the shell directs the nanomaterials towards negatively charged bacterial cells. These functionalized nanomaterials show antibacterial activity against *E. coli*, depending on the level of surface functionalization (5 μM concentration of AuNPs[Tyr-PTA-Lys] caused 80% bacterial death) (Daima *et al.*, 2013). Similarly, a stable POM surface corona was achieved on Ag NPs, using zwitterionic tyrosine as a pH-switchable reducing and capping agent. In this case, the antimicrobial activity of the multifunctional NPs was also confirmed towards Gram-positive *Staphylococcus albus*, while being biocompatible with epithelial mammalian cells (Daima *et al.*, 2014).

In deoxygenized solutions, reduced HPBs act as reductants for noble metal ions to produce metal NPs, which then benefit from POM stabilization (Kogan *et al.*, 2002; Troupis *et al.*, 2002; Weinstock, 1998). This behavior can be exploited to grow antibacterial POM-NPs within suitable materials. (POM/PEI–Ag)$_n$ composite films were created through linear Langmuir-Blodgett (LbL) self-assembly, involving the alternate deposition of $[BW_{12}O_{40}]^{5-}$ and cationic PEI–Ag$^+$ complex, followed by photochemical reduction *in-situ* to yield Ag NPs. In this particular example, the POM moiety acts as both reducing agent and assembly reagent. The resulting hybrid film completely inhibited *E. coli* growth (4×10^4 CFU mL^{-1}) (Gao *et al.*, 2011). An even more elegant approach is based on the simultaneous use of POMs as oxidizing agents, to activate monomers such as aniline, pyrrole or thiophene, and of the corresponding HPBs as reductants for metal ions (Ag, Au) in order to obtain POM-NPs embedded in conductive polymers (Kishore *et al.*, 2008).

Chitosan (CS) and its derivatives have been exploited as active matrixes for POM immobilization. At slightly acidic pH (pH < pKa(CS) = 6.28), POMs can foster ionic gelation of CS upon association with the positively charged amino groups at the C-2 positions (Meißner *et al.*, 2006). In addition, oxygen atoms on the POM surface can accept hydrogen bonds from $-OH$ and $-NH_3^+$ groups: these multiple interactions result in a non-covalent cross-linking of the CS chains, reducing water solubility while improving mechanical properties. Hybrid organic-inorganic multilayer films (Feng *et al.*, 2006; Pamin *et al.*, 2009; Yamada and Maeda, 2009), colloidal particles (Chen *et al.*, 2006), networks (Fiorani *et al.*, 2014) and capsules (Geisberger *et al.*, 2011; Menon *et al.*, 2011) were thus produced. For example, $H_5PMo_{10}V_2O_{40}$ was used to form nanoaggregates with diameters of 100–160 nm, featuring a substructure of entangled ribbons (Fig. 9.8). These composites were prepared at pH = 6, with 10% (w/w) POM. As demonstrated by UV-visible spectroscopy, CS acted as an effective scavenger of POM from the aqueous solution. Analysis of ζ-potential showed only minor changes compared to CS (around +25 mV), suggesting that the POMs were fully encapsulated into the NPs, with small effects on the surface charge density of the particles.

The nanoparticles were tested against *E. coli* (10^6 CFU mL^{-1}), and showed the synergistic activity of the heteropolyacid $H_5PMo_{10}V_2O_{40}$ (0.6 mg mL^{-1}) and CS. A likely explanation is that CS mediates membrane distortion and facilitates POM internalization within the cell, where it can affect electron transfers. Separate CS and vanadomolybdate, indeed, caused only six- to sevenfold declines in bacteria count. Upon lyophilization of the colloidal mixture, a highly porous and insoluble film, characterized by a homogeneous network of fibers (with lengths up to hundreds of μm and diameters between 2 and 12 μm), retaining antibacterial activity was also obtained (Fig. 9.8) (Fiorani *et al.*, 2014).

As a further example of POM–CS interplay, De Matteis *et al.* (2014) reported the synthesis and antibacterial activity of a series of nanometer-sized capsules obtained *via* supramolecular assembly of POMs, CS and cetyltrimethyl ammonium bromide (CTAB). Three representative POMs were used throughout this study: two Keggin structures (PTA and PMA) and a Kabanos structure, $(NH_4)_{15}\{Na[(Mo_2O_4)_6(\mu_2\text{-}SO_3)_3\text{-}(\mu_2\text{-}SO_3)]_2\}5H_2O$. The nanocapsules were prepared using a micelle-based approach, where the surfactant was used as a cationic nucleating agent and to trap the POMs prior to envelopment in the structure-directing CS matrix. Comprehensive characterization of these materials suggested a metal-organic hybrid structure based on POM-decorated micellar CTAB cores interwoven with CS-coordinated POMs. The antibacterial activity of these nanoscale composite capsules was then evaluated against *E. coli*. While the free polyoxomolybdates showed no antimicrobial activity up to a concentration of 0.5 mg mL^{-1}, hybrid materials based on the corresponding POMs presented a significant antibacterial effect, especially those containing molybdates, which induced stress and damage leading to a decrease in viability of 80% at a capsule concentration of 50 μg mL^{-1} (Fig. 9.9).

As for NMOs, the embedding of POMs into polymeric matrixes take advantages of polymer processability (Carraro and Gross, 2014). Although the antimicrobial activity often remains to be proved, the resulting materials might combine in a single unit separation and disinfection

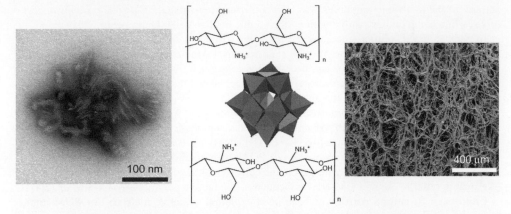

Figure 9.8.   Polyhedral representation of the molybdovanadate $[PMo_{10}V_2O_{40}]^{5-}$ embedded into CS (*center*). Color code: MO_6, dark green octahedra; P, yellow; VO_6, red octahedra. *Left*: transmission electron microscopy image of nanoparticles obtained from POM:CS=1:10 (w/w), at pH 6; *right*: scanning electron microscopy image of a film obtained from the lyophilized suspension.

capabilities. The preparation of a blend containing soluble POMs and polymeric chains is the simplest way to fabricate POM/polymer materials. Dipping or spin-coating methods were also used to obtain films based on polyvinyl alcohol (PVA), poly(ethylene glycol) (PEG), agarose, polyacrylamide and polyvinyl pyrrolidone (PVP) (Qi and Wu, 2009).

An example of bioactive material is illustrated by the incorporation of PMA, in the form of 5–60 nm aggregates, within the microchannels of colloidal Nafion. The assembly was drop cast on a glassy carbon electrode and used as an electrocatalytic film for cysteine oxidation (in phosphate buffer, pH 7). Owing to its biocidal activity (demonstrated against *Trichophyton rubrum*), the hybrid material was proposed as a solid-state antifungal coating (Swetha and Kumar, 2013).

In another case, permeable membranes based on $H_5PMo_{10}V_2O_{40}$ and a PVA/PEI blend were prepared with the aim of devising a protection against chemical and biological warfare agents. The POM was incorporated in the PVA/PEI hydrophilic matrix upon impregnation. The redox properties towards the oxidation of a model sulfide were maintained and the antibacterial activity was assessed against *E. coli*, *P. aeruginosa*, *S. aureus* and *B. subtilis* (10^7 CFU mL^{-1}). A MIC of 0.02–2 µg mL^{-1} was measured for the hybrid containing 10% POM (Wu *et al.*, 2009).

Photoactive POMs could be promising as antimicrobial agents under UV irradiation. Photo-catalytic membranes were obtained upon association of $[W_{10}O_{32}]^{4-}$, as hydrophobic tetrabutyl ammonium salt, with different polymeric films, such as polydimethylsiloxane (PDMS) or poly(vinylidene fluoride) (PVDF). The PVDF membrane, in particular, retained a suitable morphology for water filtration, with a macroporous layer as opposed to a dense skin layer (Fig. 9.10). The supported POM was used to achieve the photodegradation of alcoholic substrates under irradiation at $\lambda > 345$ nm (Bonchio *et al.*, 2003; Fontananova *et al.*, 2006).

Stronger interaction between a POM and a polymeric domain can be obtained by using polyca-tionic polymers, as in the case of polyallylammonium hydrochloride, which was used, for example, to create multilayered polyelectrolyte capsules upon LbL assemblies with alternating POM and polystyrene sulfonate layers (del Mercato *et al.*, 2014), or by using polymerizable counterions to be cross-linked *in-situ* with a suitable monomer, such as an acrylate (De Luca *et al.*, 2014; Li *et al.*, 2005).

In another approach, POMs have been immobilized through wet impregnation on the sur-face of γ-alumina and silica (Tzirakis *et al.*, 2007). For example, a sol-gel procedure, at pH = 2, gave a stable porous photoactive material containing 30% (w/w) of $[W_{10}O_{32}]^{4-}$, characterized by micropores (7–13 Å) and mesopores (30 Å) (Molinari *et al.*, 2013b). Differ-ent silica-based structures were obtained by encapsulating a POM, $[EuP_5W_{30}O_{110}]^{12-}$, with

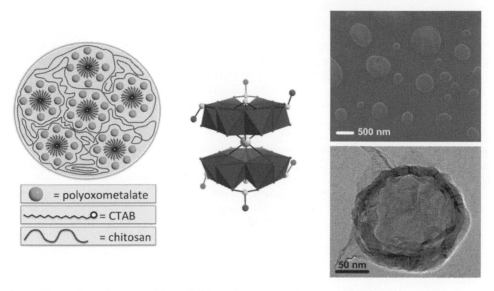

Figure 9.9. Schematic composition of chitosan/CTAB capsules with POMs (*left*). *Center*: combined ball&stick/polyhedral representation of Kabanos-like POM $\{Na[(Mo_2O_4)_6(\mu_2\text{-}SO_3)_3\text{-}(\mu_2\text{-}SO_3)]_2\}^{15-}$. Color code: MO_6, brown octahedra; S, yellow balls; Na, grey; O, red balls. *Right*: scanning electron microscopy image of composite Kabanos capsules (*top*) and bright-field transmission electron microscopy image of a discrete capsule embedded in gelatin (*bottom*). Adapted from De Matteis *et al.* (2014) with permission from The Royal Society of Chemistry.

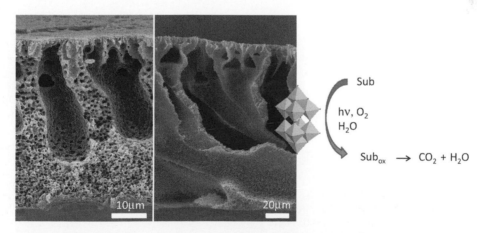

Figure 9.10. Scanning electron microscopy images of PVDF membranes without (*left*) and with (*center*) the photoactive POM $(n\text{-}Bu_4N)_4W_{10}O_{32}$, for which the reactivity is schematically represented (*right*).

a hydroxyl-terminated cationic surfactant (11-hydroxylundecyldimethyl ammonium), which was then covalently grafted into the matrix by means of a co-condensation with hydrolyzed tetraethoxylsilane. The POMs, well-dispersed within spherical structures with sizes around 150 nm, were used as reductants of metal compounds ($AgNO_3$, $HAuCl_4$ and H_2PtCl_6) for the controlled growth of NPs (Zhao *et al.*, 2010). Finally, the covalent grafting of the hybrid polyoxometalate $[AsW_9O_{33}\{P(O)(CH_2CH_2CO_2H)\}_2]^{5-}$ was obtained by formation of amide bonds with mesoporous aminopropyl-functionalized SBA-15 silica NPs (Villanneau *et al.*, 2013).

9.4   CONCLUSIONS

Nanosized metal oxides and polyoxometalates offer several advantages and opportunities for the development of novel systems for water cleaning and disinfection. Thanks to the antibacterial activity of such nanodimensional metal oxides (MOs) and the process intensification enabled by their assembly within porous supports, the integration of complementary functionalities (catalysis, disinfection and separation) is possible within a single module.

This chapter has collected selected examples of MO-based nanostructured materials with antimicrobial function, in order to provide information about suitable supports and their eventual synergistic activity. However, because many of the existing examples of MO-based antimicrobial materials deal with the preparation of coating layers or colloids, further engineering efforts are required to design systems that will operate efficiently with a continuous flow.

Due to the expanding research on MO synthesis and modification, we expect that a careful choice of materials and immobilization strategies could ensure the high stability, low leaching and long lifetime of the resulting devices. POMs, in particular, can be easily modified in terms of composition, charge, dimensions and redox potential, and although a structure–activity relationship is still lacking, all of these properties are likely to be pivotal for biological activity. In addition, they allow control of the interactions with a support matrix at a molecular level.

The photochemical production of reactive oxygen species is another promising strategy that should be implemented, since it does not involve metal ion release and can also be useful in the minimization of biofouling.

To foster innovation in this field, we have included some examples of promising hybrid materials, whose antimicrobial activity needs to be further evaluated after optimization of their constituents.

ACKNOWLEDGEMENTS

The authors gratefully acknowledge funding from the European Union's Seventh Framework Programme (EU-FP7/2007–2013) under grant agreement 246039, from the Italian Ministry of Education MIUR (FIRB project RBAP11ETKA), and from COST actions CM1003 "Biological oxidation reactions – mechanisms and design of new catalysts" and CM1203 "Polyoxometalate Chemistry for Molecular Nanoscience (PoCheMoN)" We thank Dr. Omar Saoncella, Dr. Alberto Figoli, Dr. Giorgio De Luca and Prof. Sacide Alsoy Altinkaya for fruitful discussion and collaboration.

REFERENCES

Abou-Elela, S.I., Ibrahim, H.S., Kamel, M.M. & Gouda, M. (2014) Application of nanometal oxides in situ in nonwoven polyester fabric for the removal of bacterial indicators of pollution from wastewater. *ScientificWorldJournal*, 2014, 950348.

Adams, L.K., Lyon, D.Y. & Alvarez, P.J.J. (2006) Comparative eco-toxicity of nanoscale TiO_2, SiO_2, and ZnO water suspensions. *Water Research*, 40(19), 3527–3532.

Allahverdiyev, A.M., Abamor, E.S., Bagirova, M. & Rafailovich, M. (2011) Antimicrobial effects of TiO_2 and Ag_2O nanoparticles against drug-resistant bacteria and leishmania parasites. *Future Microbiology*, 6(8), 933–940.

Ambashta, R.D. & Sillanpää, M. (2010) Water purification using magnetic assistance: a review. *Journal of Hazardous Materials*, 180(1–3), 38–49.

Applerot, G., Lipovsky, A., Dror, R., Perkas, N., Nitzan, Y., Lubart, R. & Gedanken, A. (2009) Enhanced antibacterial activity of nanocrystalline ZnO due to increased ROS-mediated cell injury. *Advanced Functional Materials*, 19(6), 842–852.

Aruoja, V., Dubourguier, H.-C., Kasemets, K. & Kahru, A. (2009) Toxicity of nanoparticles of CuO, ZnO and TiO_2 to microalgae *Pseudokirchneriella subcapitata*. *Science of the Total Environment*, 407(4), 1461–1468.

Azam, A., Ahmed, A.S., Oves, M., Khan, M.S., Habib, S.S. & Memic, A. (2012) Antimicrobial activity of metal oxide nanoparticles against Gram-positive and Gram-negative bacteria: a comparative study. *International Journal of Nanomedicine*, 7, 6003–6009.

Bae, T.-H. & Tak, T.-M. (2005) Effect of TiO$_2$ nanoparticles on fouling mitigation of ultrafiltration membranes for activated sludge filtration. *Journal of Membrane Science*, 249(1–2), 1–8.

Barsukova-Stuckart, M., Piedra-Garza, L.F., Gautam, B., Alfaro-Espinoza, G., Izarova, N.V., Banerjee, A., Bassil, B.S., Ullrich, M.S., Breunig, H.J., Silvestru, C. & Kortz, U. (2012) Synthesis and biological activity of organoantimony(III)-containing heteropolytungstates. *Inorganic Chemistry*, 51(21), 12,015–12,022.

Behnajady, M.A., Modirshahla, N. & Hamzavi, R. (2006) Kinetic study on photocatalytic degradation of C.I. Acid Yellow 23 by ZnO photocatalyst. *Journal of Hazardous Materials*, 133(1–3), 226–232.

Berardi, S., Carraro, M., Sartorel, A., Modugno, G. & Bonchio, M. (2011) Hybrid polyoxometalates: merging organic and inorganic domains for enhanced catalysis and energy applications. *Israel Journal of Chemistry*, 51(2), 259–274.

Bonchio, M., Carraro, M., Scorrano, G., Fontananova, E. & Drioli, E. (2003) Heterogeneous photooxidation of alcohols in water by photocatalytic membranes incorporating decatungstate. *Advanced Synthesis & Catalysis*, 345(9–10), 1119–1126.

Bonchio, M., Carraro, M., Conte, V. & Scorrano, G. (2005) Aerobic photooxidation in water by polyoxotungstates: the case of uracil. *European Journal of Organic Chemistry*, 2005(22), 4897–4903.

Bonchio, M., Carraro, M., Farinazzo, A., Sartorel, A., Scorrano, G. & Kortz, U. (2007) Aerobic oxidation of *cis*-cyclooctene by iron-substituted polyoxotungstates: evidence for a metal initiated auto-oxidation mechanism. *Journal of Molecular Catalysis* A: *Chemical*, 262(1–2), 36–40.

Bonetta, S., Bonetta, S., Motta, F., Strini, A. & Carraro, E. (2013) Photocatalytic bacterial inactivation by TiO$_2$-coated surfaces. *AMB Express*, 3, 59.

Carraro, M. & Gross, S. (2014) Hybrid materials based on the embedding of organically modified transition metal oxoclusters or polyoxometalates into polymers for functional applications: a review. *Materials*, 7(5), 3956–3989.

Carraro, M., Sartorel, A., Toma, F.M., Puntoriero, F., Scandola, F., Campagna, S., Prato, M. & Bonchio, M. (2011) Artificial photosynthesis challenges: water oxidation at nanostructured interfaces. *Topics in Current Chemistry*, 303, 121–150.

Carraro, M., Fiorani, G., Sartorel, A. & Bonchio, M. (2014a) Polyoxometalates catalysts for sustainable oxidations and energy applications. In: Duprez, D. & Cavani, F. (eds) *Handbook of advanced Methods and Processes in Oxidation Catalysis*. Imperial College Press, London. pp. 586–630.

Carraro, M., Modugno, G., Zamolo, V. & Bonchio, M. (2014b) Interaction of hybrid polyoxometalates with biological targets. *Journal of Biological Inorganic Chemistry*, 19, S796.

Carré, G., Hamon, E. Ennahar, S., Estner, M., Lett, M.C., Horvatovich, P., Gies, J.-P., Keller, V., Keller, N. & Andre, P. (2014) TiO$_2$ photocatalysis damages lipids and proteins in *Escherichia coli*. *Applied and Environmental Microbiology*, 80(8), 2573–2581.

Cermenati, L., Pichat, P., Guillard, C. & Albini, A. (1997) Probing the TiO$_2$ photocatalytic mechanisms in water purification by use of quinoline, photo-Fenton generated OH• radicals and superoxide dismutase. *Journal of Physical Chemistry* B: *Biophysical Chemistry, Biomaterials, Liquids, and Soft Matter*, 101(14), 2650–2658.

Chang, C.-C., Lin, C.-K., Chan, C.C., Hsu, C.-S. & Chen, C.-Y. (2006) Photocatalytic properties of nanocrystalline TiO$_2$ thin film with Ag additions. *Thin Solid Films*, 494(1–2), 274–278.

Chen, S., Wu, G., Long, D. & Liu, Y. (2006) Preparation, characterization and antibacterial activity of chitosan–Ca$_3$V$_{10}$O$_{28}$ complex membrane. *Carbohydrate Polymers*, 64(1), 92–97.

Chin, S.S., Chiang, K. & Fane, A.G. (2006) The stability of polymeric membranes in a TiO$_2$ photocatalysis process. *Journal of Membrane Science*, 275(1–2), 202–211.

Cioffi, N. & Rai, M. (eds) (2012) *Nano-Antimicrobials: Progress and Prospects*. Springer, Heidelberg, Germany.

Colmenares, J.C. & Luque, R. (2014) Heterogeneous photocatalytic nanomaterials: prospects and challenges in selective transformations of biomass-derived compounds. *Chemical Society Reviews*, 43(3), 765–778.

Daima, H.K., Selvakannan, P.R., Shukla, R., Bhargava, S.K. & Bansal, V. (2013) Fine-tuning the antimicrobial profile of biocompatible gold nanoparticles by sequential surface functionalization using polyoxometalates and lysine. *PLoS ONE*, 8(10), e79676.

Daima, H.K., Selvakannan, P.R., Kandjani, A.E., Shukla, R., Bhargava, S.K. & Bansal, V. (2014) Synergistic influence of polyoxometalate surface corona towards enhancing the antibacterial performance of tyrosine-capped Ag nanoparticles. *Nanoscale*, 6(2), 758–765.

Daoud, W.A., Xin, J.H. & Zhang, Y.-H. (2005) Surface functionalization of cellulose fibers with titanium dioxide nanoparticles and their combined bactericidal activities. *Surface Science*, 599(1–3), 69–75.

De Luca, G., Bisignano, F., Figoli, A., Galiano, F., Furia, E., Mancuso, R., Saoncella, O., Carraro, M., Bonchio, M. & Gabriele, B. (2014) Bromide ion exchange with a Keggin polyoxometalate on functionalized polymeric membranes: a theoretical and experimental study. *Journal of Physical Chemistry* B: *Biophysical Chemistry, Biomaterials, Liquids, and Soft Matter*, 118(9), 2396–2404.

De Matteis, L., Mitchell, S.G. & de la Fuente, J.M. (2014) Supramolecular antimicrobial capsules assembled from polyoxometalates and chitosan. *Journal of Materials Chemistry* B: *Materials for Biology and Medicine*, 2(41), 7114–7117.

del Mercato, L.L., Carraro, M., Zizzari, A., Bianco, M., Miglietta, R., Arima, V., Viola, I., Nobile, C., Sorarù, A., Vilona, D., Gigli, G., Bonchio, M. & Rinaldi, R. (2014) Catalytic self-propulsion of supramolecular capsules powered by polyoxometalate cargos. *Chemistry – A European Journal*, 20(35), 10,910–10,914.

Dolbecq, A., Dumas, E., Mayer, C.R. & Mialane, P. (2010) Hybrid organic-inorganic polyoxometalate compounds: from structural diversity to applications. *Chemical Reviews*, 110(10), 6009–6048.

Egger, S., Lehmann, R.P., Height, M.J., Loessner, M.J. & Schuppler, M. (2009) Antimicrobial properties of a novel silver-silica nanocomposite material. *Applied and Environmental Microbiology*, 75(9), 2973–2976.

Fan, D., Hao, J. & Wei, Q. (2012) Assembly of polyoxometalate-based composite materials. *Journal of Inorganic and Organometallic Polymers and Materials*, 22(2), 301–306.

Fatta-Kassinos, D., Kalavrouziotis, I.K., Koukoulakis, P.H. & Vasquez, M.I. (2011) The risks associated with wastewater reuse and xenobiotics in the agroecological environment. *Science of the Total Environment*, 409(19), 3555–3563.

Feng, Y., Han, Z., Peng, J., Lu, J., Xue, B., Li, L., Ma, H. & Wang, E. (2006) Fabrication and characterization of multilayer films based on Keggin-type polyoxometalate and chitosan. *Materials Letters*, 60(13–14), 1588–1593.

Fiorani, G., Saoncella, O., Kaner, P., Altinkaya, S.A., Figoli, A., Bonchio, M. & Carraro, M. (2014) Chitosan-polyoxometalate nanocomposites: synthesis, characterization and application as antimicrobial agents. *Journal of Cluster Science*, 25(3), 839–854.

Fontananova, E., Drioli, E., Donato, L., Bonchio, M., Carraro, M., Gardan, M. & Scorrano, G. (2006) Hybrid photocatalytic membranes embedding decatungstate for heterogeneous photooxydation. *Desalination*, 200(1–3), 705–707.

Fu, G., Vary, P.S. & Lin, C.-T. (2005) Anatase TiO_2 nanocomposites for antimicrobial coatings. *Journal of Materials Chemistry* B: *Materials for Biology and Medicine*, 109(18), 8889–8898.

Fukuda, N. & Yamase, T. (1997) In vitro antibacterial activity of vanadate and vanadyl compounds against *Streptococcus pneumoniae*. *Biological and Pharmaceutical Bulletin*, 20(8), 927–930.

Fukuda, N., Yamase, T. & Tajima, Y. (1999) Inhibitory effect of polyoxotungstates on the production of penicillin-binding proteins and β-lactamase against methicillin-resistant *Staphylococcus aureus*. *Biological and Pharmaceutical Bulletin*, 22(5), 463–470.

Gao, S., Wu, Z., Pan, D., Lin, Z. & Cao, R. (2011) Preparation and characterization of polyoxometalate-Ag nanoparticles composite multilayer films. *Thin Solid Films*, 519(7), 2317–2322.

Geisberger, G., Paulus, S., Carraro, M., Bonchio, M. & Patzke, G.R. (2011) Synthesis, characterisation and cytotoxicity of polyoxometalate/carboxymethyl chitosan nanocomposites. *Chemistry – A European Journal*, 17(16), 4619–4625.

Geisberger, G., Gyenge, E.B., Hinger, D., Bosiger, P., Maake, C. & Patzke, G.R. (2013) Synthesis, characterization and bioimaging of fluorescent labeled polyoxometalates. *Dalton Transactions*, 42(27), 9914–9920.

Grama, L., Man, A., Muntean, D.L., Gâz Florea, S.A., Boda, F. & Curticăpean, A. (2014) Antibacterial activity of some saturated polyoxotungstates. *Romanian Journal of Laboratory Medicine*, 22(8), 111–118.

Haghighi, F., Roudbar Mohammadi, S., Mohammadi, P., Hosseinkhani, S. & Shidpour, R. (2013) Antifungal activity of TiO_2 nanoparticles and EDTA on *Candida albicans* biofilms. *Infection, Epidemiology and Medicine*, 1(1), 33–38.

Hasenknopf, B. (2005) Polyoxometalates: introduction to a class of inorganic compounds and their biomedical applications. *Frontiers in Bioscience*, 10, 275–287.

He, J., Pang, H., Wang, W., Zhang, Y., Yan, B., Li, X., Li, S. & Chen, J. (2013) Uniform $M_3PMo_{12}O_{40} \cdot nH_2O$ ($M = NH_4^+$, K^+, Cs^+) rhombic dodecahedral nanocrystals for effective antibacterial agents. *Dalton Transactions*, 42(44), 15,637–15,644.

Hiskia, A., Mylonas, A. & Papaconstantinou, E. (2001) Comparison of the photoredox properties of polyoxometallates and semiconducting particles. *Chemical Society Reviews*, 30(1), 62–69.

Homayoonfal, M., Mehrnia, M.R., Mojtahedi, Y.M. & Ismail, A.F. (2013) Effect of metal and metal oxide nanoparticle impregnation route on structure and liquid filtration performance of polymeric nanocomposite membranes: a comprehensive review. *Desalination and Water Treatment*, 51(16–18), 3295–3316.

Hua, M., Zhang, S., Pan, B., Zhang, W., Lv, L. & Zhang, Q. (2012) Heavy metal removal from water/wastewater by nanosized metal oxides: a review. *Journal of Hazardous Materials* 211/212, 317–331.

Inoue, M., Segawa, K., Matsunaga, S., Matsumoto, N., Oda, M. & Yamase, T. (2005) Antibacterial activity of highly negative charged polyoxotungstates, $K_{27}[KAs_4W_{40}O_{140}]$ and $K_{18}[KSb_9W_{21}O_{86}]$, and Keggin-structural polyoxotungstates against *Helicobacter pylori*. *Journal of Inorganic Biochemistry*, 99(5), 1023–1031.

Inoue, M., Suzuki, T., Fujita, Y., Oda, M., Matsumoto, N. & Yamase, T. (2006) Enhancement of antibacterial activity of β-lactam antibiotics by $[P_2W_{18}O_{62}]^{6-}$, $[SiMo_{12}O_{40}]^{4-}$, and $[PTi_2W_{10}O_{40}]^{7-}$ against methicillin-resistant and vancomycin-resistant *Staphylococcus aureus*. *Journal of Inorganic Biochemistry*, 100(7), 1225–1233.

Jayakumar, R., Menon, D., Manzoor, K., Nair, S.V. & Tamura, H. (2010) Biomedical applications of chitin and chitosan based nanomaterials – a short review. *Carbohydrate Polymers*, 82(2), 227–232.

Kasemets, K., Ivask, A., Dubourguier, H.-C. & Kahru, A. (2009) Toxicity of nanoparticles of ZnO, CuO and TiO_2 to yeast *Saccharomyces cerevisiae*. *Toxicology in Vitro*, 23(6), 1116–1122.

Keita, B., Liu, T. & Nadjo, L. (2009) Synthesis of remarkably stabilized metal nanostructures using polyoxometalates. *Journal of Materials Chemistry*, 19(1), 19–33.

Kim, K.D., Han, D.N., Lee, J.B. & Kim, H.T. (2006) Formation and characterization of Ag-deposited TiO_2 nanoparticles by chemical reduction method. *Scripta Materialia*, 54(2), 143–146.

Kishore, P., Viswanathan, B. & Varadarajan, T. (2008) Synthesis and characterization of metal nanoparticle embedded conducting polymer-polyoxometalate composites. *Nanoscale Research Letters*, 3(1), 14–20.

Kogan, V., Aizenshtat, Z., Popovitz-Biro, R. & Neumann, R. (2002) Carbon-carbon and carbon-nitrogen coupling reactions catalyzed by palladium nanoparticles derived from a palladium substituted Keggin-type polyoxometalate. *Organic Letters*, 4(20), 3529–3532.

Kong, Y., Pan, L., Peng, J., Xue, B., Lu, J. & Dong, B. (2007) Preparation and antibacterial activity of nanorod-amino acid polyoxometalates. *Materials Letters*, 61(11–12), 2393–2397.

Koper, O.B., Klabunde, J.S., Marchin, G.L., Klabunde, K.J., Stoimenov, P. & Bohra, L. (2002) Nanoscale powders and formulations with biocidal activity toward spores and vegetative cells of *Bacillus* species, viruses, and toxins. *Current Microbiology*, 44(1), 49–55.

Lee, E.J. & Schwab, K.J. (2005) Deficiencies in drinking water distribution systems in developing countries. *Journal of Water and Health*, 3(2), 109–127.

Lee, H.S., Im, S.J., Kim, J.H., Kim, J.J., Kim, J.P. & Min, B.R. (2008) Polyamide thin-film nanofiltration membranes containing TiO_2 nanoparticles. *Desalination*, 219(1–3), 48–56.

Leung, Y.H., Ng, A.M.C., Xu, X., Shen, Z., Gethings, L.A., Wong, M.T., Chan, C.M., Guo, M.Y., Ng, Y.H., Djurišić, A.B., Lee, P.K., Chan, W.K., Yu, L.H., Phillips, D.L., Ma, A.P. & Leung, F.C. (2014) Mechanisms of antibacterial activity of MgO: non-ROS mediated toxicity of MgO nanoparticles towards *Escherichia coli*. *Small*, 10(6), 1171–1183.

Li, H., Qi, W., Li, W., Sun, H., Bu, W. & Wu. L. (2005) A highly transparent and luminescent hybrid based on the copolymerization of surfactant-encapsulated polyoxometalate and methyl methacrylate. *Advanced Materials*, 17(22), 2688–2692.

Li, X., Zeng, G.-M., Huang, J.-H., Zhang, D.-M., Shi, L.-J., He, S.-B. & Ruan, M. (2011) Simultaneous removal of cadmium ions and phenol with MEUF using SDS and mixed surfactants. *Desalination*, 276(1–3), 136–141.

Liu, L., Bai, H., Liu, J. & Sun, D.D. (2013) Multifunctional graphene oxide-TiO_2-Ag nanocomposites for high performance water disinfection and decontamination under solar irradiation. *Journal of Hazardous Materials*, 261, 214–223.

Liu, Z., Bai, H. & Sun, D.D. (2011) Facile fabrication of porous chitosan/TiO_2/Fe_3O_4 microspheres with multifunction for water purifications. *New Journal of Chemistry*, 35(1), 137–140.

Lok, C.-N., Ho, C.-M., Chen, R., He, Q.-Y., Yu, W.-Y., Sun, H., Tam, P.K.-W., Chiu, J.-F. & Che, C.-M. (2006) Proteomic analysis of the mode of antibacterial action of silver nanoparticles. *Journal of Proteome Research*, 5(4), 916–924.

Long, D.-L., Burkholder, E. & Cronin, L. (2007) Polyoxometalate clusters, nanostructures and materials: from self assembly to designer materials and devices. *Chemical Society Reviews*, 36(1), 105–121.

Long, D.-L., Tsunashima, R. & Cronin, L. (2010) Polyoxometalates: building blocks for functional nanoscale systems. *Angewandte Chemie – International Edition*, 49(10), 1736–1758.

Loo, S.-L., Krantz, W.B., Fane, A.G., Gao, Y., Lim, T.-T. & Hu, X. (2015) Bactericidal mechanisms revealed for rapid water disinfection by superabsorbent cryogels decorated with silver nanoparticles. *Environmental Science and Technology*, 49(4), 2310–2318.

Mahapatra, O., Bhagat, M., Gopalakrishnan, C. & Arunachalam, K.D. (2008) Ultrafine dispersed CuO nanoparticles and their antibacterial activity. *Journal of Experimental Nanoscience*, 3(3), 185–193.

Mahmoodi, N.M., Arami, M., Limaee, N.Y. & Tabrizi, N.S. (2006) Kinetics of heterogeneous photocatalytic degradation of reactive dyes in an immobilized TiO_2 photocatalytic reactor. *Journal of Colloid and Interface Science*, 295(1), 159–164.

Matsunaga, T., Tomoda, R., Nakajima, T. & Wake, H. (1985) Photoelectrochemical sterilization of microbial cells by semiconductor powders. *FEMS Microbiology Letters*, 29(1–2), 211–214.

Mei, X., Wang, Z., Zheng, X., Huang, F., Ma, J., Tang, J. & Wu, Z. (2014) Soluble microbial products in membrane bioreactors in the presence of ZnO nanoparticles. *Journal of Membrane Science*, 451, 169–176.

Meißner, T., Bergmann, R., Oswald, J., Rode, K., Stephan, H., Richter, W., Zänker, H., Kraus, W., Emmerling, F. & Reck, G. (2006) Chitosan-encapsulated Keggin anion $[Ti_2W_{10}PO_{40}]^{7-}$. Synthesis, characterization and cellular uptake studies. *Transition Metal Chemistry*, 31(5), 603–610.

Menon, D., Thomas, R.T., Narayanan, S., Maya, S., Jayakumar, R., Hussain, F., Lakshmanan, V.-K. & Nair, S.V. (2011) A novel chitosan/polyoxometalate nano-complex for anti-cancer applications. *Carbohydrate Polymers*, 84(3), 887–893.

Mitchell, S.G. & de la Fuente, J.M. (2012) The synergistic behavior of polyoxometalates and metal nanoparticles: from synthetic approaches to functional nanohybrid materials. *Journal of Materials Chemistry*, 22(35), 18,091–18,100.

Moe, C.L. & Rheingans, R.D. (2006) Global challenges in water, sanitation and health. *Journal of Water and Health*, 4(Suppl 1), 41–57.

Molinari, R., Palmisano, L., Drioli, E. & Schiavello, M. (2002) Studies on various reactor configurations for coupling photocatalysis and membrane processes in water purification. *Journal of Membrane Science*, 206(1–2), 399–415.

Molinari, A., Argazzi, R. & Maldotti, A. (2013a) Photocatalysis with $Na_4W_{10}O_{32}$ in water system: formation and reactivity of OH. radicals. *Journal of Molecular Catalysis* A: *Chemical*, 372, 23–28.

Molinari, A., Maldotti, A., Bratovcic, A. & Magnacca, G. (2013b) Photocatalytic properties of sodium decatungstate supported on sol-gel silica in the oxidation of glycerol. *Catalysis Today*, 206, 46–52.

Morones, J.R., Elechiguerra, J.L., Camacho, A., Holt, K., Kouri, J.B., Ramírez, J.T. & Yacaman, M.J. (2005) The bactericidal effect of silver nanoparticles. *Nanotechnology*, 16(10), 2346–2353.

Nagarajan, P. & Rajagopalan, V. (2008) Enhanced bioactivity of ZnO nanoparticles – an antimicrobial study. *Science and Technology of Advanced Materials*, 9(3), 035004.

Negi, H., Saravanan, P.R., Agarwal, T., Zaidi, M.G.H. & Reeta Goel, R. (2013) In vitro assessment of Ag_2O nanoparticles toxicity against Gram-positive and Gram-negative bacteria. *Journal of General and Applied Microbiology*, 59(1), 83–88.

Ng, L.Y., Mohammad, A.W., Leo, C.P. & Hilal, N. (2013) Polymeric membranes incorporated with metal/metal oxide nanoparticles: a comprehensive review. *Desalination*, 308, 15–33.

O'Connor, G.A. (1996) Organic compounds in sludge-amended soils and their potential for uptake by crop plants. *Science of the Total Environment*, 185(1–3), 71–81.

Oh, S.J., Kim, N. & Lee, Y.T. (2009) Preparation and characterization of PVDF/TiO_2 organic-inorganic composite membranes for fouling resistance improvement. *Journal of Membrane Science*, 345(1–2), 13–20.

Pamin, K., Jachimska, B., Onik, K., Połtowicz, J. & Grabowski, R. (2009) Electrostatic self-assembly of polyoxometalates on chitosan as catalysts of oxidation of cyclic hydrocarbons. *Catalysis Letters*, 127(1–2), 167–174.

Pope, M.T. (1983) *Heteropoly and Isopoly Oxometalates*. Springer, New York.

Pope, M.T. & Müller, A. (1991) Polyoxometalate chemistry: an old field with new dimensions in several disciplines. *Angewandte Chemie* – International Edition, 30(1), 34–48.

Qi, W. & Wu, L. (2009) Polyoxometalate/polymer hybrid materials: fabrication and properties. *Polymer International*, 58(11), 1217–1225.

Qu, X., Alvarez, P.J.J. & Li, Q. (2013a) Applications of nanotechnology in water and wastewater treatment. *Water Research*, 47(12), 3931–3946.

Qu, X., Brame, J., Li, Q. & Alvarez, P.J.J. (2013b) Nanotechnology for a safe and sustainable water supply: enabling integrated water treatment and reuse. *Accounts of Chemical Research*, 46(3), 834–843.

Rabea, E.I., Badawy, M.E.T., Stevens, C.V., Smagghe, G. & Steurbaut, W. (2003) Chitosan as antimicrobial agent: applications and mode of action. *Biomacromolecules*, 4(6), 1457–1465.

Rai, M., Birla, S., Ingle, A.P., Gupta, I., Gade, A., Abd-Elsalam, K., Marcato, P.D. & Duran, N. (2014) Nanosilver: an inorganic nanoparticle with myriad potential applications. *Nanotechnology Reviews*, 3(3), 281–309.

Rhule, J.T., Hill, C.L., Judd, D.A. & Schinazi, R.F. (1998) Polyoxometalates in medicine. *Chemical Reviews*, 98(1), 327–358.

Rizzello, L. & Pompa, P.P. (2014) Nanosilver-based antibacterial drugs and devices: mechanisms, methodological drawbacks, and guidelines. *Chemical Society Reviews*, 43(5), 1501–1518.

Roy, A.S., Parveen, A., Koppalkar, A.R. & Prasad, M. (2010) Effect of nano – titanium dioxide with different antibiotics against methicillin-resistant *Staphylococcus aureus*. *Journal of Biomaterials and Nanobiotechnology*, 1, 37–41.

Sami, P., Anand, T., Premanathan, M. & Rajasekaran, K. (2010) Vanadium(V)-substituted Keggin-type heteropolyoxotungstophosphates as electron transfer and antimicrobial agents: oxidation of glutathione and sensitization of MRSA towards β-lactam antibiotics. *Transition Metal Chemistry*, 35(8), 1019–1025.

Santhosh, C., Kollu, P., Doshi, S., Sharma, M., Bahadur, D., Vanchinathan, M.T., Saravanan, P., Kim, B.S. & Grace, A.N. (2014) Adsorption, photodegradation and antibacterial study of graphene-Fe_3O_4 nanocomposite for multipurpose water purification application. *RSC Advances*, 4(54), 28,300–28,308.

Sanuja, S., Agalya, A. & Umapathy, M.J. (2015) Synthesis and characterization of zinc oxide-neem oil-chitosan bionanocomposite for food packaging application. *International Journal of Biological Macromolecules*, 74, 76–84.

Sarkar, S., Guibal, E., Quignard, F. & SenGupta, A.K. (2012) Polymer-supported metals and metal oxide nanoparticles: synthesis, characterization, and applications. *Journal of Nanoparticle Research*, 14(2), 1–24.

Savage, N. & Diallo, M.S. (2005) Nanomaterials and water purification: opportunities and challenges. *Journal of Nanoparticle Research*, 7(4–5), 331–342.

Seyed Dorraji, M.S., Mirmohseni, A., Carraro, M., Gross, S., Simone, S., Tasselli, F. & Figoli, A. (2015) Fenton-like catalytic activity of wet-spun chitosan hollow fibers loaded with Fe_3O_4 nanoparticles: batch and continuous flow investigations. *Journal of Molecular Catalysis* A: *Chemical*, 398, 353–357.

Shahid, M., Sharma, P.K., Anjuli, Chibber, S. & Siddiqi, Z.A. (2014) Isolation of a decavanadate cluster [$H_2V_{10}O_{28}$][4-picH]$_4 \cdot 2H_2O$ (4-pic = 4-picoline): crystal structure, electrochemical characterization, genotoxic and antimicrobial studies. *Journal of Cluster Science*, 25(5), 1435–1447.

Sherwood, M.B. & De Beer, E.J. (1947) Discrepancy in analysis of penicillin in blood by the Oxford cup method as revealed by the paper disc technique. *Science*, 105(2729), 414.

Sondi, I. & Salopek-Sondi, B. (2004) Silver nanoparticles as antimicrobial agent: a case study on *E. coli* as a model for Gram-negative bacteria. *Journal of Colloid and Interface Science*, 275(1), 177–182.

Song, Y.-F. & Tsunashima, R. (2012) Recent advances on polyoxometalate-based molecular and composite materials. *Chemical Society Reviews*, 41(22), 7384–7402.

Stephan, H., Kubeil, M., Emmerling, F. & Müller, C.E. (2013) Polyoxometalates as versatile enzyme inhibitors. *European Journal of Inorganic Chemistry*, 2013(10–11), 1585–1594.

Stoimenov, P.K., Klinger, R.L., Marchin, G.L. & Klabunde, K.J. (2002) Metal oxide nanoparticles as bactericidal agents. *Langmuir*, 18(17), 6679–6686.

Swetha, P. & Kumar, A.S. (2013) Phosphomolybdic acid nano-aggregates immobilized nafion membrane modified electrode for selective cysteine electrocatalytic oxidation and anti-dermatophytic activity. *Electrochimica Acta*, 98, 54–65.

Tang, H., Zhang, P., Kieft, T.L., Ryan, S.J., Baker, S.M., Wiesmann, W.P. & Rogelj, S. (2010) Antibacterial action of a novel functionalized chitosan-arginine against Gram-negative bacteria. *Acta Biomaterialia*, 6(7), 2562–2571.

Touati, D. (2000) Iron and oxidative stress in bacteria. *Archives of Biochemistry and Biophysics*, 373, 1–6.

Tripathi, S., Mehrotra, G.K. & Dutta, P.K. (2011) Chitosan-silver oxide nanocomposite film: preparation and antimicrobial activity. *Bulletin of Materials Science*, 34(1), 29–35.

Troupis, A., Hiskia, A. & Papaconstantinou, E. (2002) Synthesis of metal nanoparticles by using polyoxometalates as photocatalysts and stabilizers. *Angewandte Chemie – International Edition*, 41(11), 1911–1914.

Tzirakis, M.D., Lykakis, I.N., Panagiotou, G.D., Bourikas, K., Lycourghiotis, A., Kordulis, C. & Orfanopoulos, M. (2007) Decatungstate catalyst supported on silica and γ-alumina: efficient photocatalytic oxidation of benzyl alcohols. *Journal of Catalysis*, 252(2), 178–189.

Tzirakis, M.D., Lykakis, I.N. & Orfanopoulos, M. (2009) Decatungstate as an efficient photocatalyst in organic chemistry. *Chemical Society Reviews*, 38(9), 2609–2621.

Villanneau, R., Marzouk, A., Wang, Y., Djamaa, A.B., Laugel, G., Proust, A. & Launay, F. (2013) Covalent grafting of organic-inorganic polyoxometalates hybrids onto mesoporous SBA-15: a key step for new anchored homogeneous catalysts. *Inorganic Chemistry*, 52(6), 2958–2965.

Vörösmarty, C.J., Green, P., Salisbury, J. & Lammers, R.B. (2000) Global water resources: vulnerability from climate change and population growth. *Science*, 289(5477), 284–288.

Wan Ngah, W.S., Teong, L.C. & Hanafiah, M.A.K.M. (2011) Adsorption of dyes and heavy metal ions by chitosan composites: a review. *Carbohydrate Polymers*, 83(4), 1446–1456.

Wang, X.H., Dai, H.C. & Liu, J.F. (1999) Synthesis and characterization of organotin-substituted heteropoly tungstosilicates and their biological activity I. *Polyhedron*, 18(17), 2293–2300.

Wang, X., Liu, J. & Pope, M.T. (2003) New polyoxometalate/starch nanomaterial: synthesis, characterization and antitumoral activity. *Dalton Transactions*, 5, 957–960.

Weinstock, I.A. (1998) Homogeneous-phase electron-transfer reactions of polyoxometalates. *Chemical Reviews*, 98(1), 113–170.

Wu, K.H., Yu, P.Y., Yang, C.C., Wang, G.P. & Chao, C.M. (2009) Preparation and characterization of polyoxometalate-modified poly(vinyl alcohol)/polyethyleneimine hybrids as a chemical and biological self-detoxifying material. *Polymer Degradation and Stability*, 94(9), 1411–1418.

Yamada, M. & Maeda, A. (2009) Heteropolyacid-conjugated chitosan matrix for triphase catalyst. *Polymer*, 50(25), 6076–6082.

Yamamoto, O., Ohira, T., Alvarez, K. & Fukuda, M. (2010) Antibacterial characteristics of $CaCO_3$–MgO composites. *Materials Science and Engineering* B, 173(1–3), 208–212.

Yamase, T. (2005) Anti-tumor, -viral, and -bacterial activities of polyoxometalates for realizing an inorganic drug. *Journal of Materials Chemistry*, 15(45), 4773–4782.

Yamase, T., Fukuda, N. & Tajima, Y. (1996) Synergistic effect of polyoxotungstates in combination with β-lactam antibiotics on antibacterial activity against methicillin-resistant *Staphylococcus aureus*. *Biological and Pharmaceutical Bulletin*, 19(3), 459–465.

Yavuz, C.T., Mayo, J.T., Yu, W.W., Prakash, A., Falkner, J.C., Yean, S., Cong, L., Shipley, H.J., Kan, A., Tomson, M., Natelson, D. & Colvin, V.L. (2006) Low-field magnetic separation of monodisperse Fe_3O_4 nanocrystals. *Science*, 314(5801), 964–967.

Zhao, Y., Qi, W., Li, W. & Wu, L. (2010) Covalent dispersion of surfactant-encapsulated polyoxometalates and in situ incorporation of metal nanoparticles in silica spheres. *Langmuir*, 26(6), 4437–4442.

CHAPTER 10

Atomic-force microscopy investigations of filtration membranes

Daniel Johnson & Nidal Hilal

10.1 INTRODUCTION

In this chapter we will discuss contributions which can be made to the study and characterization of filtration membrane surfaces using the technique of atomic-force microscopy (AFM), using specific examples. We will begin by giving an overview of the AFM technique itself, explaining the basic mode of operation, followed by a brief discussion of the contributions described in the literature. We will then concentrate on three examples of work which we have carried out recently using AFM to characterize various polymer membranes under different conditions. These examples are: the study of the effects of environment and imaging mode on the observations made of polymeric nanofiltration membranes using AFM; and the study of inorganic scaling on membranes used for seawater membrane distillation; the study of humic acid adhesion forces with novel polymer membranes under different conditions, including those simulating membrane bioreactor treatment of dye effluent.

10.2 ATOMIC-FORCE MICROSCOPY

AFM is a high-resolution imaging technique which was originally developed in the 1980s to overcome certain limitations found with scanning tunneling microscopy (STM) (Binnig *et al.*, 1986). Unlike STM, AFM measurements do not have to be carried out in a vacuum and work just as well with electrically non-conductive samples as with conductive ones. Measurements can be made in liquid as well as gaseous environments, enabling experiments to take place in a wide range of conditions, including those which simulate natural and industrial environments. In addition, interaction forces between the probe and the surfaces of interest can be measured, allowing a range of quantitative measurements to be made, including those of long-range and adhesive forces, and permitting investigations of the mechanical properties of those surfaces.

The basic setup of AFM is shown in Figure 10.1. At its heart the atomic-force microscope has a probe containing a sharp scanning tip mounted on a flexible micro-cantilever arm. Detection of the flexing of the cantilever is achieved by an optical lever system consisting of a beam of laser light which is reflected off the upper surface of the cantilever onto a photodetector. Raster-scanning of the sharp imaging tip across a surface and simultaneous monitoring of the deflection of the cantilever arm allows a three-dimensional image of the surface to be constructed at high resolution, which under certain circumstances can be at the sub-nanometer range.

10.2.1 *Imaging modes*

A large number of different AFM imaging modes are used, which can be divided into 'static' techniques, such as contact mode, or 'dynamic' techniques, including tapping and non-contact modes. We will concentrate on the three most basic modes here. More detailed descriptions can be found elsewhere (Butt *et al.*, 2005; Johnson *et al.*, 2009a).

Contact mode is where the scanning tip maintains constant contact with the sample's surface, operating in a repulsive force regime. As the probe is raster-scanned across the surface it will

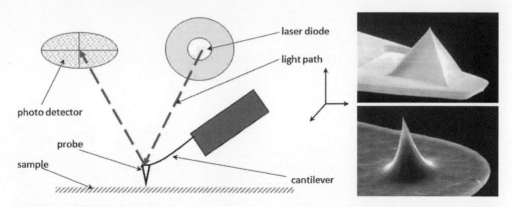

Figure 10.1.   Schematic of the basic setup of an AFM (*left*). *Right*: SEM images of the sharp imaging tips of two different probes; the heights of the imaging tips are approximately 3 μm.

encounter features of varying heights, causing the cantilever arm to flex. A feedback loop in the control electronics causes the probe to be moved up and down by adjusting a piezoelectric crystal in the scanner, maintaining a constant deflection of the cantilever arm at a level set by the user. By monitoring the change in height relative to the x, y position of the probe, a three-dimensional image of the surface can be built up. This is contact mode operation with a constant force. Alternatively, when imaging a surface which is relatively smooth, with low-lying features, contact mode can be operated with varying force. Here the feedback mechanisms are turned off and the deflection of the cantilever is instead used to monitor changes in surface height. Contact mode is the simplest mode and is relatively quick and easy to accomplish.

One of the most used modes of operation is tapping mode (also referred to as intermittent contact, amplitude-modulation mode or AC-mode scanning), developed to overcome certain limitations of contact mode (Hansma *et al.*, 1993; Zhong *et al.*, 1993). Here the cantilever is oscillated at close to its resonant frequency, allowing it to come into intermittent contact with the surface of interest. As it scans over the surface, features of greater and lesser height cause its amplitude to decrease or increase, respectively. A feedback mechanism is employed to adjust the height of the probe to maintain a constant amplitude. This height signal is then used to construct the surface topography in a similar way to contact mode. Whilst generally slower than contact mode, reduced normal and lateral force interactions give this mode an advantage over contact mode when imaging soft or rough surfaces.

Finally, in non-contact (or frequency-modulation) mode the cantilever is again oscillated at close to its resonant frequency, but at a much smaller amplitude than in tapping mode. As the probe approaches the sample's surface, long-range attractive interactions, including electrostatic and van der Waals forces, occur between atoms on the opposing surfaces. These interactions lead to a detectable change in the oscillation frequency, causing a change in the phase between the driving and oscillating frequencies. Feedback loops are then employed to maintain a constant distance between the probe and surface, without making contact with the surface (Lüthi *et al.*, 1994). This method minimizes the interaction forces between probe and sample, leading to a decreased area of interaction and a potentially better x, y resolution than for the other modes. In practice, this mode is more difficult to use with finesse than the other, simpler modes, but under appropriate conditions is able to achieve true atomic resolution.

10.2.2   *AFM as a force sensor*

One of the major applications of AFM is as a quantitative sensor of interaction forces between the probe and the surface of interest. Because, under normal operating conditions, the degree of cantilever flex has a linear relationship with the force applied, and hence follows Hooke's Law,

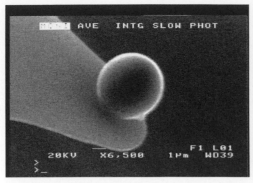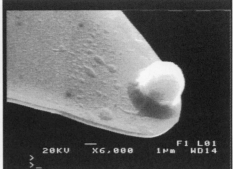

Figure 10.2. Images of a colloidal probe (*left*) and a cell probe (*right*). The colloidal probe was created using a silica microsphere and the cell probe by using the yeast *Saccharomyces cerevisiae* (Bowen *et al.*, 1998a, 1998b).

the direct measurement of interaction forces with a cantilever whose mechanical stiffness has been determined is straightforward. The probe tip simply needs to be brought into and out of contact with a point on the sample surface and the deflection recorded as a function of probe height. The academic literature is full of examples of such measurements, including the measurement of nanomechanical material properties (Domke and Radmach, 1998; Kaliapan and Capella, 2005; Weisenhorn *et al.*, 1992), surface adhesion forces (Kappl and Butt, 2002; Roberts, 2005; Weisenhorn *et al.*, 1992), long-range interaction forces (Burnham *et al.*, 1990; Gillies *et al.*, 2005), interaction forces between particles and interfaces (Johnson *et al.*, 2009b) and mechanical properties and bond strengths of polymers and biomolecules (Allen *et al.*, 1997; Best *et al.*, 2003; Rief *et al.*, 1997). By coating the probe tip, or replacing it with a particle or cell, a large number of different interactions may be measured under different operating conditions that mimic natural or industrial conditions of interest to the investigator. The deflection is typically converted to force in nanonewtons (nN) by using the optical lever sensitivity (obtained from the slope of the contact region when measurements are taken against a hard surface) and the spring constant of the lever. The height displacement of the probe can be converted into probe-sample separation distance by subtracting cantilever deflection from the height and then setting the region of hard contact as zero, although the determination of actual zero separation may be non-trivial.

10.2.3 Colloidal probes

By replacing the sharp scanning tip with a particle or cell of interest, a colloidal or cell probe can be produced. In the case of particles, if a microsphere is used the simplified geometry, compared with the shape of a sharp scanning tip, makes it easy to scale the forces by the radius of the particle. In addition, it is relatively straightforward to coat or otherwise functionalize the microspheres with a substance of interest. Alternatively, particles of substances, such as calcium carbonate or drug crystals (Al-Anezi *et al.*, 2008; Davies *et al.*, 2005), may be used to study their adhesive properties. Figure 10.2 shows scanning electron microscopy (SEM) images of two examples: on the left is a colloidal probe formed by using a silica microsphere (Bowen *et al.*, 1998a), whilst on the right is a yeast cell (*Saccharomyces cerevisiae*) mounted on the end of the cantilever arm for use as a cell probe (Bowen *et al.*, 1998b).

Colloidal probes were first utilized for force interaction measurements by Ducker and colleagues (Ducker *et al.*, 1991; Ducker and Senden, 1992), who used a 3.5 μm diameter silica sphere to measure its interactions with a polystyrene surface as a function of solution ionic strength and pH. By measuring the long-range DLVO forces under different conditions it is possible to estimate the ability of a membrane surface to reject attachment of a colloidal foulant, whilst measurement of adhesion forces can give information about the strength of foulant-membrane attachments. This is of particular use when modifying membrane surfaces to resist fouling. From adhesion measurements it can be seen whether a membrane is likely to be resistant to initial foulant attachment using only a small sample of membrane, obviating the need for fabrication of larger amounts of membrane which may be expensive. For example, Hilal *et al.* (2003) measured adhesion forces for poly(ether sulfone) (PES) membranes, both unmodified and modified with quaternized N, N-dimethylaminoethyl methacrylate (qDMAEMA). When using a silica sphere as the probe, it was found that the modification reduced the measured adhesion force from 39.5 to 24.8 mN m^{-1}, suggesting that the modified membrane would be more resistant to fouling by silicate particulates.

By chemically modifying the surfaces of colloidal microspheres, probes can be made which are either coated with the foulant of interest, or which simulate the behavior of colloid and biocolloid foulants. A carboxylated latex particle probe was used by Herzberg *et al.* (2009) to simulate inter-actions between a bacterial cell and reverse osmosis membranes fouled by extracellular polymeric substances (EPS), in the presence and absence of calcium carbonate in the feed solution. During approach, only repulsive forces were measured with an unfouled membrane, both in the presence and absence of Ca^{2+} ions, which was found to be due to electrostatic double layer repulsion. When EPS fouling was present, an attractive "jump-in" of the probe to the membrane surface was observed when Ca^{2+} was added to the solution. These jump-in events occur when the gradient of attractive forces become greater than the restoring force of the AFM cantilever (Butt *et al.*, 2005), causing the probe to make rapid contact with the sample. It was concluded that this was due to binding of Ca^{2+} to carboxylate groups both on the surface of the probe and in the fouling layer (Herzberg *et al.*, 2009). These observations are similar to measurements reported by Li and Elimelech (2004), who measured interactions between carboxylated particle probes and mem-branes fouled with humic acid. When calcium ions were added, the measured adhesion force was seen to increase dramatically, whether the EPS fouling layer was present or not, although it was particularly marked when an EPS layer was present. For some measurements when EPS was present in solution, no measurable adhesion was seen. This was attributed to interaction of the Ca^{2+} ions with carboxylate groups in the EPS layer. The same type of probe was also used to demonstrate a reduction in adhesion forces for a poly(vinylidene fluoride) (PVDF) membrane when modified with a layer of grafted copolymer, with no attractive forces observed when the membrane was immersed in 100 mM NaCl solution (Herzberg *et al.*, 2011). Similar work by Bernstein *et al.* (2011) also used a carboxylated probe to assess the fouling resistance of reverse osmosis membranes with surfaces modified by graft polymerization. Interaction forces between the probe and the membrane surface were found to be dominated by electrostatic double layer forces, as seen when the solution pH was changed altering the charged state of the interacting sur-faces. Adhesion forces were largely a result of electrostatic mechanisms, hydrophobic/hydrophilic interaction and possible hydrogen bonding, with the explanations being different for the different membranes, demonstrating the complexity of foulant-membrane interaction mechanisms.

Evans *et al.* (2009) used the colloidal probe technique to assess the efficacy of membrane cleaning on fouling caused by the polyphenols present in black tea. Using a silica microsphere coated with a model polyphenol, adhesion force measurements were carried out with regenerated cellulose ultrafiltration membranes. It was found that adhesion forces were greater for virgin and fouled-then-cleaned membranes than for fouled membranes. Pre-treatment of the membrane using sodium hydroxide reduced the adhesion forces by comparison with the virgin and cleaned membranes, indicating the benefits of membrane pre-treatment prior to filtration use.

10.3   IMAGING NANOFILTRATION MEMBRANES: EFFECT OF IMAGING MODE AND ENVIRONMENT

The atomic-force microscope has become a useful tool for the study of membrane surface mor-phology. One of the major advantages AFM has over other high-resolution imaging techniques is the ability it gives to perform experiments and make observations in both ambient air and liquid

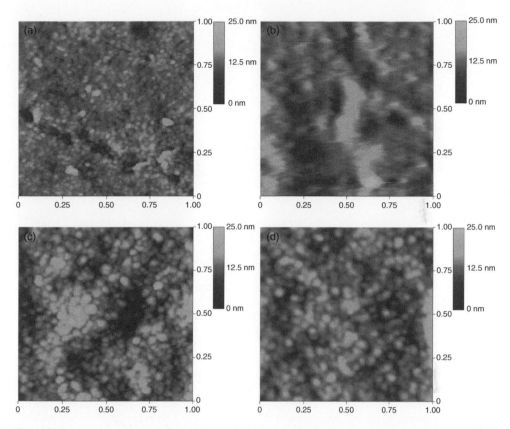

Figure 10.3.   1 × 1 μm images of membranes taken in air: (a) PES 20%, PVP 20%, tapping mode; (b) PES 20%, PVP 20%, contact mode; (c) NF270, tapping mode; (d) NF270, contact mode.

environments. The various imaging modes all come with advantages and disadvantages. In this section, we describe a study (Johnson *et al.*, 2012) to compare the two most commonly utilized imaging modes in both air and water when examining two different nanofiltration membranes, to identify the strengths and weaknesses of different approaches when applied to the study of typical polymeric nanofiltration membranes. For this work, two different membranes were chosen. Firstly, a laboratory-fabricated membrane, made using the phase inversion method with 20% w/w PES and 20% polyvinylpyrrolidone (PVP) in the casting solution, was examined. The second membrane was a commercially available Filmtech NF270 polyamide thin-film composite nanofiltration membrane (Dow Chemicals). Contact angles were determined for both membranes using the sessile drop technique, being 72.61° and 21.89° for the laboratory-made and commercial membranes, respectively.

10.3.1   *AFM imaging*

Both membranes were scanned in air under ambient conditions and in high-purity water (18 MΩ cm), using both tapping and contact modes, with image sizes of 0.25, 1.0, 5.0 and 100 μm. For clarity, only the 1 μm images are included here, with representative images taken in air shown in Figure 10.3. For the PES 20%, PVP 20% membrane, whilst the tapping-mode image (Fig. 10.3a) shows a granular appearance, this detail is almost completely absent in the contact-mode image (Fig. 10.3b), which bears similarities to the ripple patterns observed due to

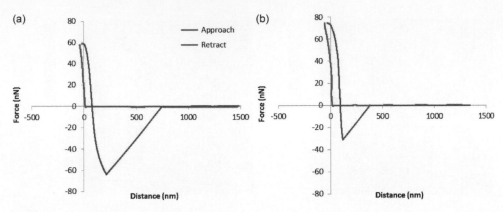

Figure 10.4.   Typical force measurements between contact probe and membrane surfaces in air for: (a) PES 20%, PVP 20% membrane; (b) NF270 membrane.

nanowear previously reported for other polymer surfaces under a number of conditions (Berger *et al.*, 2007; Filippov *et al.*, 2011; Leung and Goh, 1992; Schmidt *et al.*, 1999, 2003). This is most likely to be due to higher interaction forces being experienced between scanning tip and sample during contact-mode imaging, by comparison with tapping mode, inducing ordering of the upper surface of the membrane (Bowen and Hilal, 2009; Filippov *et al.*, 2011; Knoll *et al.*, 2010; Leung and Goh, 1992).

The appearance of the commercial NF270 membrane is similar to that of the PES 20%, PVP 20% membrane, although the lateral dimensions of the surface features appear larger in the 1 × 1 μm scans for NF270 than for the laboratory-fabricated membrane. Of note is the comparison between the tapping- and contact-mode images in Figures 10.3c and 10.3d, which show very little difference in image quality by comparison with the comparable pair of images obtained with the PES 20%, PVP 20% membrane. Scanning conditions, including the loading force, were kept as closely matched for the two membranes as was practicable, so are unlikely to be the cause of this difference, which could possibly be caused by a difference in the resistance to friction between the two membranes.

Force measurements were carried out in air to assess whether there was any difference in surface stiffness between the two membranes. Cantilevers had their spring constants calculated using the reference cantilever method (Gibson *et al.*, 1996; Torii *et al.*, 1996). Five measurements were made for each sample, at different parts of the sample surface. The optical lever sensitivity of the probes was calculated from force curves acquired from a clean silicon surface. The force measurements for the PES 20%, PVP 20% and NF270 membranes are shown in Figures 10.4a and 10.4b, respectively. Adhesion forces were assessed from the force at the point at which the tip separated from the surface when retracting (minimum value for the blue trace). The adhesion force measured in air for the fabricated membrane (61.1 nN, s.d. = 4.42) was approximately twice that measured for the commercial one, NF270 (27.5 nN, s.d. = 3.70). The greater adhesive forces experienced between the probe and the fabricated membrane lead to a greater total load force when imaging the PES 20%, PVP 20% membrane. For contact-mode imaging, the set point was set at 0.5 V, which corresponded to an imaging force of 3.9 nN, much less than the adhesion forces measured for either of the membranes.

By using the optical lever sensitivity value used when pressing the same lever with the same laser alignment against a non-deforming silicon surface, the deflection of the cantilever can be subtracted from the distance axis of the force curves. This allows calculation of the indentation of the probe into the sample surface. Plots of indentation *versus* applied force for the two membranes are shown in Figure 10.5. At forces up until approximately 10 nN the profiles of the two membranes are similar. Above this value a kink in the profile of the NF270 membrane causes it to deviate from that of the PES 20% PVP 20% membrane. The similarity of the profiles at the low forces

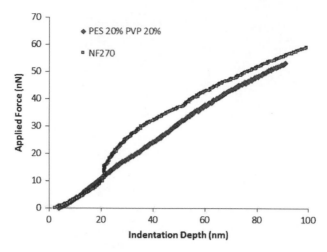

Figure 10.5.   Applied force *versus* indentation depth for nano-indentation measurements of PES 20%, PVP 20% and NF270 polymer membranes.

comparable to those experienced during imaging suggests that the stiffness of the membrane surfaces does not account for the differences in ability to image the two membrane surfaces in contact mode.

Images obtained in a liquid environment are shown in Figure 10.6. In Figure 10.6a, a tapping-mode image is shown for the PES 20%, PVP 20% membrane. This shows globular features not seen in either the tapping-mode image in air (Fig. 10.3a) or the contact-mode image in liquid (Fig. 10.6b) of the same scan area. These features are relatively circular in shape, resembling spherical caps, and are approximately 0.1 μm across. They are much larger than the granular features observed in air for this membrane and appear to be rounded features on top of the flat surface of the membrane, rather than part of the membrane material itself. One possible explanation for these features is the presence of small pockets of air adhering to the surface. Such features have been previously described in the literature on a range of hydrophobic surfaces and termed 'nanobubbles' (Borkent *et al.*, 2010; Hampton and Nguyen, 2009; Ishida *et al.*, 2000; Ishida and Higashitani, 2006; Liu *et al.*, 2008). Nanobubbles have been extensively described, with a number of existing reviews in the literature (Alheshibri *et al.*, 2016; Attard, 2003; Christenson and Claesson, 2001; Meyer *et al.*, 2006). Their presence on a hydrophobic surface and their absence on the same surfaces when imaged in air or in contact mode are both indications that the features observed on the PES 20%, PVP 20% membrane may be explained by the presence of nanobubbles. Similar features are absent when imaging the hydrophilic NF270 membrane in water under identical conditions (Fig. 10.6c) Although rounded granular features are present on NF270, they are identical to features seen for this membrane in air, are still visible in contact mode in water (Fig. 10.6d), and appear to be part of the fabric of the membrane itself.

The equation of a circle was compared to the cross section of one of the putative nanobubbles using a least squares fitting approach (Fig. 10.7b). The close fit ($R^2 = 0.89$) shows that the bubble is in the shape of a spherical cap (Borkent *et al.*, 2010). From this fit, the contact angle of the bubble (angle in liquid and external to the bubble) may be calculated as approximately 165°. This is more than twice the measured macroscopic contact angle of 72.6°. The reason for this difference is unknown, but a study by Borkent *et al.* (2010) noted that all previous investigations of nanobubbles using AFM had reported contact angles in the region of 150° to 170°, independent of the macroscopic contact angles for the materials examined. It was suggested that the reason for the discrepancy is the sensitivity of the method to organic contamination at the sample surface. Whilst the authors of the study described in this section (Johnson *et al.*, 2012) were careful to keep all surfaces and probes clean and free from contamination, the air-water interface readily

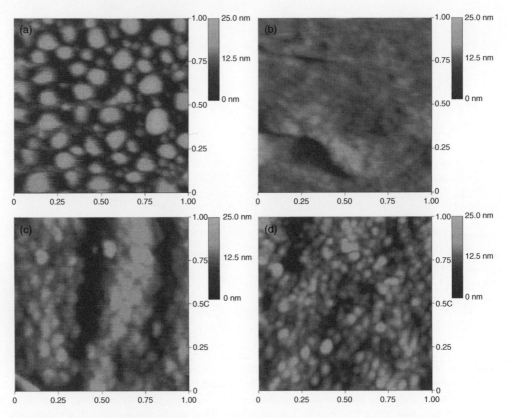

Figure 10.6. $1 \times 1\,\mu\text{m}$ images of membranes taken in high-purity water: (a) PES 20%, PVP 20%, tapping mode; (b) PES 20%, PVP 20%, contact mode; (c) NF270, tapping mode; (d) NF270, contact mode.

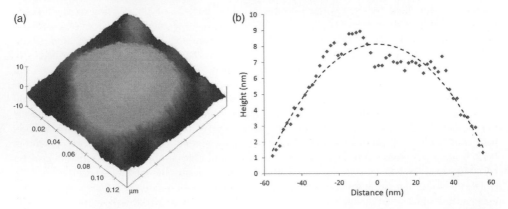

Figure 10.7.   (a) Perspective view of zoomed-in area of PES 20%, PVP 20% membrane as imaged in tapping mode in water, the image showing one of the putative nanobubbles; (b) Cross section of same nanobubble showing actual height data (red diamonds). The dashed line shows the fit of the equation of a circle to the nanobubble profile ($R^2 = 0.89$).

collects contaminants, so this cannot be ruled out entirely. In addition, mechanical deformation of the air bubbles may occur during scanning, which will affect their observational profile.

The existence of nanobubbles on hydrophobic surfaces has been used to explain the presence of the long-range ($>10\,\text{nm}$) hydrophobic force (Considine *et al.*, 1999; Mahnke *et al.*, 1999;

Zhang *et al.*, 2004). These interactions have implications in a number of situations, including the fouling of filtration membranes by organic particulates and in the separation of minerals using froth flotation (Hampton and Nguyen, 2009; Johnson *et al.*, 2006, 2009b). These forces are probably better described as capillary forces (Hampton and Nguyen, 2010) rather than true hydrophobic forces, which occur at shorter ranges and have been attributed to interactions between overlapping solvation zones (Eriksson *et al.*, 1989; Hampton and Nguyen, 2010; Israelachvili, 1992).

10.3.2 *Surface roughness measurements*

Numerous roughness parameters may be derived from surface topographies obtained by AFM, typically by using the instrument's onboard software. Two of the most common parameters quoted in the literature are the roughness average (R_a) and the root mean square (RMS) roughness (R_q). R_a is the arithmetic mean of the surface height deviations from the mean plane of the image and may be calculated by the following formula:

$$R_a = \frac{1}{n} \sum_{i=1}^{n} |Z_i| \tag{10.1}$$

where Z_i is the height for any given pixel, and n is the number of pixels present in the image.

RMS roughness can be calculated by the following equation:

$$RMS, R_q = \sqrt{\frac{\sum (Z_i - Z_{ave})^2}{n}} \tag{10.2}$$

where Z_{ave} is the arithmetic mean of the height values for all the pixels in the image. The RMS roughness is the standard deviation of the pixel height data.

The percentage surface area difference is the ratio (as a percentage) of the surface area of the image to the area of the two-dimensional projected plane of the image. This is directly comparable with Wenzel's roughness value, which can be related to the effect of roughness on contact angle hysteresis (Kamusewitz and Possart, 2003; Marmur, 2003, 2004; Whyman *et al.*, 2008). The fourth roughness measurement parameter applied to the images obtained in this study is the fractal dimension, which has been previously reported in the literature to be scale invariant (Wong *et al.*, 2009; Wyart *et al.*, 2008).

Figure 10.8 presents the roughness data for these four different roughness parameters, plotted as a function of the scan size. Each value plotted is the mean of three different measurements. For the RMS and R_a data (Figs. 10.8a and 10.8b), it is clear that the calculated roughness is a function of the scan size, in accordance with the literature (Boussu *et al.*, 2005, 2006). Both these roughness parameters are statistical descriptions of the variation in pixel height values about the mean. As a greater area is scanned, larger surface features are liable to be included in the image, leading to a broader distribution of pixel heights. In both cases, the increase in measured roughness with scan size is slightly greater for the PES 20%, PVP 20% membrane than for NF270. The trend is reversed for the percentage surface area difference observations (Fig. 10.8c), with the calculated roughness value decreasing as image coverage increases, although the magnitude of the effect is much less pronounced than that in the RMS and R_a parameters (note the smaller scale of the *y* axis). For the fractal dimension values (Fig. 10.8d), there is little or no apparent effect of image size on the calculated roughness, suggesting that it is a much more appropriate measure when comparing images taken at different scan sizes.

10.4 INVESTIGATION OF INORGANIC SCALING ON PVDF AND PTFE DISTILLATION MEMBRANES

When desalinating brine and brackish water the primary scalants are dissolved mineral salts, including Ca^{2+}, Mg^{2+} and Si^{4+} ions. The effect of the drying out of membranes on their scaling was

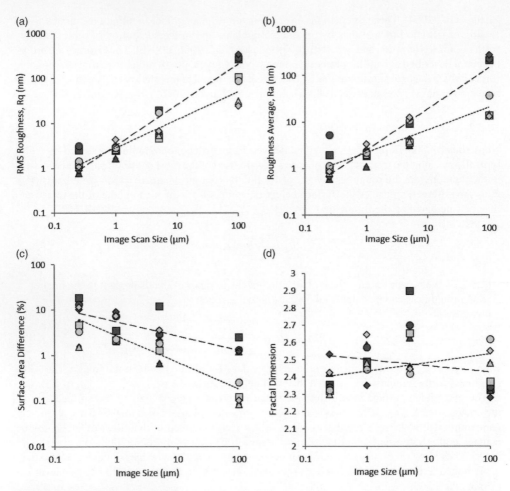

Figure 10.8.   Change in calculated roughness parameters for AFM images according to scan size (image edge length): (a) RMS; (b) R_a; (c) percentage surface area difference; (d) fractal dimension. PES 20%, PVP 20% membrane data is in purple, NF270 membrane in yellow. Symbols represent: □ contact mode in air; △ contact mode in water; ○ tapping mode in air; ◇ tapping mode in water. Trend lines are presented for PES 20%, PVP 20% (dashed line) and NF270 (dotted line).

studied by Guillen-Burrieza *et al.* (2013), in conditions which simulated membrane distillation of seawater. To increase understanding of the surface phenomena related to the salt scaling of these membranes, both imaging and force measurement investigations were carried out, amongst other techniques.

Four different commercial flat-sheet microfiltration membranes were studied, including two polytetrafluoroethylene (PTFE) and two PVDF membranes. The two PTFE membranes differed in that the active layers of PTFE-1 and PTFE-2 were designed for high and standard flux, respectively. All images taken were of a $1 \times 1\,\mu m$ size (Fig. 10.9). Image sizes greater than this were not used, due to the height variations in larger images being close to or greater than the maximum z-axis range of the instrument (approximately $+/-5\,\mu m$ from the centre point).

AFM analyses were performed on membranes exposed to seawater for two weeks and compared to their as-received counterparts. The RMS roughness values obtained from these images are summarized in Figure 10.10. The measured surface roughness of the two clean (unfouled) PTFE membranes was much greater than that of the two PVDF membranes, with the RMS values being

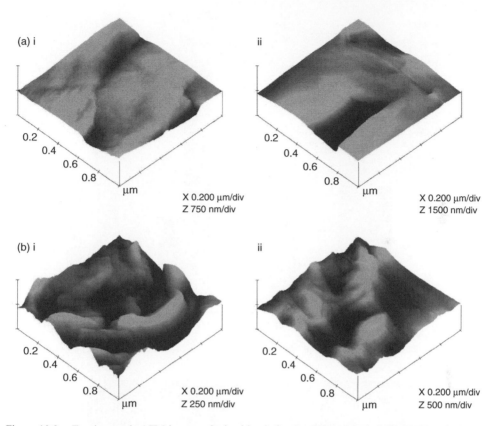

Figure 10.9. Tapping-mode AFM images obtained in air for clean (i) and fouled (ii) PVDF membranes: (a) PVDF-1; (b) PVDF-2.

approximately twice as high. This was also seen in the greater range of height observed for the PTFE membranes than the PVDF ones. This reflects the different structures of the two membrane types, as seen in the AFM images, with the PTFE membranes having a much more open and less dense structure than the PVDF ones. It should also be noted that the RMS roughness values for both membranes were relatively high when compared to nanofiltration and other types of polymeric membrane (Al Malek *et al.*, 2012; Al-Rashdi *et al.*, 2013; Johnson *et al.*, 2012).

For the PVDF membranes, an increase in RMS roughness was observed after membrane fouling, suggesting that deposited salt crystals (scale) were sitting on the surface and leading to an increase in the surface roughness of the membranes. For both of the PTFE samples this was reversed, with the surface roughness seen to decrease after fouling. This suggests that, to some extent, the fouling layer on the PTFE membranes filled in the gaps between the more open structural elements of the PTFE membranes, leading to a decrease in overall surface roughness. The higher roughness of the fouled PVDF membranes compared with the PTFE ones may also be due to the thicker fouling layer which was observed on the PVDF membrane by SEM (Guillen-Burrieza *et al.*, 2013).

Measurements of the adhesion force between a $CaCO_3$ probe and the membrane surfaces were made in seawater at ambient laboratory temperatures ($\sim$22°C). The seawater was filtered with a 0.1 μm syringe filter to remove particulates prior to measurement. $CaCO_3$ crystals were immobilized on the end of a tip-less AFM cantilever using the same method as described by Al-Anezi *et al.* (2009): using a micromanipulator setup, a tip-less AFM probe was dipped into a thin layer of glue at one end of a microscope slide; the probe was then positioned on top of a

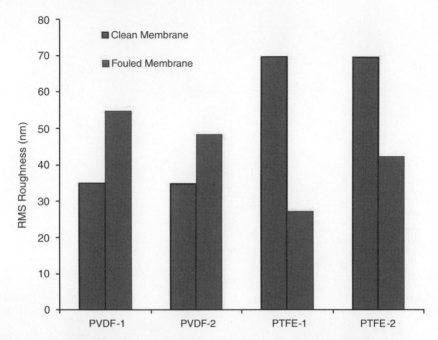

Figure 10.10.   RMS roughness of scans in air of as-received and fouled (two-week seawater exposure) membranes.

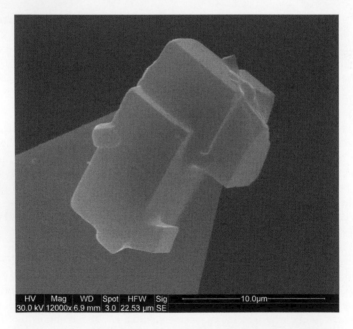

Figure 10.11.   SEM image of a typical $CaCO_3$ probe prepared for force measurements.

single $CaCO_3$ crystal on the same glass slide and the glue allowed to set firmly before starting measurements. An SEM image of a particle probe consisting of a single $CaCO_3$ crystal is shown, subsequent to observations being taken, in Figure 10.11. Around 50 probe-membrane interaction measurements were taken for each membrane sample and the adhesion forces were recorded.

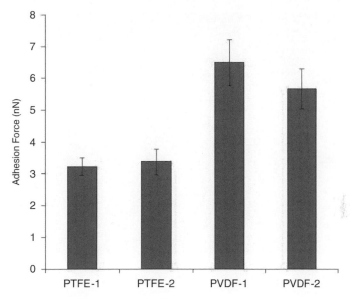

Figure 10.12. Summary of adhesion forces between a $CaCO_3$ particle probe and different membrane surfaces in seawater. Error bars denote standard error of the mean for the measurements.

The measured mean adhesion forces between the $CaCO_3$ probe and the membranes appeared to be higher for the PVDF membranes than for the PTFE ones (Fig. 10.12). When comparing the RMS roughness values with the adhesion forces, it is clear that there is an inverse correlation between the roughness of the initial, unfouled membrane and the measured adhesion force, with rougher membranes showing a lower adhesion between the $CaCO_3$ probe and the membrane surface. It is likely that the increased roughness causes a decrease in the area of contact between the probe and the membrane – in other words, the smoother surface of the PVDF membrane allows for a greater proportion of the crystal surface to make contact with the membrane, allowing for a more stable attachment to occur. As the membranes were measured as having a similar water contact angle prior to fouling, this is a more attractive explanation for the measured difference in adhesion forces, as this rules out differences in wetting as an explanation. However, it does not completely rule out differences in chemical interaction between membranes and probe as a factor in the observed difference in adhesion forces. Contact angle is generally accepted to be affected by surface roughness as well as by chemical interactions (Galiano *et al.*, 2015; Kamusewitz and Possart, 2003; Tamai and Aritani, 1971). The similar contact angles observed for each membrane, despite having markedly different roughness values, points to the likelihood of different chemical interactions between each membrane type and the surrounding water.

It was concluded that an improved attachment of microcrystals to PVDF membrane surfaces, when compared to PTFE membranes, was likely to produce better conditions for the initiation of fouling by mineral salts. Microcrystals attached to the membrane surface are likely to then provide nucleation sites for further crystal growth.

10.5 ADHESION FORCE MEASUREMENT BETWEEN HUMIC ACID AND POLYMER MEMBRANES

The fouling of membrane surfaces by substances present in feed water is a significant cause of increased operational costs and power consumption for membrane separation processes. Novel polymeric membrane surfaces are being developed by numerous research groups to counter foulant

S4800 5.0kV 9.1mm x6.00k SE(M)   5.00um

Figure 10.13.   SEM image of a humic acid-coated latex microsphere mounted on the end of an AFM cantilever.

attachment and build-up. One such group of membranes under development are those which contain an active surface created using polymeric bicontinuous microemulsions (PBMs) (Galiano *et al.*, 2015), as described in Chapter 1. In this section, we describe the use of colloidal probes functionalized with humic acid (HA) to simulate organic foulants and investigate the adhesion behavior of various PBM-coated membrane preparations, both in high-purity water and model textile dye wastewater (MTDW) (Johnson *et al.*, 2015).

The commercial membranes used were PES ultrafiltration membranes (150 kDa) manufactured by Microdyn-Nadir (Wiesbaden, Germany). Other membranes were prepared by production of a PBM as a top layer on the commercial PES membrane. The fabrication of these PBM membranes is described elsewhere (Galiano *et al.*, 2015).

Amine-functionalized polystyrene latex spheres were further functionalized with a coating of HA, following a protocol based upon that used by Koopal *et al.* (1998) for the attachment of HA to aminopropyl silica surfaces. Mean particle size was measured as 5.60 μm diameter (standard deviation 0.67 μm), as verified from SEM imaging. Verification of HA functionalization of the microspheres was made by attenuated total reflectance Fourier transform infra-red spectroscopy (ATR-FTIR). Comparisons were made to the spectra obtained from HA, unfunctionalized polystyrene microspheres and functionalized microspheres. In order to carry out measurements which closely mimic the operating environment in which membranes designed to be low-fouling in a membrane-based reactor system are likely to be used, a model textile dye wastewater (MTDW) was used, based on chemical components typically used in the textile dyeing industry (Alaton *et al.*, 2002; Deowan *et al.*, 2012; Işik and Sponza, 2008; Körbahti and Tanyolaç, 2009). A representative HA-functionalized colloidal probe is shown in Figure 10.13.

Figure 10.14 shows representative AFM tapping-mode images of the commercial PES membrane and four prepared PBM membranes. RMS roughness values obtained from the 1×1 and 5×5 μm scans are shown in Figure 10.15, plotted against the membrane fabrication temperature. As can be seen, roughness values are higher for the larger five-micron size scans than for the smaller one-micron size scans, a feature noted previously as a common issue when determining surface roughness (see Section 10.3.2 and Johnson *et al.*, 2012). For the 1×1 μm scan sizes, a strong and statistically significant correlation ($R^2 = 0.95$, $p = 0.024$) was noted between surface roughness and fabrication temperature, with the smoothest sample (B) being fabricated at the highest temperature (35°C), which may have been due to a lower viscosity of polymer solution at this temperature. However, when examining roughness for larger sample areas (5×5 μm),

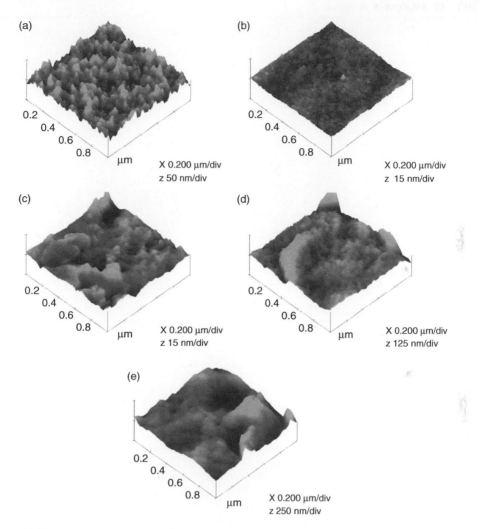

Figure 10.14. $1 \times 1\,\mu m$ AFM scans of membrane surfaces (z range scaled to allow clear view of surface detail): (a) commercial PES membrane (z range $= 50$ nm/division); (b) PBM membrane A ($z = 15$ nm/div); (c) PBM membrane B ($z = 15$ nm/div); (d) PBM membrane C ($z = 125$ nm/div); (e) PBM membrane D ($z = 250$ nm/div).

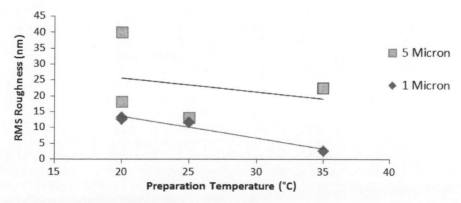

Figure 10.15. Relationship between PBM-membrane fabrication temperature and surface roughness in $1 \times 1\,\mu m$ and $5 \times 5\,\mu m$ AFM scans.

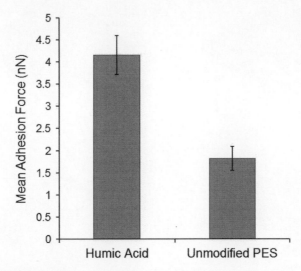

Figure 10.16.   Mean adhesion forces between HA-coated probe and commercial PES membrane surface, fouled with a layer of HA ('Humic Acid') and clean ('Unmodified PES').

no statistically significant correlation was observed ($R^2 = 0.07$, $p = 0.735$), suggesting that the fabrication temperature only affects the size of surface features at the nanoscale.

Initial force measurements were made in water between the HA-coated colloidal probe and the unmodified commercial PES membrane used as the scaffold for creation of the PBM membranes. Typically, unless otherwise stated, 40 measurements were taken at different interaction points on the membrane surface before averaging. All measurements were carried out in a liquid environment consisting of high-purity water (resistivity of 18.2 MΩ cm) or MTDW.

Figure 10.16 shows the adhesion forces between the HA probe and the unmodified PES membrane surface, and between the HA probe and the same type of membrane with a continuous overlayer of HA, formed after filtration of HA suspensions. The mean observed adhesion force for the HA-fouled membrane was higher than that of the unfouled membrane. This suggests that once initial fouling with HA has occurred at the surface of the membrane it will facilitate further fouling, due to the greater affinity of HA for HA than for the membrane. The implication is that for a membrane or any other surface to be resistant to HA fouling, the membrane/surface will have to completely reject all HA particulates in suspension.

When considering the number of force measurements to be acquired for each sample, if each such measurement is made at a different point on the membrane surface the time taken can increase to between 30 minutes and an hour. In addition, when first immersed in liquid, the deflection baseline measurement drifts due to bending of the AFM cantilever, making it usual to leave the probe to equilibrate in the solution for up to an hour before commencing measurement. For this reason, it was decided to assess whether there was a time-dependent effect on force curves taken in MTDW when measurements were taken regularly over a period of three hours. Figure 10.17 shows the effect of time on measurements taken with PBM membrane A. First, the colloidal probe was immersed in high-purity water and allowed to equilibrate for approximately one hour. The solution surrounding the probe was then replaced with several rinses of MTDW before taking measurements. Ten measurements were taken half-hourly for a total of three hours.

The trend line for the time-dependent measurements taken in the model wastewater has a negligible negative slope of -0.15 nN h^{-1}. This shows that observations are not time-dependent, as well as demonstrating the robustness of the probes. If deposition of chemical species were to occur on the surface of the probe and membrane, it is likely to occur immediately on contact. Any further deposition over several hours does not appear to affect the magnitude of the adhesion

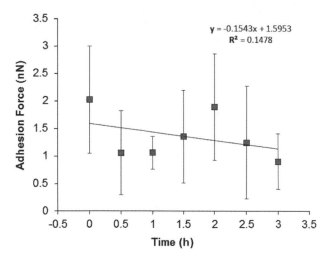

Figure 10.17.   Time dependency of adhesion forces when measured in MTDW.

forces measured to a significant degree. It is worth noting that deposition of dye molecules must occur, due to the change in color of the membrane from white to blue, seen after its immersion in the model wastewater. This was not removed by rinsing with clean water.

Figure 10.18a shows mean adhesion forces measured between the functionalized colloidal probe and various membrane surfaces in high-purity water, including both the commercial PES membrane and PBM-coated surfaces. Forces were normalized by dividing the force value by the microsphere radius, to scale the data according to the area of interaction between the probe and the surface (Butt *et al.*, 2005), with forces being presented in units of $mN\,m^{-1}$. The unmodified PES membrane showed a mean adhesion force of $0.4\,mN\,m^{-1}$. A ranking of the membranes in terms of adhesion forces is shown in Table 10.1, in which the membrane with the lowest adhesion force is assigned a value of 1. In high-purity water, the lowest adhesion force was seen with the commercial PES membrane, while membrane sample B had a measured adhesion with the HA-coated probe much greater than that of any other sample. Sample B was prepared with the same composition as samples A and C, the only difference being that it had the highest preparation temperature ($35°C$). Repeat measurements with different probes consistently showed this sample to have much greater adhesion forces in relation to HA. A correlation ($R^2 = 0.83$) was observed between preparation temperature and HA-membrane adhesion forces measured in pure water.

Figure 10.18b shows the adhesion forces measured between a HA-coated colloidal probe and membrane surfaces in a MTDW environment. The adhesion force trend is markedly changed from that seen in high-purity water, giving a different ranking in terms of adhesion (see Table 10.1). Here, membrane samples B and C both had mean adhesion forces below that measured for the unmodified PES membrane, with the latter membrane and membrane sample D having a much greater measured adhesion (of the order of 270–300% larger). Only the measurements for sample A remained broadly similar in both environments, with a two-sample T-test showing the differences before and after MTDW exposure for this membrane sample not being statistically significant ($p = 0.504$). For each of the other membranes, the differences were found to be significant, with p values $<<0.01$. By comparing the data in Figures 10.18a and 10.18b, it can be seen that only membrane sample C gave low measured adhesion values in both environments. Also of note is that sample B, which was the worst performer in water, was the second-best performing membrane in MTDW in terms of lowest measured adhesion force.

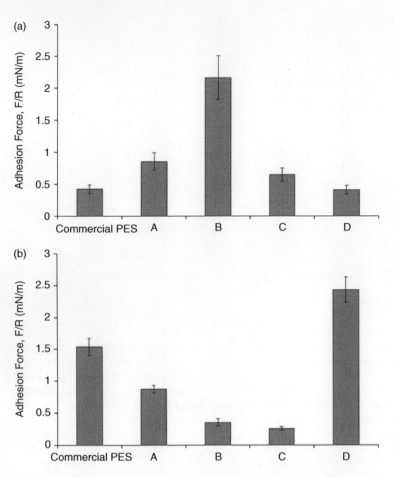

Figure 10.18. Normalized mean adhesion forces for unmodified commercial PES membrane and four PBM-modified PES membranes in: (a) high-purity water; (b) model textile dye wastewater.

Table 10.1. Ranking of membranes according to adhesion measurements carried out in high-purity water and model textile dye wastewater (MTDW), with a ranking of 1 indicating the lowest adhesion and 5 the highest adhesion.

Membrane sample	Adhesion ranking – high-purity water	Adhesion ranking – MTDW
Commercial PES	1	4
A (PMB-coated)	4	3
B (PMB-coated)	5	2
C (PMB-coated)	3	1
D (PMB-coated)	2	5

In order to better understand the adhesion force trends, water contact angle measurements were taken for clean membranes and also for membranes which had been immersed in MTDW. These measurements are summarized in Figure 10.19, which shows that the clean membranes show differing degrees of wettability, with PBM-membrane sample B ($CA = 95.3°$) being the

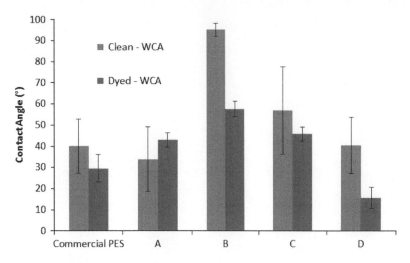

Figure 10.19.   Water contact angle (*WCA*) measurements for clean membranes and membranes dyed following exposure to MTDW.

only membrane to show a considerable degree of hydrophobicity. The water contact angle was modified in all of the membrane samples after exposure to MTDW, with all showing a decrease in contact angle, with the exception of sample A. All changes in contact angle were determined as statistically significant ($p < 0.05$).

A degree of correlation with the measured adhesion forces was seen for all of the water contact angles, both before and after exposure to MTDW. Figure 10.20 shows plots of this relationship before (Fig. 10.20a) and after (Fig. 10.20b) exposure to MTDW. For the clean membrane, $R^2 = 0.81$ and $p = 0.037$; for the membrane exposed to MTDW, $R^2 = 0.91$ and $p = 0.012$. From this high degree of correlation, it appears that the wettability of the membrane is likely to be a major factor in determining the adhesion forces between foulants and these membrane surfaces. It is interesting to observe that the direction of the relationship is inverted after exposure of the membranes to MTDW: in the clean membranes an increase in hydrophobicity of the surface favors increased adhesion with HA, whereas after exposure to MTDW, it is membrane surfaces which are more hydrophilic that show the most adhesion. Contact angle measurements on the HA used to functionalize the probes before and after overnight exposure to MTDW showed that the HA itself also undergoes surface modification, with a reduction in contact angle demonstrating a change from a hydrophobic material ($CA = 129°$) to a moderately hydrophilic material ($CA = 29°$). This demonstrated that the change in the adhesion behavior of the membranes in clean water and MTDW is due to modification of both the probe and the membrane surface. In addition, a strong correlation was also noted between the contact angle and the preparation temperature ($R^2 = 0.99$, $p = 0.005$), as illustrated by Figure 10.20c.

Analysis of surface functional groups for the membranes before and after exposure to MTDW was carried out using ATR-FTIR (Johnson *et al.*, 2015). Examination of the subtracted spectra suggested a possible reason for the change in contact angle of the membranes after exposure to MTDW. It was previously noted that PBM membrane A showed an increase in contact angle, with all other samples showing a decrease. In all membranes, a possible increase in alcohols and carboxylate groups was seen at the surface. PBM membranes B and C showed an increase in the number of hydrophilic C-O groups on their surfaces. However, for PBM membrane A, a reduction in the number of hydrophilic C-O and C=O groups at the surface was seen, explaining its slightly increased hydrophobicity.

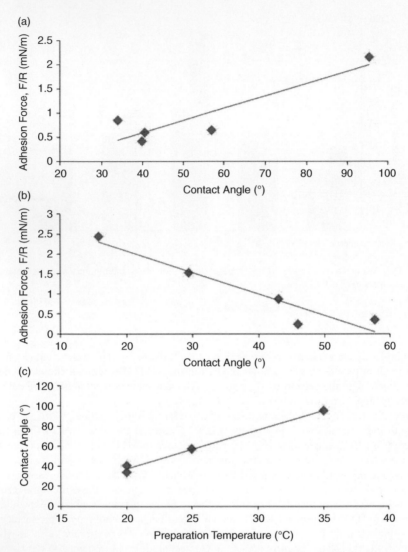

Figure 10.20.   (a) Plot of adhesion forces measured between HA probe and clean membranes *vs.* contact angle, in water ($R^2 = 0.81$, $p = 0.037$); (b) plot of adhesion forces *vs.* contact angle of surfaces exposed to MTDW, in MTDW ($R^2 = 0.91$, $p = 0.012$); (c) plot of contact angle *vs.* membrane fabrication temperature, for clean membranes ($R^2 = 0.99$, $p = 0.005$).

10.6   CONCLUSIONS AND FUTURE DIRECTIONS

In this chapter we have discussed, in general terms, the techniques of atomic-force microscopy and the contributions that have been made to the development of polymer filtration membranes, before giving detailed accounts of some recently published experiments. AFM has a number of advantages over other high-resolution imaging techniques, such as the ability to image in liquid and ambient air conditions, direct three-dimensional interaction with the sample surface, and the ability to 'feel' interaction forces between the probe and the sample surface. This allows the measurement of surface morphological parameters, such as roughness, as well as foulant-membrane interaction forces in both air and environments which mimic conditions in which the membranes are expected to operate, in terms of factors such as pH and ionic strength and composition, as well as the effect of chemical additives.

The AFM technique is now 30 years old and becoming a mature technology, especially in terms of instrument design, with a large number of manufacturers now providing a wide range of commercially available models, and imaging modes covering many applications. In terms of using AFM as a tool to aid in the development of polymer membranes and the assessment of surface fouling, this instrument will continue to be a useful complement to the suite of characterization techniques available to membrane technologists. However, there is always room for further development. As has been demonstrated here, and in numerous studies described in the scientific literature, surface functionalization of AFM probes allows access to information about the interactions between foulants and surfaces. Development of surface chemical modifications into ones mimicking more complex chemistry could provide much useful information about the mechanisms of biofouling of membrane surfaces, and aid in the development of novel biofouling-resistant surfaces.

REFERENCES

Al-Anezi, K., Johnson, D.J. & Hilal, N. (2008) An atomic force microscope study of calcium carbonate adhesion to desalination process equipment: effect of anti-scale agent. *Desalination*, 220(1–3), 359–370.

Alaton, I.A., Balcioglu, I.A. & Bahnemann, D.W. (2002) Advanced oxidation of a reactive dyebath effluent: comparison of O_3 H_2O_2/UV-C and TiO_2/UV-A process. *Water Research*, 36, 1143–1154.

Alheshibri, M., Qian, J., Jehannin, M. & Craig, V.S.J. (2016) A history of nanobubbles *Langmuir*, 32, 11,086–11,100.

Allen, S., Chen, X., Davies, J., Davies, M.C., Dawkes, A.C., Edwards, J.C., Roberts, C.J., Sefton, J., Tendler, S.J.B. & Williams, P.M. (1997) Detection of antigen-antibody binding events with the atomic force microscope. *Biochemistry*, 36, 7457–7463.

Al-Malek, S.A., Abu Seman, M.N., Johnson, D.J. & Hilal, N. (2012) Formation and characterization of polyethersulfone membranes using different concentrations of polyvinylpyrrolidone. *Desalination*, 288, 31–39.

Al-Rashdi, B.A.M., Johnson, D.J. & Hilal, N. (2013) Removal of heavy metal ions by nanofiltration. *Desalination*, 315, 2–17.

Attard, P. (2003) Nanobubbles and the hydrophobic attraction. *Advances in Colloid and Interface Science*, 104, 75–91.

Berger, R., Cheng, Y., Förch, R., Gotsmann, B., Gutmann, J.S., Pakula, T., Rietzler, U., Shärtl, W., Schmidt, M., Strack, A., Windeln, J. & Butt, H.-J. (2007) Nanowear on polymer films of different architecture. *Langmuir*, 23, 3150–3156.

Bernstein, R., Belfer, S. & Freger, V. (2011) Bacterial attachment to RO membranes surface-modified by concentration-polarization-enhanced graft polymerization. *Environmental Science and Technology*, 45, 5973–5980.

Best, R.B., Brockwell, D.J., Toca-Herarra, J.L., Blake, A.W., Smith, D.A., Radford, S.E. & Clarke, J. (2003) Force mode atomic force microscopy as a tool for protein folding studies. *Analytica Chimica Acta*, 479, 87–105.

Binnig, G., Quate, C.F. & Gerber, C. (1986) Atomic force microscope. *Physical Review Letters*, 56(9), 930–933.

Borkent, B.M., de Beer, S., Mugele, F. & Lohse, D. (2010) On the shape of surface nanobubbles. *Langmuir*, 26(1), 260–268.

Boussu, K., Van der Bruggen, B., Volodin, A., Snauwaert, J., Van Haesendonck, C. & Vandecasteele, C. (2005) Roughness and hydrophobicity studies of nanofiltration membranes using different modes of AFM. *Journal of Colloid and Interface Science*, 286, 632–638.

Boussu, K., Van der Bruggen, B., Volodin, A., Van Haesendonck, C., Delcour, J.A., Van der Meeren, P. & Vandecasteele, C. (2006) Characterization of commercial nanofiltration membranes and comparison with self-made polyethersulfone membranes. *Desalination*, 191, 245–253.

Bowen, W.R. & Hilal, N. (eds) (2009) *Atomic Force Microscopy in Process Engineering*. Butterworth-Heinemann, Oxford.

Bowen, W.R., Hilal, N., Lovitt, R. & Wright, C.J. (1998a) Direct measurements of interaction between adsorbed protein layers using an atomic force microscope. *Journal of Colloid and Interface Science*, 197, 348–352.

Bowen, W.R., Hilal, N., Lovitt, R. & Wright, C.J. (1998b) Direct measurement of the force of adhesion of a single biological cell using an atomic force microscope. *Colloids and Surfaces* A: *Physicochemical and Engineering Aspects*, 136, 231–234.

Burnham, N.A., Dominguez, D.D., Mowery, R.L. & Colton, R.J. (1990) Probing the surface forces of monolayer films with an atomic-force microscope. *Physical Review Letters*, 64(16), 1931–1934.

Butt, H.-J., Cappella, B. & Kappl, M. (2005) Force measurements with the atomic force microscope: technique, interpretation and applications. *Surface Science Reports*, 59, 1–152.

Christenson, H.K. & Claesson, P.M. (2001) Direct measurements of the force between hydrophobic surfaces in water. *Advances in Colloid and Interface Science*, 91, 391–436.

Considine, R.F., Hayes, R.A. & Horn, R.G. (1999) Forces measured between latex spheres in aqueous electrolyte: non-DLVO behavior and sensitivity to dissolved gas. *Langmuir*, 15, 1657–1659.

Davies, M., Brindley, A., Chen, X., Marlow, M., Doughty, S.W., Shrubb, I. & Roberts, C.J. (2005) Characterization of drug particle surface energetics and Young's modulus by atomic force microscopy and inverse gas chromatography. *Pharmaceutical Research*, 22(7), 1158–1166.

Deowan, S.A., Wagner, B., Aresipathi, C., Hoinkis, J., Figoli, A. & Drioli, E. (2012) Treatment of model textile dye wastewater (MTDW) towards developing novel submerged membrane bioreactor process. *Procedia Engineering*, 44, 1768–1771.

Domke, J. & Radmach, M. (1998) Measuring the elastic properties of thin polymer films with the atomic force microscope. *Langmuir*, 14, 3320–3325.

Ducker, W.A. & Senden, T.J. (1992) Measurement of forces in liquids using a force microscope. *Langmuir*, 8, 1831–1836.

Ducker, W.A., Senden, T.J. & Pashley, R.M. (1991) Direct measurement of colloidal forces using an atomic force microscope. *Nature*, 353, 239–241.

Eriksson, J.C., Ljunggren, S. & Claesson, P.M. (1989) A phenomenological theory of long-range hydrophobic attraction forces based on a square-gradient variational approach. *Journal of the Chemical Society Faraday Transactions II*, 85, 163–176.

Evans, P.J., Bird, M.R., Rogers, D. & Wright, C.J. (2009) Measurement of polyphenol-membrane interaction forces during the ultrafiltration of black tea liquor. *Colloids and* Surfaces A: *Physicochemical and Engineering Aspects*, 335(1–3), 148–153.

Filippov, A.E., Popov, V.L. & Urbakh, M. (2011) Mechanism of wear and ripple formation induced by the mechanical action of an atomic force microscope tip. *Physical Review Letters*, 106, 025502.

Galiano, F., Figoli, A., Johnson, D., Hilal, N., Altinkaya, S. & Hoinkis, J. (2015) A step forward to a more efficient wastewater treatment by membrane surface modification via polymerizable bicontinuous microemulsion. *Journal of Membrane Science*, 482: 103–114.

Gibson, C.T., Watson, G.S. & Myra, S. (1996) Determination of the spring constants of probes for force microscopy/spectroscopy. *Nanotechnology*, 7, 259–262.

Gillies, G., Kappl, M. & Butt, H.-J. (2005) Surface and capillary forces encountered by zinc sulfide microspheres in aqueous electrolyte. *Langmuir*, 21, 5882–5886.

Guillen-Burrieza, E., Thomas, R., Mansoor, B., Johnson, D., Hilal, N. & Arafat, H. (2013) Effect of dry-out on the fouling of PVDF and PTFE membranes under conditions simulating intermittent seawater membrane distillation (SWMD). *Journal of Membrane Science*, 438, 126–139.

Hampton, M.A. & Nguyen, A.V. (2009) Accumulation of dissolved gases at hydrophobic surfaces in water and sodium chloride solutions: implication for coal flotation. *Minerals Engineering*, 22, 786–792.

Hampton, M.A. & Nguyen, A.V. (2010) Nanobubbles and the nanobubble bridging capillary force. *Advances in Colloid and Interface Science*, 154, 30–55.

Hansma, H.G., Sinsheimer, R.L., Groppe, J., Bruice, T.C., Elings, V., Gurley, G., Bezanilla, M., Mastrangelo, I.A., Hough, P.V.C. & Hansma, P.K. (1993) Recent advances in atomic force microscopy of DNA. *Scanning*, 15(5), 296–299.

Herzberg, M., Kang, S. & Elimelech, M. (2009) Role of extracellular polymeric substances in biofouling of reverse osmosis membranes. *Environmental Science and Technology*, 43, 4393–4398.

Herzberg, M., Sweity, A., Brami, M., Kaufman, Y., Freger, V. & Oron, G. (2011) Surface properties and reduced biofouling of graft-copolymers that possess oppositely charged groups. *Biomacromolecules*, 12, 1169–1177.

Hilal, N., Al-Khatib, L., Atkin, B.P., Kochkodan, V. & Potapchenko, N. (2003) Photochemical modification of membrane surfaces for (bio)fouling reduction: a nanoscale study using AFM. *Desalination*, 158, 65–72.

Ishida, N. & Higashitani, K. (2006) Interaction forces between chemically modified hydrophobic surfaces evaluated by the AFM – the role of nanoscopic bubbles in the interactions. *Minerals Engineering*, 19, 719–725.

Ishida, N., Inoue, T., Miyahara, M. & Higashitani, K. (2000) Nano bubbles on a hydrophobic surface in water observed by tapping-mode atomic force microscopy. *Langmuir*, 16, 6377–6380.

Işik, M. & Sponza, D.T. (2008) Anaerobic/aerobic treatment of a simulated textile wastewater. *Separation and Purification Technology*, 60, 64–72.

Israelachvili, J. (1992) *Intermolecular and Surface Forces*. Academic Press, London.

Johnson, D.J., Miles, N.J. & Hilal, N. (2006) Quantification of particle bubble interactions. *Advances in Colloid and Interface Science*, 127(2), 67–81.

Johnson, D.J., Hilal, N. & Bowen, W.R. (2009a) Basic principles of atomic force microscopy. In: Bowen, W.R. & Hilal, N. (eds) *Atomic Force Microscopy in Process Engineering: An Introduction to AFM for improved Processes and Products*. Butterworth-Heinemann, Oxford. pp. 1–30.

Johnson, D.J., Hilal, N., Waters, K., Hadler, K. & Cilliers, J. (2009b) Measurements of interactions between particles and charged microbubbles using a combined micro and macroscopic strategy. *Langmuir*, 25(9), 4880–4885.

Johnson, D.J., Al-Malek, S.A., Al-Rashdi, B.A.M. & Hilal, N. (2012) Atomic force microscopy of nanofiltration membranes: effect of imaging mode and environment. *Journal of Membrane Science*, 389, 486–498.

Johnson, D.J., Galiano, F., Deowan, S.A., Hoinkis, J., Figoli, A., Scurr, D.J. & Hilal, N. (2015) Adhesion forces between humic acid functionalized colloidal probes and polymer membranes to assess fouling potential. *Journal of Membrane Science*, 484, 35–46.

Kaliapan, S.K. & Capella, B. (2005) Temperature dependent elastic-plastic behavior of polystyrene studied using AFM force-distance curves. *Polymer*, 46, 11,416–11,423.

Kamusewitz, H. & Possart, W. (2003) Wetting and scanning force microscopy on rough polymer surfaces: Wenzel's roughness factor and the thermodynamic contact angle. *Applied Physics* A: *Materials Science and Processing*, 76, 899–902.

Kappl, M. & Butt, H.-J. (2002) The colloidal probe technique and its application to adhesion force measurements. *Particle and Particle Systems Characterization*, 19, 129–143.

Knoll, A., Rothuizen, H., Gotsmann, B. & Duerig, U. (2010) Wear-less floating contact imaging of polymer surfaces. *Nanotechnology*, 21, 185701.

Koopal, L.K., Yang, Y., Minnaard, A.J., Theunissen, P.L.M. & Riemsdijk, W.H. (1998) Chemical imobilisation of humic acid on silica. *Colloids and Surfaces* A: *Physicochemical and Engineering Aspects*, 141, 385–395.

Körbahti, B.K. & Tanyolaç, A. (2009) Continuous electrochemical treatment of simulated industrial textile wastewater from industrial components in a tubular reactor. *Journal of Hazardous Materials*, 170, 771–778.

Leung, O.M. & Goh, M.C. (1992) Oriental ordering of polymers by atomic force microscope tip-surface interaction. *Science*, 255(5040), 64–66.

Li, Q. & Elimelech, M. (2004) Organic fouling and chemical cleaning. *Environmental Science and Technology*, 38, 4683–4693.

Liu, G., Wu, Z. & Craig, V.S.J. (2008) Cleaning of protein-coated surfaces using nanobubbles: an investigation using a quartz crystal microbalance. *Journal of Physical Chemistry* C, 112, 16,748–16,753.

Lüthi, R., Meyer, E.E., Howald, L., Haefke, H., Asnelmetti, D., Dteier, M., Rüetschi, M., Bonner, T., Overney, R.M., Frommer, J. & Güntherodt, H.-J. (1994) Progress in non-contact atomic force microscopy. *Journal of Vacuum Science and Technology* B, 12(3), 1673–1676.

Mahnke, J., Stearnes, J., Hayes, R.A., Fornasiero, D. & Ralston, J. (1999) The influence of dissolved gas on the interactions between surfaces of different hydrophobicity in aqueous media. *Physical Chemistry Chemical Physics*, 1, 2793–2798.

Marmur, A. (2003) Wetting on hydrophobic rough surfaces: to be heterogeneous or not to be? *Langmuir,* 19, 8343–8348.

Marmur, A. (2004) Adhesion and wetting in an aqueous environment: theoretical assessment of sensitivity to the solid surface energy. *Langmuir*, 20, 1317–1320.

Meyer, E.E., Rosenberg, K.J. & Israelachvili, J. (2006) Recent progress in understanding hydrophobic interactions. *PNAS*, 103(43), 15,739–15,746.

Rief, M., Gautel, M., Oesterhelt, F., Fernandez, J.M. & Gaub, H.E. (1997) Reversible unfolding of individual titin immunoglobulin domains by AFM. *Science*, 276, 1109–1113.

Roberts, C.J. (2005) What can we learn from atomic force microscopy adhesion forces with single drug particles? *European Journal of Pharmaceutical Sciences*, 24, 153–157.

Schmidt, R.H., Haugstad, G. & Gladfelter, W.L. (1999) Correlation of nanowear patterns to viscoelastic response in a thin polystyrene melt. *Langmuir*, 15, 317–321.

Schmidt, R.H., Haugstad, G. & Gladfelter, W.L. (2003) Scan-induced patterning in glassy polymer films: using scanning induced force microscopy to study plastic deformation at the nanometer length scale. *Langmuir*, 19, 898–909.

Tamai, Y. & Aratani, K. (1972) Experimental study of relation between contact angle and surface-roughness. *Journal of Physical Chemistry*, 76(22), 3267–3271.

Torii, A., Sasaki, M., Hane, K. & Okuma, S. (1996) A method for determining the spring constant of cantilevers for atomic force microscopy. *Measurement Science and Technology*, 7, 179–189.

Weisenhorn, A.L., Maivald, P., Butt, H.-J. & Hansma, H.G. (1992) Measuring adhesion, attraction, and repulsion between surfaces in liquids with an atomic-force microscope. *Physical Review* B: *Condensed Matter and Materials Physics*, 45(19), 11,226–11,233.

Whyman, G., Bormashenko, E. & Stein, T. (2008) The rigorous derivation of Young, Cassie-Baxter and Wenzel equations and the analysis of the contact angle hysteresis phenomenon. *Chemical Physics Letters*, 450, 355–359.

Wong, P.C.Y., Kwon, Y.-N. & Criddle, C.S. (2009) Use of atomic force microscopy and fractal geometry to characterize the roughness of nano-, micro-, and ultrafiltration membranes. *Journal of Membrane Science*, 340, 117–132.

Wyart, Y., Georges, G., Deumié, C., Amra, C. & Moulin, P. (2008) Membrane characterization by microscopic methods: multiscale structure. *Journal of Membrane Science*, 315, 82–92.

Zhang, X.H., Zhang, X.D., Lou, S.T., Zhang, Z.X., Sun, J.L. & Hu, J. (2004) Degassing and temperature effects on the formation of nanobubbles at the mica/water interface. *Langmuir*, 20, 3813–3815.

Zhong, Q., Inniss, D., Kjoller, K. & Ellings, V. (1993) Fractured polymer silica fiber surface studied by tapping mode atomic force microscopy. *Surface Science*, 290(1–2), L688–L692.

CHAPTER 11

Molecular simulations of water and ion transport through nanoporous membranes

Richard Renou, Minxia Ding, Haochen Zhu, Aziz Ghoufi & Anthony Szymczyk

11.1 INTRODUCTION

Water is vital for all forms of life and for all types of industrial development. Although it covers around 70% of the earth's surface, only 3% of the earth's water is fresh water and about 99% of this latter is locked in polar ice and not readily accessible as groundwater so that less than 0.1% of the global water resource is available for people and ecosystems (Drioli and Macedonio, 2012). The availability of drinking water has nowadays become a worldwide problem due to the continuous growth in water demand that has not been balanced by an adequate replenishment. Furthermore, water sources are suffering ever more frequently from a worsening of their quality due to the indiscriminate discharge of both domestic and industrial effluents without adequate treatment (Macedonio *et al.*, 2012). The UN predicts that by 2025, two-thirds of the world's population will live in areas of significant water stress, lacking sufficient safe water for drinking, industry or agriculture (United Nations, 2010). The lack of fresh water is further aggravated by factors such as pollution and the inequality of its distribution. People's access to drinking water is therefore a major challenge for the coming decades, not only for developing countries but also for the industrialized states (Shannon *et al.*, 2008). Another important and emerging issue is the removal of contaminants increasingly identified in water streams, such as hydrophilic organic compounds, disinfection byproducts, pharmaceutical compounds and also many different ions originating from electronic products which often end up in landfills, thus contaminating land, water and air (Macedonio *et al.*, 2012).

The only methods by which to increase water supply beyond what is available from the hydrological cycle are desalination and water reuse (Elimelech and Phillip, 2011). Moreover, ensuring future safe worldwide water supply creates demands today for advanced and environmentally acceptable processes that enable the preservation of water and reduction in the amounts consumed.

Membrane separation processes are already recognized worldwide as promising tools for addressing the global issues of water shortage and water pollution as part of a strategy of process intensification (i.e. the development of methods aimed at decreasing raw material utilization, energy consumption, equipment size and waste generation (Drioli *et al.*, 2011)). Indeed, these techniques are energy-efficient (no phase change is required to operate the separation), environmentally friendly (in the sense that they require little or no addition of chemicals), modular and compact.

Among the various kinds of membrane processes, reverse osmosis (RO) and nanofiltration (NF) are particularly well-suited for desalination. RO is typically used to separate out dissolved salts and small organic molecules. Its applications range from the production of ultra-pure water for semiconductor and pharmaceutical use, to the desalination of seawater for drinking water production and the purification of industrial wastewater (Drioli and Macedonio, 2012). RO has gained in popularity during the last two decades and is currently the world's leading desalination mechanism, representing about 60% of the installed desalination capacity. NF emerged in the late 1980s, filling the gap between ultrafiltration (UF) and RO. NF rejects smaller solutes than those removed by UF and has a higher rate of filtrate flow than RO at the same pressure, leading to smaller systems with the same production capacity. Due to its lower operating pressures, NF reduces energy consumption when compared to RO. NF membranes are typically characterized by lower rejection of monovalent ions than RO membranes, but maintain high rejection of divalent ions. NF membranes have been employed in pretreatment operations in both thermal and membrane seawater desalination processes, and used for softening brackish water and seawater, as well as in membrane-mediated wastewater reclamation (Drioli and Macedonio, 2012).

On a number of occasions, conventional commercial separation processes in industry have already been converted to membrane separation processes with significant reductions in cost, energy and environmental impact (Drioli *et al.*, 2011). Nevertheless, current applications of NF/RO remain small in number compared to the potential applications that still await. The major reason is that the physical phenomena involved in separations occurring at the nanoscale (as in both NF and RO processes) are not yet fully understood and, for all the current descriptions of transport through these systems, the approach remains more descriptive than truly predictive due to the use of a variety of empirical fitting parameters in the transport models used (Bowen and Welfoot, 2005). The complexity of the separations performed *via* NF and RO membranes results from the combination of nanosized pores or cavities (free volumes) with electrically charged materials (the mechanism of membrane charge formation includes dissociation of surface functional groups and/or adsorption of charged species from the solution onto the membrane surface), which enables the occurrence of phenomena that are not possible at larger scales.

There have been considerable efforts in the last decades to develop transport models suited to NF and RO (Bowen and Mukhtar, 1996; Dresner, 1972; Lonsdale *et al.*, 1965; Palmeri *et al.*, 1999; Spiegler and Kedem, 1966; Szymczyk and Fievet, 2005; Wang *et al.*, 1995). It is true that continuum-based theories and macroscopic models have shown quite impressive capabilities, but mainly as a qualitative description of transport. In order to better capture the physics involved in these membranes with nanometric (or even sub-nanometric) void spaces, and then go one step further for an understanding of transport through NF and RO membranes, we must use a more detailed description of these systems by considering microscopic modeling tools.

Molecular simulations have proved to be efficient techniques for the rationalization of microscopic phenomena occurring in nanometric-sized objects (Balme *et al.*, 2011; Cohen-Tanugi and Grossman, 2012; Kulik *et al.*, 2012; Sala *et al.*, 2012). Indeed, they offer unique possibilities to connect some macroscopic properties to a microscopic description of the physical phenomena involved in nanoconfined phases.

Molecular dynamics (MD) is a particularly promising molecular simulation technique for the investigation of transport phenomena through nanoporous membranes. Basically, an MD simulation consists of determining the trajectories of a system of particles interacting with each other *via* a specified force field (empirical parameters). From the trajectories of the different particles, it is possible to determine some microscopic properties of the system but also, thanks to statistical mechanics, to evaluate some of its macroscopic properties. This chapter provides a short illustration of the potentialities of MD simulations in the investigation of water and ion transport through model nanoporous membranes.

11.2   MOLECULAR DYNAMICS SIMULATIONS

MD simulation is a computational technique in which atoms are 'moved' by solving numerically the equations of motion of classical mechanics (Newton's equations) for a set of molecules (Allen and Tildesley, 1987). The first proper MD simulations were reported more than fifty years ago by Alder and Wainwright (1957), who studied the dynamics of an assembly of hard spheres. Nowadays, MD simulation has become one of the most powerful tools in engineering and science for understanding the behavior of fluids and materials at the atomistic level.

11.2.1   *Fundamentals of molecular dynamics*

Any natural or synthetic system is composed of molecules consisting of atoms whose internal structure is made up of nuclei and electrons. The behavior of these entities is normally described by relativistic quantum mechanics based on the following postulates:

- *De Broglie hypothesis*: any particle of mass m and velocitiy $\mathbf{v}$ is associated with a wave phenomenon characterized by a wavelength.
- *Born postulate*: a complete description of the behavior of an electron is given by a mathematical function, the wave function Ψ, which is defined by the Schrödinger equation.
- *Heisenberg uncertainty principle*: this asserts that it is impossible to determine simultaneously both the position and the momentum of a particle with an accuracy greater than a fundamental limit.
- *Schrödinger postulate*: the Schrödinger equation describes how the quantum state of a physical system changes in time. It is as central to quantum mechanics as Newton's laws are to classical mechanics. In contrast to Newton's equations, its solution no longer provides unique, absolute trajectories, positions and velocities of particles, but only probabilistic statements about the positions and velocities of the particles.

The use of the postulates outlined above is justified because the logical development of mathematical formalism leads to calculation results consistent with experimental observations. Nevertheless, quantum mechanics quickly reaches its limits, especially in terms of the number of calculations required, which increases exponentially with system size (i.e. the number of electrons), thus limiting the scope of investigation to very small systems (typically from a few tens to about a hundred atoms).

In MD simulations, the problem is simplified by considering the Born-Oppenheimer approximation. The mass of electrons is much smaller than that of the nuclei. Therefore, the nuclei move with such a comparatively low amplitude that all of the electronic integrations that have kinetic terms can be approximated to zero. In an MD simulation, we are therefore interested only in the dynamics of the nuclei. The Born-Oppenheimer approximation allows the description of a system by the position $\mathbf{r}$ and momentum $\mathbf{p}$ of atoms. If one considers the atoms as punctual points defined by $(\mathbf{r}, \mathbf{p})$, then we can apply Newtonian physical principles, notably the equation of motion (Newton's second law):

$$m\mathbf{a} = \sum \mathbf{F} = -\sum \nabla_r U_\mathrm{p} = \frac{\mathrm{d}\mathbf{p}}{\mathrm{d}t} = m\frac{\mathrm{d}^2\mathbf{r}}{\mathrm{d}t^2} \tag{11.1}$$

where m, $\mathbf{a}$ and $\Sigma \mathbf{F}$ are the mass of the atom, its acceleration and the sum of forces acting on this atom, respectively, and U_p is the potential energy of the system.

Newton's equations of motion relate the derivative of the potential energy to the changes in position as a function of time. Equation (11.1) shows explicitly that knowledge of the potential energy allows the trajectory of an atom to be described because two successive integrations allow calculation of the velocity and the position of this atom. The equations of motion are deterministic, that is, the positions and the velocities at time zero determine the positions and velocities at all other times. Therefore, to calculate a trajectory, one only needs to know the initial positions of the atoms and the initial distribution of their velocities.

The potential energy is a function of the atomic positions of all atoms in the system. Due to the complicated nature of this function (see Section 11.2.4), there is no analytical solution to Equation (11.1), which must therefore be solved numerically (Allen and Tildesley, 1987). The MD method generates the successive configurations of the system for time intervals that are around 10^{-15} seconds (a typical time step in classical MD simulations). For each time step, the forces acting on the different atoms are computed and combined with the current positions and velocities to generate new positions and velocities.

It is necessary to combine Newton's dynamics, statistical mechanics and the ergodic principle (which states that the time averages of functions that depend on coordinates and momenta of all the particles of a system in the phase space $(\mathbf{r}, \mathbf{p})$ are equal to the statistical averages) in order to obtain the thermodynamic properties of the system, its structural properties, etc. Although MD is computationally demanding, the rapid development of computing power over past decades now makes it possible to perform MD simulations with larger molecular systems (up to $\sim 10^6$ atoms) on timescales ranging from tens of picoseconds to a few hundred nanoseconds.

11.2.2   *Statistical ensembles*

An MD simulation generates information at the microscopic level (the atomic positions and the velocities of the different atoms in the system). The conversion of this microscopic information to macroscopic observables such as pressure, heat capacity, etc., can be achieved thanks to statistical mechanics. The macroscopic quantities have different forms of expression depending on the specific statistical ensemble (in which a number of the system parameters are set). The most common statistical ensembles considered in molecular simulations are:

- *Microcanonical ensembles (NVE)*: a system with a fixed number of particles N, a fixed volume V, and a fixed energy E. It corresponds to an isolated system.
- *Canonical ensembles (NVT)*: a system with a fixed number of particles N, a fixed volume V, and a constant temperature T.
- *Grand canonical ensembles (μVT)*: a system with a constant chemical potential μ, a fixed volume V, and a constant temperature T. The system is an extension of the canonical ensemble, allowing fluctuation in the number of particles.
- *Isothermal-isobaric ensembles (NPT)*: a system with a fixed number of particles N, a fixed pressure P, and a constant temperature T. This ensemble plays an important role in chemistry, as many chemical processes are carried out under conditions of constant pressure.

11.2.3   *Periodic boundary conditions*

Because of the high computational cost of MD simulations, the sides of the simulation box inside which atoms move are typically limited to a few tens of Angströms in size. In order to avoid problems associated with boundary effects caused by the limited size of the simulation box and to make the system equivalent to an infinite medium, periodic boundary conditions are applied. This consist of replicating the original box (without walls at its boundaries) in the three spatial dimensions to form an infinite lattice. In this way, as an atom moves outside the original box, one of its periodic images simultaneously enters through the opposite face of the box.

11.2.4   *Force field*

The force field is the set of functions and parameters needed to compute the potential energy (U_P) of the system. Force-field functions and parameter sets are derived from both experiments and high-level quantum mechanics calculations. Different force fields are available, such as AMBER, CHARMM and COMPASS. The AMBER force field, for example, encapsulates both bonded terms (for atoms that are linked by covalent bonds), and non-bonded terms (describing short-range van der Waals interactions and long-range electrostatic interactions):

$$U_\mathrm{p} = U_\mathrm{bond} + U_\mathrm{angle} + U_\mathrm{torsion} + U_\mathrm{vdW} + U_\mathrm{elec} \tag{11.2}$$

where U_bond describes explicit bonds between atoms and contains intramolecular interactions, U_angle is the valence angle potential, U_torsion is the dihedral angle potential describing the interaction arising from torsional forces in molecules, U_vdW accounts for van der Waals interactions and U_elec describes electrostatic interactions.

Different functional forms have been proposed for these potentials. The ones used to perform the simulations that will be shown in this chapter are described below.

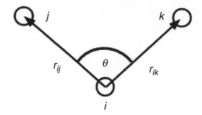

Figure 11.1.   The valence angle for three atoms (i, j and k) and its associated vectors (Smith *et al.*, 2008).

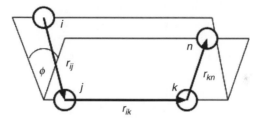

Figure 11.2.   The dihedral angle and associated vectors (Smith *et al.*, 2008).

11.2.4.1   *Harmonic bond potential*

The harmonic bond potential describes explicit bonds between specified atoms. It is a function of the interatomic distance. This potential is based on a simplification of the development of Hooke's law. The potential energy associated with small bond stretches about the equilibrium bond length r_0 is:

$$U_{\text{bond}} = \sum_{\text{bond}} k_{\text{bond}}(r_{ij} - r_0)^2 \tag{11.3}$$

where r_{ij} is the distance between atoms i and j, and k_{bond} is the force constant that quantifies the stiffness of the covalent bond.

11.2.4.2   *Harmonic valence angle potential*

This potential is a harmonic function of the angle between three atoms i, j and k (see Fig. 11.1):

$$U_{\text{angle}} = \sum_{\text{angle}} k_{\theta}(\theta_{jik} - \theta_0)^2 \tag{11.4}$$

where k_{θ} is the angular force constant, θ_{jik} is the angle between bond vectors $\mathbf{r}_{ij}$ and $\mathbf{r}_{ik}$, and θ_0 is the equilibrium angle value.

11.2.4.3   *Dihedral angle potential*

The dihedral angle potential describes the interaction arising from torsional forces in molecules. It requires the specification of four atomic positions:

$$U_{\text{torsion}} = \sum_{\text{dihedral}} \sum_{n} A[1 + \cos(\varpi\phi - \delta)] \tag{11.5}$$

where A, ϖ and δ represent the barrier torsion, frequency and phase, respectively, and ϕ denotes the angle between the plane containing the atoms i, j and k and that containing atoms j, k and n (see Fig. 11.2).

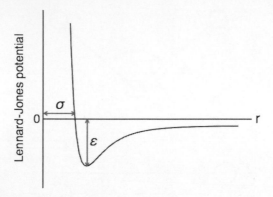

Figure 11.3.   Illustration of a typical Lennard-Jones 12-6 potential.

11.2.4.4   *Lennard-Jones (van der Waals) potential*

The Lennard-Jones potential (U_{LJ}) is a mathematical approximation to describe van der Waals interactions between two non-bonded atoms. A common form of this potential, known as 12-6 potential, is expressed as:

$$U_{LJ} = \sum_{i<j} 4\varepsilon_{ij} \left[\left(\frac{\sigma_{ij}}{r_{ij}} \right)^{12} - \left(\frac{\sigma_{ij}}{r_{ij}} \right)^{6} \right] \tag{11.6}$$

where r_{ij} is the distance between atoms i and j, σ_{ij} is the distance between i and j for which the potential is zero (it gives a measurement of how close two non-bonded atoms can get and is thus referred to as the van der Waals radius) and ε_{ij} is the depth of the potential well (see Fig. 11.3) and is therefore a measure of how strongly atoms i and j attract each other.

When atoms i and j are different chemical species, ε_{ij} and σ_{ij} are usually estimated by the Lorentz-Berthelot mixing rules:

$$\varepsilon_{ij} = \sqrt{\varepsilon_{ii}\varepsilon_{jj}} \tag{11.7}$$

$$\sigma_{ij} = \frac{\sigma_{ii} + \sigma_{jj}}{2} \tag{11.8}$$

In Equation (11.6), the r^{-12} term is the repulsive part of the Lennard-Jones potential (resulting from overlapping electron orbitals) and is dominant at short separation distances, while the r^{-6} term represents the attractive part of the potential (van der Waals or dispersion forces) and is dominant at larger distances. Because the attractive interaction decreases rapidly with separation distance, prohibitively expensive calculations can be avoided by applying a cut-off distance r_{cut}, that is, van der Waals interactions are computed only for atom pairs separated by a distance not exceeding r_{cut} (note that because of periodic boundary conditions the cut-off distance must be less than half the size of the simulation box).

11.2.4.5   *Electrostatic potential*

The electrostatic potential (U_{elec}) is an effective pair potential that describes the interaction between two electrically charged particles q_i and q_j. It can be expressed by Coulomb's law:

$$U_{elec} = \sum_{i<j} \frac{q_i q_j}{4\pi \varepsilon_0 r_{ij}} \tag{11.9}$$

where ε_0 is the vacuum permittivity.

Long-range charge-charge interactions are difficult to handle in computer simulations because of the limited size of the simulation boxes. Equation (11.9) is, therefore, not practical in molecular simulations and several alternative methods, such as the Ewald summation (Ewald, 1921), the particle mesh Ewald method (Darden *et al.*, 1993) or the smooth particle mesh Ewald method (Essmann *et al.*, 1995), have been developed to optimize the computation of electrostatic interactions.

11.3   MD SIMULATIONS OF TRANSPORT THROUGH MODEL NANOPORES

All of the simulations described below were performed in the canonical statistical ensemble (NVT) at a temperature T of 300 K (unless otherwise specified). The temperature was kept constant by using a Berendsen thermostat (Allen and Tildesley, 1987). The van der Waals interactions were truncated at 12 Å (cut-off radius) and the long-range electrostatic interactions were computed from the Ewald summation method. The equations of motion were integrated using the velocity Verlet algorithm with a time step of 2 fs.

Model silica nanopores with different radii and surface chemistries were considered. Each pore was generated from an equilibrated structure of amorphous silica provided by Vink and Barkema (2003) through which a cylindrical cavity was further carved according to the procedure proposed by Bródka and Zerda (1996).

Water molecules were described by the rigid and non-polarizable TIP4P/2005 model (Abascal and Vega, 2005).

The different force field parameters can be found in Abascal and Vega (2005), Balbuena *et al.* (1998), Renou *et al.* (2013) and Ghoufi *et al.* (2013).

Data analysis was performed for the last 10–20 ns after 5 ns of equilibration. All simulations were carried out with the DL-POLY package (Forester and Smith, 2004).

11.3.1   *MD simulations at equilibrium*

The results described in this section were obtained from simulations performed at equilibrium, that is, no external field was applied to the system.

11.3.1.1   *Structure of aqueous solutions confined in nanoporous membranes*

Because MD simulations allow determination of the time evolution of atomic positions, they are well-suited to giving information about the molecular structure of fluids confined in nanopores and nanocavities.

Figure 11.4 shows the radial profile of the density of water confined in nanopores of two different radii (6 and 12 Å). The surface of both nanopores is hydrophilic, with an identical density of silanol groups set to 7.5 nm^{-2} (a snapshot of the narrowest pore is provided in Fig. 11.5).

For both pore radii, Figure 11.4 shows an increase in the water density (with respect to the bulk density, $\sim$1000 kg m^{-3} at $P = 0.1$ MPa and $T = 300$ K) in the vicinity of the pore surface as a result of excluded-volume effects (Morineau and Alba-Simionesco, 2003). The density profiles exhibit oscillations with a period of about 3 Å, which basically corresponds to the effective diameter of a water molecule. This indicates a layered organization of water molecules inside the pores (see also Fig. 11.6). Moving away from the interface, the local density decreases and the layered structuration of the water tends to vanish in the central part of the wider pore (i.e. R12 in Fig. 11.4) because the water molecules no longer "feel" the solid surface. Because of the more severe confinement occurring inside the pore of radius 6 Å, the bulk density of water is not recovered even at the center of the pore.

Such a layered structuration of water molecules in the vicinity of a solid surface has been highlighted with other kinds of solids (Argyris *et al.*, 2011; Gallo *et al.*, 2002), and it can be seen as a general property of any liquid adjacent to any solid surface (Ghoufi *et al.*, 2013; Lyklema *et al.*, 1998). As an illustration, Figure 11.7 shows that a similar layered organization of water

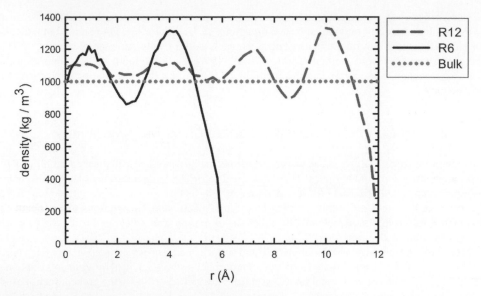

Figure 11.4. Radial variations of water density inside nanopores of radius $R = 6$ Å (R6) and $R = 12$ Å (R12). The position $r = 0$ Å corresponds to the center of the pore.

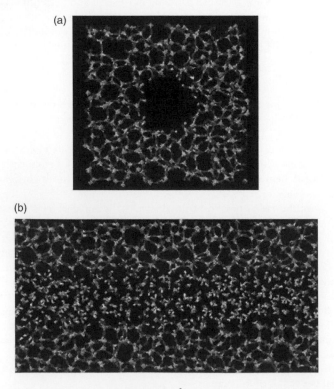

Figure 11.5. (a) Cross section of the nanopore of 6 Å in radius (silicon, oxygen and hydrogen atoms are shown in yellow, red and white, respectively); (b) longitudinal view of the nanopore filled with water molecules. The number of water molecules to be inserted in the nanopore was determined by a grand-canonical Monte Carlo sampling simulation. The dimensions of the simulation box are $36 \times 36 \times 72$ Å.

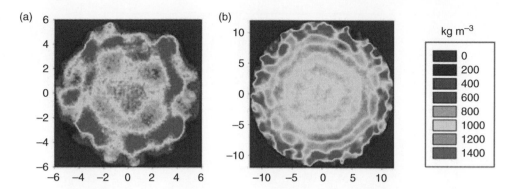

Figure 11.6. 2D density profiles according to the xy section of the nanopore. The figure shows the layered organization of water molecules inside nanopores of radius: (a) 6 Å; (b) 12 Å.

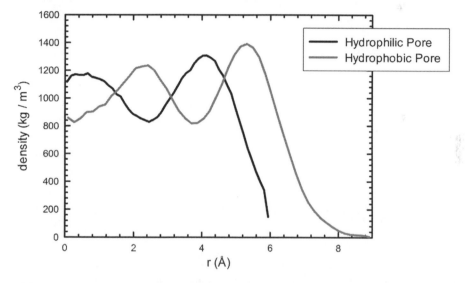

Figure 11.7. Radial variations of water density inside nanopores with hydrophilic and hydrophobic surface properties. The hydrophobic nanopore was generated from a nanopore of radius 12 Å by replacing silanol surface groups with trimethylsilane ($-Si(Me)_3$) groups. The position $r = 0$ Å corresponds to the center of the pore.

molecules occurs inside a nanopore with hydrophobic surface properties (a water-depleted region is found in the close vicinity of the pore surface because of the absence of interactions between water molecules and the hydrophobic surface).

The analysis of radial distribution functions (RDFs) is commonly performed in molecular simulations in order to obtain information about the local structural organization around a given atom. The RDF $g_{ij}(r)$ gives the probability of finding a pair of atoms i and j a distance r apart (j around i), relative to the probability expected for a completely random distribution at the same density (Allen and Tildesley, 1987). In bulk phase, the RDF is given by:

$$g_{bulk}(r) = \frac{<\rho(r)>}{\rho} = \frac{<n(r)>}{4\pi r^2 dr \rho} \tag{11.10}$$

where $<\rho(r)>$ is the average of the local volume density within a spherical shell located at a distance between r and $r + dr$ from the center of mass of the central particle, $<n(r)>$ is the average local number of particles within this shell, and ρ is the volume density of the bulk phase.

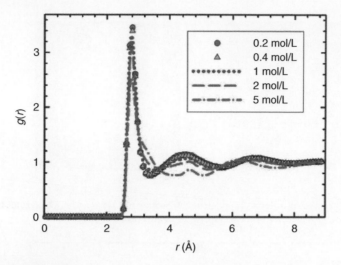

Figure 11.8.   Radial distribution function between the oxygen atoms of water molecules for sodium chloride solutions of various concentrations confined inside a nanopore of radius 6 Å.

Inside the nanopore, the RDF has to be corrected in order to take into account the presence of the surface. For a cylindrical pore, the RDF corrected for the so-called excluded-volume effect takes the form (Morineau and Alba-Simionesco, 2003):

$$g_{pore}(r) = \frac{g_{bulk}(r)}{\dfrac{1}{r^2}\displaystyle\int_{r'=0}^{r}\left(\dfrac{2}{\pi}\,a\cos\left(\dfrac{r'}{2R}\right) - \dfrac{r'}{2R}\sqrt{1-\left(\dfrac{r'}{2R}\right)^2}\right)\dfrac{r'\,dr'}{\sqrt{1-\left(\dfrac{r'}{r}\right)^2}}} \tag{11.11}$$

Figure 11.8 shows the RDF between the oxygen atoms of water molecules for sodium chloride solutions of various concentrations confined inside a hydrophilic nanopore (7.5 silanol surface groups per nm^2). The presence of ions has a substantial impact on the local organization around the water molecules. Notably, the second peak (corresponding to the second hydration shell), located at $r = 4.4$ Å, tends to vanish as the salt concentration increases.

Modifications of the local organization of water molecules can also be tracked through the inspection of the tetrahedrality of the hydrogen-bond (HB) network of water. In molecular simulations, HBs are usually described on the basis of the following geometric criteria (Padró *et al.*, 1997) (see Fig. 11.9):

- The distance between the donor and acceptor oxygen atoms (R_{OO}) has to be less than 3.5 Å.
- The distance between the acceptor oxygen atom and the hydrogen atom (R_{OH}) has to be less than 2.5 Å.
- The angle between the intramolecular O–H bond and the line passing through the donor and acceptor oxygen atoms (θ) has to be smaller than 30°.

The degree of tetrahedrality of the HB network can be quantified by the so-called tetrahedral-order parameter q (Errington and Debenedetti, 2001):

$$q = 1 - \frac{3}{8}\sum_{i=1}^{3}\sum_{j=i+1}^{4}\left(\cos\phi_{ij} + \frac{1}{3}\right)^2 \tag{11.12}$$

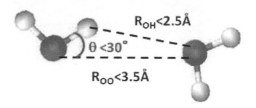

Figure 11.9.   The three geometric criteria defining a hydrogen bond between two water molecules (oxygen and hydrogen atoms are shown in red and white, respectively).

where ϕ_{ij} is the angle formed by the lines joining the oxygen atom of a given water molecule and the oxygen atoms of the nearest neighboring water molecules i and j (only the four closest neighbors are considered).

The average value of q varies between 0 (ideal gas) and 1 (perfect tetrahedral arrangement) (Galamba, 2013). The distributions of the tetrahedral-order parameter for pure water in bulk phase and inside nanopores of two different radii (6 and 12 Å) are shown in Figure 11.10. Two different zones have been considered, namely the central part of the pore and the interfacial region, that is, the first water layer located in the vicinity of the pore surface (see Fig. 11.4). In bulk phase water, the most probable value of q is found to be 0.78, in good agreement with the value of 0.8 reported by Overduin and Patey (2012). The distributions of q in the central part of both pores are close to that in bulk phase (see Fig. 11.10a), indicating that the local arrangement of water molecules is not substantially modified. On the other hand, Figure 11.10b shows that the presence of the pore surface leads to a strong distortion of the pseudo-tetrahedral HB network because the most probable values of q in the interfacial region are found to be 0.70 and 0.52 in pores of radius 12 and 6 Å, respectively.

Surface chemistry also exerts a profound influence on the orientation of water molecules inside nanopores. Let us define θ as the angle formed by the dipole moment vector of water molecules and the surface normal vector (see Fig. 11.11). Figure 11.12 shows the distribution of the angle θ for neat water confined in silica nanopores with various negative surface charge densities (the negative charge density on the surface of the different nanopores was generated by removing some protons from the surface silanol groups). As in Figure 11.10, results are shown for both the central part of the pore and the interfacial region corresponding to the first water layer located in the vicinity of the pore surface. For the uncharged nanopore, all angles are sampled in the central part of the pore (Fig. 11.12a), which means that there are no preferential orientations of water molecules with respect to the pore surface. Close to the surface (Fig. 11.12b), water molecules are slightly more oriented as a result of HBs formed between water molecules and the curved hydrophilic surface. Adding negatively charged surface groups (SiO$^-$) has a significant impact on the orientation of the water molecules, as revealed by the sharper angle-distribution profiles shown in Figure 11.12. The most probable value of θ increases with the degree of deprotonation of the surface and it tends progressively to 180°. This indicates that water molecules rotate under the influence of the electric field generated by the charged surface groups, with their oxygen atom pointing towards the pore center and their hydrogen atoms orientated towards the surface (obviously, the impact of the surface charge density is stronger in the interfacial region).

11.3.1.2   *Dielectric properties of electrolyte solutions confined in nanopores*

Gaining insight into the dielectric properties of confined solutions is of the utmost interest in NF/RO (Szymczyk and Fievet, 2005; Yaroshchuk, 2000). Notably, Szymczyk *et al.* (2007) have explained the behavior of a polyamide membrane with respect to four asymmetric electrolytes with different double-charged cations (Cu^{2+}, Zn^{2+}, Ni^{2+} and Ca^{2+}) by a combination of dielectric exclusion and electrostatic interaction.

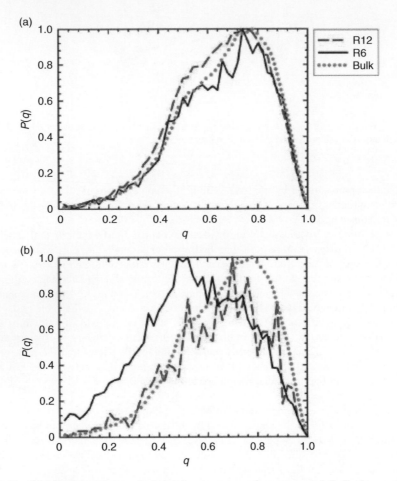

Figure 11.10.   Distribution of the tetrahedral-order parameter q for neat water in bulk phase and confined inside nanopores of radius $R = 6$ Å (R6) and $R = 12$ Å (R12): (a) central part of the pore; (b) interfacial region (i.e. within the first water layer in the vicinity of the pore surface, see Fig. 11.4).

Any ion has an electrostatic free energy, even if it does not interact with other ions. For an ion in vacuum, this free energy is called self-energy, and in any other medium it is termed Born energy. It corresponds to the work required to charge the ion in its environment (Israelachvili, 1985). When applied to membrane separations, the Born model states that the work required to transfer a mole of ion i from a solution outside the membrane into the membrane pores is given by:

$$\Delta G_{i,\text{Born}} = \frac{N_A (z_i e)^2}{8 \pi \varepsilon_0 r_i} \left(\frac{1}{\varepsilon_p} - \frac{1}{\varepsilon_{\text{bulk}}} \right) \tag{11.13}$$

where N_A is the Avogadro number, e is the elementary charge, z_i is the valence of ion i, r_i is the radius of ion i, ε_p is the effective dielectric constant of the pore-filling solution and $\varepsilon_{\text{bulk}}$ is the dielectric constant of the solution outside the membrane.

According to Equation (11.11), ion transfer from the bulk phase into the membrane pores is thermodynamically unfavorable (i.e. $\Delta G_{i,\text{Born}} > 0$) if the effective dielectric constant of the pore-filling solution is lower than that of the bulk phase. In other words, transport theory predicts that desalination performance is enhanced if $\varepsilon_p < \varepsilon_{\text{bulk}}$ (an illustration is provided in Figure 11.13,

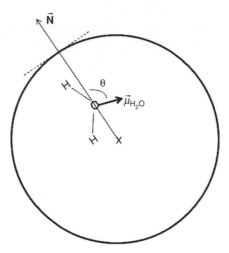

Figure 11.11. Definition of the angle θ between the dipole moment vector of a water molecule (μ_{H2O}) and the surface normal vector (N).

for which the theoretical salt rejection was computed from the continuum-based SEDE model developed by Szymczyk and Fievet (2005)).

The lack of knowledge about the dielectric constant of confined liquids strongly limits our ability to model liquid-membrane interactions and, more generally, our understanding of the behavior of confined fluids. In MD simulations, the dielectric constant of confined water can be computed from the fluctuations of the dipole moments (Ghoufi *et al.*, 2012). MD simulations have recently been used to investigate the dielectric properties of confined water within carbon nanotubes (Lin *et al.*, 2009; Mikami *et al.*, 2009; Qi *et al.*, 2013), graphene nanosheets (Bonthuis *et al.*, 2011; Zhang *et al.*, 2013) and hydrophilic silica membranes (Ghoufi *et al.*, 2012; Renou *et al.*, 2013; Zhu *et al.*, 2012, 2013). Inside cylindrical nanopores, simulations have, notably, shown a significant dielectric anisotropy, the radial component of the dielectric constant being much smaller than the axial one, which can be significantly higher than the bulk value, as shown in Figure 11.14. The axial and radial components of the dielectric constant were computed as follows (Ghoufi *et al.*, 2012):

$$\varepsilon_\parallel(r) = 1 + \frac{1}{\varepsilon_0 k_B T} \left(<P_\parallel(r) \cdot M_\parallel> - <P_\parallel(r)> \cdot <M_\parallel> \right) \tag{11.14}$$

$$\varepsilon_\perp(r) = 1 + \frac{1}{\varepsilon_0 k_B T} \left(<P_\perp(r) \cdot M_\perp> - <P_\perp(r)> \cdot <M_\perp> \right) \tag{11.15}$$

where $P(r)$ is the local polarization density and $M = \int P(r) dr$

The strong dielectric anisotropy highlighted in Figure 11.14 is modulated by the presence of ions inside the nanopore. Indeed, as shown in Figure 11.15, the dielectric anisotropy is reduced as the ion concentration inside the pore increases.

11.3.1.3 *Dynamics of confined water*

In computer simulations, transport coefficients may be calculated from equilibrium correlation functions, by means of Einstein relations or by conducting a suitable non-equilibrium simulation (Allen and Tildesley, 1987). The self-diffusion coefficient (D_s) of solvent molecules (or solutes)

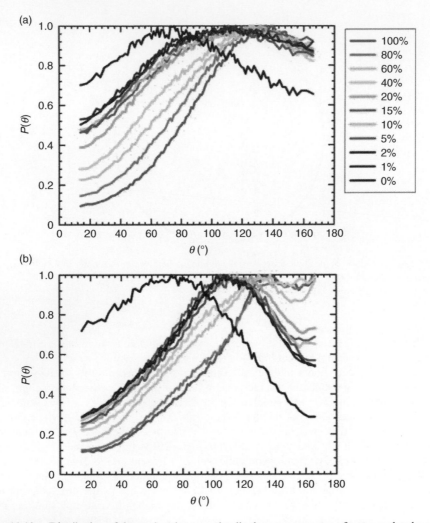

Figure 11.12.   Distribution of the angle θ between the dipole moment vector of water molecules and the surface normal vector for neat water confined inside uncharged and negatively charged nanopores (radius $r = 6$ Å) with various degrees of deprotonation (indicated in the figure): (a) orientation of water molecules in the central part of the pore; (b) orientation of water molecules in the interfacial region (i.e. within the first water layer in the vicinity of the pore surface, see Fig. 11.4).

is frequently computed from MD simulations performed at equilibrium from the time evolution of their mean square displacement according to the Einstein relation:

$$D_S = \frac{1}{2\Omega} \lim_{t \to \infty} \frac{dMSD}{dt} \tag{11.16}$$

where Ω is the dimensionality of the system in which diffusion is considered (i.e. $\Omega = 1$ if diffusion is considered to be in a single direction, $\Omega = 2$ for diffusion in a plane, and $\Omega = 3$ for the overall diffusion through the sample volume) and MSD stands for the mean square displacement, defined as:

$$MSD = \left\langle \frac{1}{N} \sum_{i=1}^{N} (\mathbf{r}_i(t) - \mathbf{r}_i(0))^2 \right\rangle \tag{11.17}$$

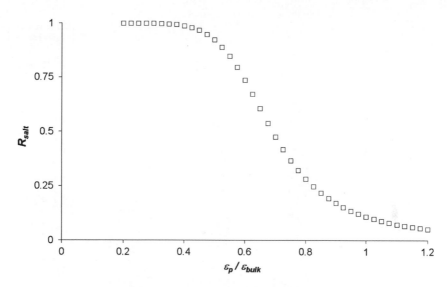

Figure 11.13.   Rejection rate of a 0.01 M NaCl solution by an NF membrane with pores of radius 6 Å as a function of $\varepsilon_p/\varepsilon_{bulk}$ (simulations were performed using the SEDE model (Szymczyk and Fievet, 2005) with $X = -0.1$ M, $\Delta z/A_k = 1\,\mu$m and $J_V = 10^{-5}\,\mathrm{m\,s^{-1}}$, where X, $\Delta z/A_k$ and J_V represent the volume-charge density of the membrane, its thickness-to-porosity ratio and the transmembrane volume flux, respectively).

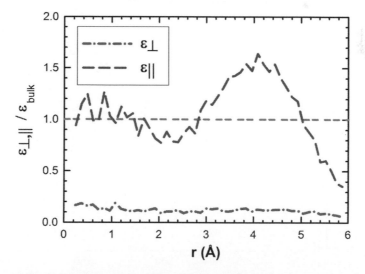

Figure 11.14.   Profile of the radial ($\varepsilon_\perp$) and axial ($\varepsilon_\parallel$) components of the dielectric constant of water confined in a nanopore of radius 6 Å (data are normalized with respect to the bulk dielectric constant of water, ε_{bulk}). The position $r = 0$ Å corresponds to the center of the pore.

where N represents the number of diffusing solvent molecules (or solutes) and the brackets denote a statistical average over the different configurations.

Figure 11.16 shows the variation of the *MSD* of water molecules *versus* time for pure water in bulk phase and inside a nanopore of radius 12 Å. From Equation (11.16), the self-diffusion coefficient of water in bulk phase is found to be $2.30 \times 10^{-9}\,\mathrm{m^2\,s^{-1}}$, which is in excellent agreement with the experimental value ($2.27 \times 10^{-9}\,\mathrm{m^2\,s^{-1}}$) (Abascal and Vega, 2005). The self-diffusion coefficient of water inside the nanopore is much smaller than the bulk value and is found to be

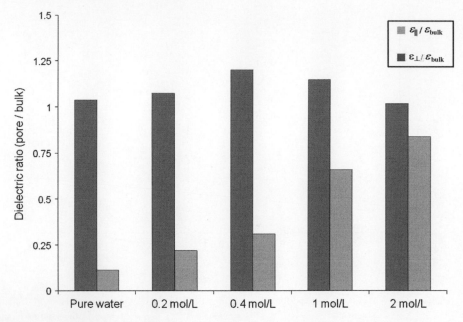

Figure 11.15.   Averages of the radial ($\varepsilon_\perp$) and axial ($\varepsilon_\parallel$) components of the dielectric constant of sodium chloride solutions of varying concentration confined in a nanopore of radius 6 Å (data are normalized with respect to the bulk dielectric constant of water, ε_{bulk}).

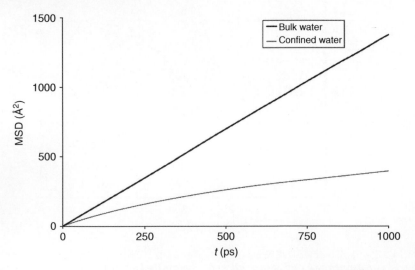

Figure 11.16.   Mean square displacement (MSD) *vs.* time for bulk water and water confined in a nanopore of radius 12 Å.

$4.20 \times 10^{-10}\,\mathrm{m^2\,s^{-1}}$, that is, it is reduced by more than a factor of five with respect to the bulk self-diffusivity.

The rotational dynamics of water molecules can be explored from an analysis of the autocorrelation function of the dipole moments, $\Phi(t)$, defined (Guardia and Marti, 2004) as:

$$\Phi(t) = \frac{\langle \vec{\mu}(t) \cdot \vec{\mu}(0) \rangle}{\langle \mu(0)^2 \rangle} \qquad (11.18)$$

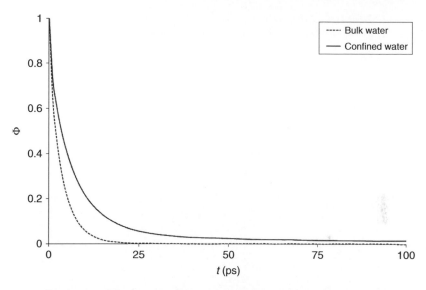

Figure 11.17.   Time variation of the autocorrelation function of the dipole moments for bulk water and water confined in a silica nanopore of radius 12 Å.

Table 11.1.   Self-diffusion coefficient ($D_{S,pore}$) and relaxation time (τ_{pore}) of water molecules inside different silica nanopores.

Water molecule situation	$D_{S,pore}$ [m^2 s^{-1}]	τ_{pore} [ps]
In bulk phase	2.30×10^{-9}	3.1
In a hydrophilic pore of radius 12 Å	4.20×10^{-10}	6.2
In a hydrophilic pore of radius 6 Å	3.15×10^{-10}	7.8
In a hydrophobic pore of radius 9 Å	7.16×10^{-10}	7.6

Figure 11.17 shows the time variation of the autocorrelation function of the dipole moments for pure water in bulk phase and inside the same 12 Å radius nanopore as considered above (and in Fig. 1.16). As can be seen, the dipole moments of the water molecules relax much more slowly inside the nanopore than in the bulk phase. The so-called relaxation time, τ, can be determined by assuming a Debye exponential behavior for $\Phi(t)$ (Marti *et al.*, 1994):

$$\Phi(t) = \exp(-t/\tau) \tag{11.19}$$

Both the translational (D_S) and rotational (τ) dynamics of water slow down as the degree of confinement is increased, as shown in Table 11.1. Interestingly, the hydrophobic/hydrophilic properties of the pore surface strongly impact the translational dynamics of water (D_S is higher inside a hydrophobic nanopore than inside a hydrophilic pore because of the absence of interactions between water molecules and the hydrophobic surface), while the rotational dynamics of the water molecules are virtually unaffected (see Table 11.1). On the other hand, the presence of ions strongly affects both the diffusion and the relaxation of water molecules (see Table 11.2), the impact of the ion concentration on water dynamics being more pronounced under nanoconfinement than in bulk phase (see columns 3 and 5 in Table 11.2).

11.3.2   *Non-equilibrium MD simulations*

Different methods have been proposed in order to model a pressure-driven flow through nanopores in MD simulations (Huang *et al.*, 2006; Zhu *et al.*, 2004). It requires the application of an external

Table 11.2.   Influence of the concentration of sodium chloride on the dynamics of water confined in a silica nanopore of radius 6 Å.

NaCl [mol L^{-1}]	$D_{S,pore}$ [m^2 s^{-1}]	$D_{S,pore}/D_{S,bulk}$	τ [ps]	τ_{pore}/τ_{bulk}
0	3.15×10^{-10}	0.14	7.8	1.7
0.2	2.28×10^{-10}	0.10	8.6	1.8
0.4	1.78×10^{-10}	0.08	9.2	1.8
1	1.60×10^{-10}	0.08	11.2	2
2	1.36×10^{-10}	0.09	15.5	2.2
5	3.22×10^{-11}	0.04	–	–

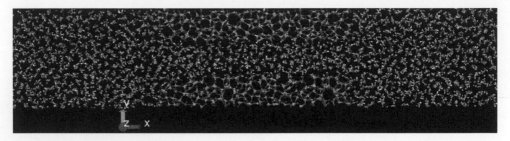

Figure 11.18.   Longitudinal view of a nanopore of radius 6 Å enclosed by external reservoirs filled with water molecules.

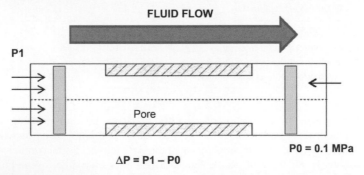

Figure 11.19.   Schematic illustration of the mobile walls method to allow NEMD simulation of a pressure-driven flow through a nanoporous system. The pressure P_1 exerted on the left-hand mobile wall is greater than the pressure P_0 exerted on the right-hand mobile wall, creating a pressure gradient through the nanopore that is responsible for the fluid flow.

field to the system, and the associated simulations are known as non-equilibrium molecular dynamics (NEMD) simulations. In NEMD simulations, the nanopore has to be enclosed by two reservoirs, which play the roles of the feed and permeate compartments (see illustration in Fig. 11.18).

One efficient method of simulating pressure-driven transport through nanopores consists of adding two mobile walls, which act as pistons, into the simulation box (see Fig. 11.19). Basically, a constant force is applied on the different atoms of each wall, this force being different for the two walls so that a pressure gradient is established through the simulation box (Huang *et al.*, 2006). This approach has been successfully applied to generate a pressure-driven flow through different kinds of model nanotubes (Frentrup *et al.*, 2011; Goldsmith and Martens, 2010; Wang *et al.*, 2013).

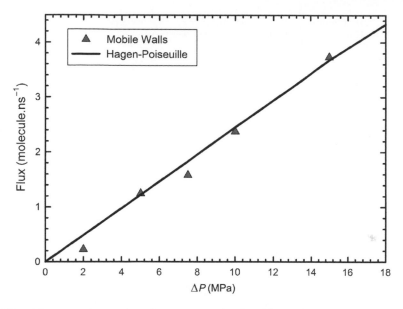

Figure 11.20.   Flow rate of water through a nanopore of radius 6 Å as a function of the applied-pressure difference, determined from NEMD simulations (triangular symbols) and from the Hagen-Poiseuille equation (solid line).

Figure 11.20 shows the flux of water molecules through a nanopore of 6 Å in radius, determined from five NEMD simulations performed for a range of applied-pressure differences (ΔP). Water flux has also been computed from the Hagen-Poiseuille (HP) equation, which reads as follows for a single cylindrical pore:

$$J_{HP} = \frac{R^2}{8\eta L_{pore}} \Delta P \qquad (11.20)$$

where J_{HP} is the solvent volume flux, η is the solvent viscosity, and R and L_{pore} are the pore radius and length, respectively.

The water flux inferred from NEMD simulations is found to vary linearly with the difference in applied pressure, as predicted by the theory (Hagen-Poiseuille equation). Interestingly, the results inferred from molecular simulations are in good quantitative agreement with those predicted by the continuum-based theory, even though the silica nanopore under consideration has a radius as small as 6 Å.

The high computational cost of molecular simulations strongly limits the size of the systems as well as the timescale that can be investigated with these microscopic modeling tools. In NEMD simulations, a direct consequence is that the external force applied on the system in order to induce transport must often be much higher than the actual force applied in experiments. This is especially true when investigating ion transport through a nanopore because the number of ions entering the nanopore is very limited, which significantly increases the uncertainty surrounding the estimates because of poor statistics.

As an example, Figure 11.21 shows the axial profiles of ion pairs obtained for various electrolytes (molar solutions of NaCl, NaI and $MgCl_2$) from NEMD simulations conducted with a pressure difference of 100 MPa. For the mono-monovalent electrolytes (NaCl and NaI), almost no ion pairs are formed, while for the asymmetric electrolyte ($MgCl_2$), ions remain paired inside and outside the nanopore (each Mg^{2+} ion is surrounded by two Cl^- ions). An additional simulation was performed with the NaCl solution by applying a pressure difference of 5 MPa (much

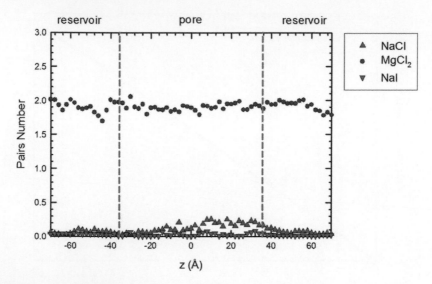

Figure 11.21.   Axial profile of the number of ion pairs in molar solutions of NaCl, NaI and MgCl$_2$ flowing through a nanopore of radius 6 Å under a pressure difference of 100 MPa. The vertical lines indicate the interfaces between the external reservoirs and the nanopore.

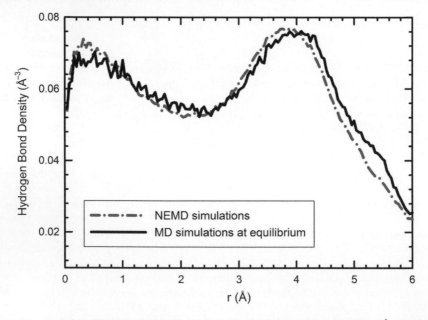

Figure 11.22.   Radial profile of the hydrogen-bond density inside a nanopore of radius 6 Å obtained from simulations performed at equilibrium (Np_zT statistical ensemble) under atmospheric pressure and from NEMD simulations with an applied-pressure difference of 100 MPa.

closer to experimental conditions) but it was not possible to obtain reliable results since only one cation and one anion crossed the nanopore after a simulation time of 20 ns (for the system under consideration, simulating 20 ns takes approximately two months using 16 CPUs of computing capacity).

Because much higher pressure differences than those encountered in experimental conditions are necessary to obtain reliable results from NEMD simulations in a reasonable timeframe, it is

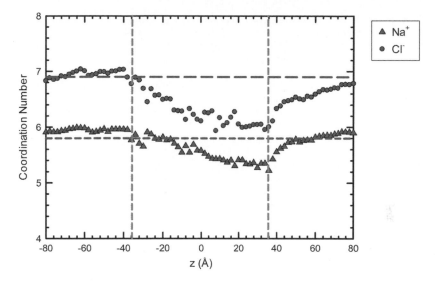

Figure 11.23. Axial profile of the coordination number of Na$^+$ and Cl$^-$ ions inferred from an NEMD simulation of a molar solution of NaCl flowing through a nanopore of radius 6 Å under a pressure difference of 100 MPa. The vertical lines indicate the interfaces between the external reservoirs and the nanopore, and the horizontal lines represent the respective coordination numbers of Na$^+$ and Cl$^-$ ions in bulk phase.

essential to assess the impact of such pressures on the physical properties of the fluid. This is illustrated in Figure 11.22, which compares the density of hydrogen bonds formed between water molecules inside the nanopore in simulations performed at equilibrium and atmospheric pressure, and in NEMD simulations carried out by applying a pressure difference of 100 MPa. As can be seen, both profiles are very similar, thus indicating that the structural properties of water are not substantially impacted by the high pressure applied in the NEMD simulation.

The axial profiles of the coordination numbers of Na$^+$ and Cl$^-$ ions determined from NEMD simulation are shown in Figure 11.23. The simulation highlights a partial dehydration of both ions as they enter the nanopore, with a progressive decrease in the coordination number as ions move towards the pore end. Just before exiting the pore, both Na$^+$ and Cl$^-$ ions have lost approximately one water molecule in their solvation shell, which is recovered as the ions reach the low-pressure external reservoir.

REFERENCES

Abascal, J.L.F. & Vega, C. (2005) A general purpose model for the condensed phases of water: TIP4P/2005. *Physical Review* B: *Condensed Matter and Materials Physics*, 123, 234505.

Alder, B.J. & Wainwright, T.E. (1957) Phase transition for a hard sphere system. *Journal of Chemical Physics*, 27, 1208–1209.

Allen, M.P. & Tildesley, D.J. (1987) *Computer Simulation of Liquids*. Oxford University Press, New York.

Argyris, D., Ho, T., Cole, D.R. & Striolo, A. (2011) Molecular dynamics studies of interfacial water at the alumina surface. *Journal of Physical Chemistry* C, 115, 2038–2046.

Balbuena, P.B., Johnston, K.P., Rossky, P.J. & Hyun, J.-K. (1998) Aqueous ion transport properties and water reorientation dynamics from ambient to supercritical conditions. *Journal of Physical Chemistry* B, 102, 3806–3814.

Balme, S., Janot, J.M., Berardo, L., Henn, F., Bonhenry, D., Kraszewski, S., Picaud, F. & Ramseyer, C. (2011) New bioinspired membrane made of a biological ion channel confined into the cylindrical nanopore of a solid-state polymer. *Nano Letters*, 11, 712–716.

Bonthuis, D.J., Gekle, S. & Netz, R.R. (2011) Dielectric profile of interfacial water and its effect on double-layer capacitance. *Physical Review Letters*, 107, 166102.

Bowen, W.R. & Mukhtar, H.R. (1996) Characterisation and prediction of separation performance of nanofiltration membranes. *Journal of Membrane Science*, 112, 263–274.

Bowen, W.R. & Welfoot, J.S. (2005) Modelling the performance of nanofiltration membranes. In: Schäfer, A.I., Fane, A.G. & Waite, T.D. (eds) *Nanofiltration – Principles and Applications*. Elsevier, Oxford. pp. 119–146.

Bródka, A. & Zerda, T.W. (1996) Properties of liquid acetone in silica pores: molecular dynamics simulation. *Journal of Chemical Physics*, 104, 6319–6326.

Cohen-Tanugi, D. & Grossman, J.C. (2012) Water desalination across nanoporous graphene. *Nano Letters*, 12, 3602–3608.

Darden, T., York, D. & Pedersen, L. (1993) Particle mesh Ewald: an $N \cdot \log(N)$ method for Ewald sums in large systems. *Journal of Chemical Physics*, 98, 10,089–10,092.

Dresner, L. (1972) Stability of the extended Nernst-Planck equations in the description of hyperfiltration through ion-exchange membranes. *Journal of Physical Chemistry*, 76, 2256–2267.

Drioli, E. & Macedonio, F. (2012) Membrane engineering for water engineering. *Industrial & Engineering Chemistry Research*, 51, 10,051–10,056.

Drioli, E., Stankiewicz, A.I. & Macedonio, F. (2011) Membrane engineering in process intensification – an overview. *Journal of Membrane Science*, 380, 1–8.

Elimelech, M. & Phillip, W.A. (2011) The future of seawater desalination: energy, technology, and the environment. *Science*, 333, 712–717.

Errington, J.R. & Debenedetti, P.G. (2001) Relationship between structural order and the anomalies of liquid water. *Nature*, 409, 318–321.

Essmann, U., Perera, L., Berkowitz, M.L., Darden, T., Lee, H. & Pedersen, L. (1995) A smooth particle mesh Ewald method. *Journal of Physical Chemistry*, 103, 8577–8593.

Ewald, P. (1921) Die Berechnung optischer und elektrostatischer Gitterpotentiale. *Annals of Physics*, 369, 253–287.

Forester, T.R. & Smith, W. (2004) DLPOLY, CCP5 Program Library. STFC Daresbury Laboratory, Warrington, UK.

Frentrup, H., Avendaño, C., Horsch, M., Salih, A. & Müller, E. A. (2011) Transport diffusivities of fluids in nanopores by non-equilibrium molecular dynamics simulation. *Molecular Simulation*, 38, 540–553.

Galamba, N. (2013) Water's structure around hydrophobic solutes and the iceberg model. *Journal of Physical Chemistry* B, 117, 2153–2159.

Gallo, P., Ricci, M.A. & Rovere, M. (2002) Layer analysis of the structure of water confined in vycor glass. *Journal of Chemical Physics*, 116, 342–346.

Ghoufi, A., Szymczyk, A., Renou, R. & Ding, M. (2012) Calculation of local dielectric permittivity of confined liquids from spatial dipolar correlations. *EPL (Europhysics Letters)*, 99, 37008.

Ghoufi, A., Hureau, I., Morineau, D., Renou, R. & Szymczyk, A. (2013) Confinement of tert-butanol nanoclusters in hydrophilic and hydrophobic silica nanopores. *Journal of Physical Chemistry* C, 117, 15,203–15,212.

Goldsmith, J. & Martens, C.C. (2010) Molecular dynamics simulation of salt rejection in model surface-modified nanopores. *Journal of Physical Chemistry*, 1, 528–535.

Guardia, E. & Marti, J. (2004) Density and temperature effects on the orientational and dielectric properties of supercritical water. *Physical Review* E, 69, 011502.

Huang, C., Nandakumar, K., Choi, P.Y.K. & Kostiuk, L.W. (2006) Molecular dynamics simulation of a pressure-driven liquid transport process in a cylindrical nanopore using two self-adjusting plates. *Journal of Physical Chemistry*, 124, 234701.

Israelachvili, J.N. (1985) *Intermolecular and Surface Forces*. Academic Press, London.

Kulik, H.J., Schwegler, E. & Galli, G. (2012) Probing the structure of salt water under confinement with first-principles molecular dynamics and theoretical X-ray absorption spectroscopy. *Journal of Physical Chemistry Letters*, 3, 2653–2658.

Lin, Y., Shiomi, J., Maruyama, S. & Amberg, G. (2009) Dielectric relaxation of water inside a single-walled carbon nanotube. *Physical Review* B, 80, 045419.

Lonsdale, H.K., Merten, U. & Riley, R.L. (1965) Transport properties of cellulose acetate osmotic membranes. *Journal of Applied Polymer Science*, 9, 1341–1362.

Lyklema, J., Rovillard, S. & De Coninck, J. (1998) Electrokinetics: the properties of the stagnant layer unraveled. *Langmuir*, 14, 5659–5663.

Macedonio, F., Drioli, E., Gusev, A.A., Bardow, A., Semiat, R. & Kurihara, M. (2012) Efficient technologies for worldwide clean water supply. *Chemical Engineering and Processing*, 51, 2–17.

Marti, J., Guardia, E. & Padro, J.A. (1994) Dielectric properties and infrared spectra of liquid water: influence of the dynamic cross correlations. *Journal of Physical Chemistry*, 101, 10,883–10,891.

Mikami, F., Matsuda, K., Kataura, H. & Maniwa, Y. (2009) Dielectric properties of water inside single-walled carbon nanotubes. *ACS Nano*, 3, 1279–1287.

Morineau, D. & Alba-Simionesco, C. (2003) Liquids in confined geometry: how to connect changes in the structure factor to modifications of local order. *Journal of Physical Chemistry*, 118, 9389–9400.

Overduin, S.D. & Patey, G.N. (2012) Understanding the structure factor and isothermal compressibility of ambient water in terms of local structural environments. *Journal of Physical Chemistry* B, 116, 12,014–12,020.

Padró, J., Saiz, L. & Guàrdia, E. (1997) Hydrogen bonding in liquid alcohols: a computer simulation study. *Journal of Molecular Structure*, 416, 243–248.

Palmeri, J., Blanc, P., Larbot, A. & David, P. (1999) Theory of pressure-driven transport of neutral solutes and ions in porous ceramic nanofiltration membranes. *Journal of Membrane Science*, 160, 141–170.

Qi, W., Chen, J., Lei, X., Song, B. & Fang, H. (2013) Anisotropic dielectric relaxation of the water confined in nanotubes for terahertz spectroscopy studied by molecular dynamics simulations. *Journal of Physical Chemistry* B, 117, 7967–7971.

Renou, R., Ghoufi, A., Szymczyk, A., Zhu, H., Neyt, J.-C. & Malfreyt, P. (2013) Nanoconfined electrolyte solutions in porous hydrophilic silica membranes. *Journal of Physical Chemistry* C, 117, 11,017–11,027.

Sala, J., Guardia, E. & Marti, J. (2012) Specific ion effects in aqueous eletrolyte solutions confined within graphene sheets at the nanometric scale. *Physical Chemistry Chemical Physics*, 14, 10,799–10,808.

Shannon, M.A., Bohn, P.W., Elimelech, M., Georgiadis, J.G., Marinas, B.J. & Mayes, A.M. (2008) Science and technology for water purification in the coming decades. *Nature*, 452, 301–310.

Smith, W., Forester, T.R., Todorov, I.T. & Leslie, M. (2008) The DL_POLY_2 User Manual. CCLRC Daresbury Laboratory, Warrington, UK.

Spiegler, K.S. & Kedem, O. (1966) Thermodynamics of hyperfiltration (reverse osmosis): criteria for efficient membranes. *Desalination*, 1, 311–326.

Szymczyk, A. & Fievet, P. (2005) Investigating transport properties of nanofiltration membranes by means of a steric, electric and dielectric exclusion model. *Journal of Membrane Science*, 252, 77–88.

Szymczyk, A., Fatin-Rouge, N., Fievet, P., Ramseyer, C. & Vidonne, A. (2007) Identification of dielectric effects in nanofiltration of metallic salts. *Journal of Membrane Science*, 287, 102–110.

United Nations. (2010) The Millennium Development Goals Report. Available from: http://www.un.org/millenniumgoals/ [accessed December 2016].

Vink, R.L.C. & Barkema, G.T. (2003) Large well-relaxed models of vitreous silica, coordination numbers, and entropy. *Physical Review* B, 67, 245201.

Wang, L., Dumont, R.S. & Dickson, J.M. (2013) Nonequilibrium molecular dynamics simulation of pressure-driven water transport through modified CNT membranes. *Journal of Physical Chemistry*, 138, 124701.

Wang, X.L., Tsuru, T., Nakao, S.I. & Kimura, S. (1995) Electrolyte transport through nanofiltration membranes by the space-charge model and the comparison with Teorell-Meyer-Sievers model. *Journal of Membrane Science*, 103, 117–133.

Yaroshchuk, A.E. (2000) Dielectric exclusion of ions from membranes. *Advances in Colloid and Interface Science*, 85, 193–230.

Zhang, C., Gygi, F. & Galli, G. (2013) Strongly anisotropic dielectric relaxation of water at the nanoscale. *Journal of Physical Chemistry Letters*, 4, 2477–2481.

Zhu, F., Tajkhorshid, E. & Schulten, K. (2004) Theory and simulation of water permeation in aquaporin-1. *Biophysical Journal*, 86, 50–57.

Zhu, H., Ghoufi, A., Szymczyk, A., Balannec, B. & Morineau, D. (2012) Anomalous dielectric behavior of nanoconfined electrolytic solutions. *Physical Review Letters*, 109, 107801.

Zhu, H., Ghoufi, A., Szymczyk, A., Balannec, B. & Morineau, D. (2013) Reply to comment on "Anomalous dielectric behavior of nanoconfined electrolytic solutions". *Physical Review Letters*, 111, 089802.

CHAPTER 12

Use of carbon nanotubes in polymer membranes for wastewater purification: An *ab initio* theoretical study

Giorgio De Luca & Federica Bisignano

12.1 INTRODUCTION

The removal of pollutants or the recovery of valuable molecules with low molecular weight is a current challenge in wastewater treatment. Membrane processes are, as of now, considered to be a viable option to solve these issues. In terms of separation in membranes, the rejection of organic matter and water permeability are two fundamental, coupled aspects to optimize. In general, an increase in the former comes at the expense of the latter; in fact, a high level of rejection is often a cause of membrane fouling, which in turn dramatically reduces water permeability. Thus, membrane fouling constitutes one of the major drawbacks in membrane separation processes, which, unfortunately, is intrinsically connected to the rejection of solutes. In addition to these challenges, two other related aspects have to be considered to obtain efficient membranes for water purification: resistance and lifetime, related to chemical, biological or mechanical degradation, swelling or embrittlement.

   Thus, the choice of innovative materials (e.g. nanostructures) that show high molecular rejection and low propensity to fouling (thereby maintaining high water permeability) should take these aspects into account, as well as material processability and cost. Composite membranes formed by adding nanostructures such as carbon nanotubes (CNTs) or nanoparticles are receiving huge attention as the properties of these structures can improve the efficiency of the membrane in terms of permeability and selectivity, resistance and, ultimately, prevention of fouling (Cao and Wang, 2011; White House, 2000). Computational modeling can greatly accelerate the choice of appropriate nanostructure. In addition, the description at nanoscale level of the intermolecular interactions that control rejection and water permeability is crucial because it allows correlations to be drawn between nanoscale and macroscale properties. In this context, *ab initio* modeling is very useful for the design and/or choice of innovative nanostructures, since it allows accurate assessment of key nanoscale features which are very difficult to obtain experimentally.

   In this chapter, *ab initio* modeling will be illustrated in order to investigate particular properties of some promising nanostructures, such as CNTs studied in the context of the BioNexGen project (www.bionexgen.eu) with the aim of proposing novel mixed matrix polymer membranes. Indeed, nanostructures such as CNTs or graphene could confer an increase in membrane efficiency in terms of both selectivity and permeability; moreover, they meet the other requirements that have been listed above too. Specifically, the rejection by multi-walled CNTs (MW-NTs) of low molecular weight organic solutes coming from industrial wastewater, such as cosmetics and textiles, and olive oils is investigated in depth, together with their water permeability. To start with, two key properties of a nano-enhanced membrane are investigated: firstly, the rejection of low molecular weight compounds by non-functionalized NTs, in order to predict the optimal nanotube sizes able to reject the target solutes; secondly, the permselective properties of a CNT composite membrane are studied in order to predict the permeability enhancement as a function of the density and inlet diameters of vertically aligned nanotubes. Finally, effective functionalization models are analyzed in order to maximize solute selectivity while retaining higher permeability.

12.2   THEORETICAL BACKGROUND AND ALGORITHMS

Quantum chemical (QC) approaches coupled with bespoke algorithms were used to achieve the proposed objectives. The geometries of the nanostructures under consideration and the related physico-chemical properties controlling their selective features, such as binding energy or steric hindrance, were all assessed by using *ab initio* approaches in the framework of density-functional theory (DFT) (Parr and Yang, 1989). A QC approach, although it may require a rather significant computational effort, helps to estimate accurately some properties that it is currently impossible to assess experimentally. Moreover, its definitely nanostructure represents the most reliable method since it does not make use of any adjustable parameters. It is also worth pointing out that the *ab initio* results are homogeneous and hence overcome the common inconvenience of handling data obtained under different experimental conditions or parameters, which makes comparison difficult.

Furthermore, the proposed in-house algorithms also do not make use of adjustable parameters, being based only on the QC geometries of the nanostructures. In particular, general algorithms were developed, leading to the potential for a wide range of application, such that the methodologies illustrated here could be applied to other nano-architectures.

12.2.1   *Theoretical background*

The choice of a theoretical and computational procedure depends strictly on the target property. For example, binding energies, molecular electrostatic properties such as partial charges and electron density distribution, dipoles and orbital configurations, as well as the breaking or/and formation of interactions, must necessarily be evaluated by QC calculations. However, QC methods are not applicable to the geometry optimization of macromolecules (or related properties), that is, systems containing thousands of atoms. Thus, each property requires the use of an appropriate computational procedure determined both by the accuracy required and the time needed to obtain the desired results.

Moreover, when QC procedures are used, attention has to be paid to the structural models since they dramatically affect the computational time required for the calculations. The structural models, in turn, are also dependent on the properties to be assessed (De Luca, 2013). For example, if non-covalent binding energies have to be calculated, the associated structural model, such as a carbon nanotube with guest molecules inside, must contain a sufficient number of atoms to obtain reliable energies. In this chapter, the QC calculations were carried out on systems having large numbers of atoms (up to 700), and although high computational time was required, accurate energies associated with the non-covalent interactions and molecular structures involved in the target systems were achieved. The non-covalent interactions play an important role and control many nanostructure system features. However, for flexible molecules, molecular mechanics calculations were also carried out in order to have an accurate description of the conformational space of these systems. A brief introduction to DFT, used in the QC calculations, now follows.

12.2.1.1   *Density-functional theory*

DFT is a relatively new quantum theory that replaces the complicated N-electron wave function, $\Psi(x_1, x_2, ..., x_N)$, and the associated Schrödinger equation with the much simpler electron density, $\rho(\mathbf{r})$, and its associated Kohn-Sham equations.

For an isolated N-electron atomic or molecular system in the Born-Oppenheimer non-relativistic approximation, the time-independent Schrödinger equation is given by:

$$\hat{\mathbf{H}}|\Psi\rangle = \mathbf{E}|\Psi\rangle \tag{12.1}$$

where $\mathbf{E}$ is the diagonal matrix of the eigenvalues, $|\Psi\rangle$ is the vectorial representation of the N-electron wave function and $\hat{\mathbf{H}}$ is the Hamiltonian operator, defined by the kinetic energy operator, electron-nuclei attraction and Coulomb electron-electron repulsion operators. When a quantum system is in the state Ψ, which may or may not satisfy Equation (12.1), the average of many measurements of the total energy is given by the relationship:

$$E[\Psi] = \frac{\langle \Psi | \hat{H} | \Psi \rangle}{\langle \Psi | \Psi \rangle} \tag{12.2}$$

In the Hartree-Fock approach, the N-electron wave function is approximated by a Slater determinant of N single-electron orbitals, that is, depending on the single-electron position. The energy of the ground state is then found by minimizing the functional (12.2) subject to the orthonormalization conditions of the single-electron orbitals and in agreement with the variational principle (Szabo and Ostlund, 1994).

The foundation of DFT rests on two theorems, by Hohenberg and Kohn, that legitimize the use of electron density $\rho(\mathbf{r})$ as a basic variable in place of the N-electron wave function, Ψ. The first theorem states that the external potential $v(\mathbf{r})$ is determined, within a trivial additive constant, by the electron density $\rho(\mathbf{r})$. In the second theorem, the energy variational principle is provided: for a trial density $\tilde{\rho}(\mathbf{r})$, such that $\tilde{\rho}(\mathbf{r}) \geq 0$ and $\int \tilde{\rho} d\mathbf{r} = N$, $E_0 \leq E_v[\tilde{\rho}]$. The $E_v[\rho(\mathbf{r})]$ is the total DFT energy functional, expressed as the sum of the kinetic energy, the external potential $v(\mathbf{r})$ and the electron-electron potential energy:

$$E_v[\rho] = T[\rho] + V_{ne}[\rho] + V_{ee}[\rho]$$

$$= F_{HK}[\rho] + \int \rho(\mathbf{r}) v(\mathbf{r}) d\mathbf{r} \tag{12.3}$$

$$F_{HK}[\rho] = T[\rho] + V_{ee}[\rho] \tag{12.4}$$

F_{HK} is called the universal functional and it depends only on the total number of electrons by means of the electron density $\rho(\mathbf{r})$. The functional $V_{ee}[\rho]$, for the electron-electron potential energy, is written as the sum of the classical electron repulsion energy, $J[\rho]$, and a non-classical term:

$$V_{ee}[\rho] = J[\rho] + \text{non-classical term} \tag{12.5}$$

It is worth noting that the external potential is defined by the electron-nuclei interactions as well as by other external electrostatic potentials. In 1965, Kohn and Sham proposed an ingenious, indirect approach to the kinetic functional, $T[\rho]$, by setting the problem in such a way as to consider an exact and known kinetic energy functional, $T_s[\rho]$, namely that of an isoelectronic non-interacting quantum system. To obtain the desired separation from $T_s[\rho]$, the universal functional, depending only on the number of electrons, was rewritten as:

$$F_{HK}[\rho] = T_s[\rho] + J[\rho] + E_{xc}[\rho] \tag{12.6}$$

with

$$E_{xc}[\rho] = T[\rho] - T_s[\rho] + V_{ee}[\rho] - J[\rho] \tag{12.7}$$

The functional $E_{xc}[\rho]$ is called the exchange-correlation energy and it includes the difference between the kinetic energy of the real isoelectronic interacting system and the kinetic energy of the analogous non-interacting system, as well as the non-classical contribution to the $V_{ee}[\rho]$ electron-electron interactions. Assuming the differentiability of the total energy functional, the variational principle (the second Hohenberg-Kohn theorem) requires that the ground-state density satisfy the stationary principle:

$$\delta \left\{ E_v[\rho] - \mu \left[\left(\int \rho(\mathbf{r}) d\mathbf{r} - N \right) \right] \right\} = 0 \tag{12.8}$$

giving the Euler-Lagrange equation:

$$\mu = \frac{\delta E_v[\rho]}{\delta\rho(\mathbf{r})} = v(\mathbf{r}) + \frac{\delta F_{HK}[\rho]}{\delta\rho(\mathbf{r})} \tag{12.9}$$

where μ is the chemical potential of the quantum system, dependent on $\rho(\mathbf{r})$. It should be noted that the universal functional, F_{HK}, depends on the total number of electrons, while $E_v[\rho]$ depends on N and on the external potential $v(\mathbf{r})$; thus the quantum system is completely described by N and $v(\mathbf{r})$ and solving the Euler-Lagrange equation (12.9) above (Parr and Yang, 1989).

A useful procedure to obtain the $\rho(\mathbf{r})$ of the ground state was, nevertheless, given by Kohn and Sham that did not directly resolve the Euler-Lagrange equation. Specifically, atomic orbitals, $\phi_j(\mathbf{r})$, were introduced into the problem and the electron density was then defined through a linear combination of these atomic orbitals, namely the basis set (Parr and Yang, 1989):

$$\rho(\mathbf{r}) = \sum_i^N |\psi_i(\mathbf{r})|^2 \tag{12.10}$$

with:

$$\psi_i(\mathbf{r}) = \sum_j^K c_j^i \phi_j(\mathbf{r}), \quad i = 1, \dots, N \tag{12.11}$$

Again, the variational principle ensures that the minimum of the functional $E_v[\rho(\phi_j(\mathbf{r}))]$ is the best approximation of the ground state energy, but this time in the space of the atomic orbital basis set, $\phi_j(\mathbf{r})$. The search for this minimum leads us to solve the non-linear Kohn-Sham orbital equations (Parr and Yang, 1989):

$$\left[-\frac{1}{2}\nabla^2 + v_{\text{eff}}(\mathbf{r}) \right] \psi_i(\mathbf{r}) = \varepsilon_i \psi_i(\mathbf{r})$$

$$v_{\text{eff}}(\mathbf{r}) = v(\mathbf{r}) + \frac{\delta J[\rho]}{\delta\rho(\mathbf{r})} + \frac{\delta E_{xc}[\rho]}{\delta\rho(\mathbf{r})}$$

$$= v(\mathbf{r}) + \int \frac{\rho(\mathbf{r}')}{|\mathbf{r} - \mathbf{r}'|} d\mathbf{r}' + v_{xc}(\mathbf{r}) \tag{12.12}$$

where $v_{\text{eff}}(\mathbf{r})$ and $v_{xc}(\mathbf{r})$ are the effective and the exchange-correlation potentials, respectively.

Each approximation of the $F_{HK}[\rho]$ functional is expressed as a function of the exchange-correlation functional that is the integral of $v_{xc}(\mathbf{r})$. The form of the exchange-correlation functional, $E_{xc}[\rho]$, continues to be upgraded, although several of its approximations have been published (Beck, 1993; Perdew et al., 1996; Xu et al., 2005). In summary, the accuracy and the computational time of a DFT calculation is affected by the choice of functional $E_{xc}[\rho]$ and the choice of an appropriate orbital basis set (Xu et al., 2005). In practical terms, a large basis set is necessary to obtain high accuracy and convergence of the Kohn-Sham equations above; however, a large basis set also requires a lengthy calculation.

Other key points that should be taken into account in a DFT calculation are as follows. In Equation (12.10), the external potential $v(\mathbf{r})$ depends on the geometrical arrangement of the nuclei, considered as fixed in the Born-Oppenheimer approximation. Thus, a specific arrangement of the nuclei defines a specific $v(\mathbf{r})$ potential. Nevertheless, a particular arrangement of nuclei does not necessarily provide the lowest total energy of the molecular system. Thus, an optimization of the molecular geometry is always necessary to find the arrangement of nuclei that corresponds to the lowest total energy, that is, a minimum in the potential energy surface. Furthermore, in addition to a global minimum, molecular structures may have several local minima (especially for flexible molecules), corresponding to the various conformational isomers. The research of these conformational isomers through the use of pure QC approaches requires high computational time, so that, in this case, other computational approaches, coupled with QC methods, should be used. The research of local and global minima for flexible nanostructures is a necessary first step in evaluating the physico-chemical properties of interest. The external potential, $v(r)$, can also be substituted by an effective potential, replacing the core electrons of the atoms with an effective electrostatic potential. In this case, only the electrons of the external orbitals are explicitly considered in the quantum calculation (Pacios and Christiansen, 1985). This procedure decreases the computational time and, at the same time, allows relativistic effects to be taken into account, which are important when transition metals are present in the nanostructures.

Ultimately, all of these points should be taken into account during a DFT calculation to optimize accuracy and computational effort (De Luca, 2013). There follows a brief description of the bespoke algorithms used to describe and, hence, to predict some key NT properties.

12.2.2 *Bespoke topological algorithms*

In order to model property-structure relationships, some bespoke algorithms have been implemented, for example, to build zigzag CNTs with the C–C distance and phenyl angle as a function of the nanotube diameters (Gulseren *et al.*, 2002; Saito *et al.*, 1998). Moreover, other algorithms have been implemented in order to evaluate the sizes of the principal molecules present in olive oil, and cosmetic, pharmaceutical and textile wastewaters. In addition, an original in-house algorithm was developed to model molecular rejection by CNTs with functionalized inlets.

12.2.2.1 *Molecular topological analysis algorithm*

Organic compound sizes have been widely discussed in the literature because they are fundamental descriptors that control molecular retention by nanoporous membranes (Kiso *et al.*, 1992; Ozaki and Li, 2002; Van der Bruggen *et al.*, 1999).

The most easily accessible parameter correlated to the size of a solute is the molecular weight. However, it may mislead in terms of molecular rejection. In fact, although empirical correlations between weight and size are available for some compounds, they do not have general validity. Thus, in various studies, molecular sizes such as molecular width, Stokes radius and molecular mean size have been shown to be better descriptors of steric hindrance than molecular weight (Kiso *et al.*, 1992; Ozaki and Li, 2002; Van der Bruggen *et al.*, 1998). Stokes radius is a commonly used descriptor for the assessment of molecular steric hindrance. However, the diffusivities necessary to calculate the Stokes radius cannot be obtained for many organic solutes or for the same experimental conditions (Kiso *et al.*, 1992). In addition, the Stokes radius is based on the assumption that solutes are spherical in shape, rigid and not hydrated, which is not always correct. Since the Stokes radius can be difficult to measure for some molecules, other molecular sizes have been developed to correctly take into account the steric hindrance and, in turn, solute rejection by a size exclusion mechanism. (Agenson *et al.*, 2003; Gulseren *et al.*, 2002; Santos *et al.*, 2006; Yangali-Quintanilla *et al.*, 2010).

When the pore size is comparable to the molecular dimension and the steric hindrance is the driving mechanism for the rejection, then the morphology of solutes is especially important. In this case, the aspect ratios of the penetrants become important descriptors. Thus, the proposed algorithm (*molecular topological analysis*) allows evaluation of the effective diameters and maximum and minimum molecular cross sections without resorting to empirical parameters, but using the geometries of solutes obtained from *ab initio* optimizations. It is worth noting that the *ab initio* calculations require only a knowledge of the chemical formula of the compounds, and they correctly take into account the non-covalent intramolecular and intermolecular interactions that often modify nanostructure geometries and hence their molecular sizes. As a result, the proposed procedure is quite general and thus has been applied to estimate the principal sizes of hydrated solutes and ions in solutions.

In detail, the algorithm or procedure is developed as follows. First, a hard sphere, having radius equal to the corresponding van der Waals radius, represents each atom of the molecule

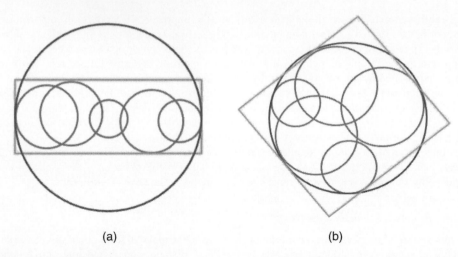

(a)   (b)

Figure 12.1. Smallest enclosing rectangle (green) and smallest enclosing circle (blue) of the atom projections in two limit cases: (a) flat linear molecule; (b) non-flat molecule.

or nanostructure that is reliably described by QC calculations. Then, three main dimensions are evaluated: the height, H, d_{min} and d_{max}. The height is defined as the distance between the two farthest atoms. A 3D, Cartesian coordinate system is fixed in such a way that the origin of the axes matches the geometric center of the molecule, while the z-axis is parallel to the straight line connecting the two most distant atoms of the target structure. Each atom of the molecule is now projected onto a plane perpendicular to the z-axis, and the optimal geometric form, including all the atom projections, having minimum area is then researched. This form yields the molecular minimum cross section (MCS). Braeken *et al.* (2005) determined this as a circle of minimum area around the projected molecule, that is, they considered a circle as the optimal geometric shape. Referred to as the 'smallest enclosing circle problem' in computational geometry, its solution is not trivial. In fact, most algorithms (Chrystal, 1885; Eliosoff and Unger, 1998; Hearn and Vijan, 1982) consider only the special case of enclosing points, namely, the case in which the radii are equal to zero, although several approaches are also proposed for the general cases in which at least one circle has non-zero radius (Xu *et al.*, 2001). Nevertheless, a circle including all of the atom projections could overestimate the MCS of the target solutes, as shown in Figure 12.1. In this figure, the smallest enclosing circle (colored blue) and the smallest enclosing rectangle (colored green) around the atom projections have been delineated. Figure 12.1a shows that the circle overestimates the MCS in the case of flat linear molecules, whereas, for a non-flat molecule (Fig. 12.1b), the minimum circle only slightly overestimates the projected atoms. However, this drawback has been considered in the proposed algorithm, and the minimum rectangle can be exploited instead of the enclosing circle, minimizing even this slight overestimation.

In more detail, the smallest enclosing rectangle is adopted in the implementation of the bespoke algorithm. In order to find the smallest enclosing rectangle, the rectangle of minimum area having sides parallel to the x- and y-axes is first computed, then a counterclockwise rotation of the Cartesian x,y-plane is carried out and another rectangle of minimum area is computed. This procedure is sequentially applied for different angles in such a way that the whole interval $[0, \pi/2]$ is scanned. Figure 12.2 illustrates the computational procedure applied. Among all the rectangles, the one of minimum area is finally found and it provides the molecular MCS, while the sides of this smallest enclosing rectangle are denoted by d_{min} and d_{max}. The overestimation of non-flat molecule cross sections is strongly minimized by this approach, and the new algorithm becomes more generalized than the published one since it allows the description of nanostructures with different forms. All the details of this novel code can be found in Bisignano (2014).

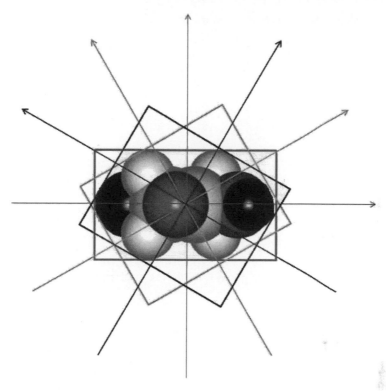

Figure 12.2.   Initial rectangle of atom projections (in red), applied $\pi/6$ rotations (in green) and $\pi/3$ rotations (in blue).

The principal sizes d_{min} and d_{max} can be used to calculate the effective diameter of the molecule. If the z-axis of the molecule is supposed to make an angle α with the membrane surface, the new projection H$'$ can be evaluated as:

$$H' = H\cos(\alpha) + d_{max}\sin(\alpha) \qquad (12.13)$$

The effective diameter, d_{eff}, can then be obtained by assuming that the probability distribution of α is equal to $\cos(\alpha)$. As a result, the effective diameter is obtained by integration of Equation (12.11):

$$d_{eff} = \frac{\pi}{4}H + \frac{1}{2}d_{max} \qquad (12.14)$$

The d_{eff} descriptor is an average size, taking into account the different orientations of a molecule when approaching a membrane surface (Braeken *et al.*, 2005). In summary, the proposed algorithm gives different topological descriptors, such as H, d_{min} and d_{max} (smallest enclosing rectangle) as well as d_{eff}, without using parameters that require the fitting of experimental data.

This algorithm was then used to calculate the main sizes of several molecules (Buekenhoudt *et al.*, 2013) in order to take into account the non-spherical character of some solvents and overcome the lack, in the literature, of their kinetic diameters (very important sizes for solvent diffusion in inorganic nanoporous membranes). The computed d_{eff} and d_{min} of some solvents, together with their experimental kinetic diameters, d_{kin} (Bowen *et al.*, 2003; Iliyas *et al.*, 2007; Shao and Huang, 2007), are listed in Table 12.1.

Interestingly, the d_{eff} calculated with the proposed algorithm is consistently greater than the experimentally determined d_{kin}; conversely, the d_{min} is found to be strikingly similar to this

Table 12.1.   Computed main molecular sizes and experimental d_{kin} for some important solvents.

Molecules	d_{eff} [nm]	d_{min} [nm]	d_{kin} [nm]
Water	0.47	0.28	0.27
Methanol	0.62	0.42	0.38
Ethanol	0.74	0.42	0.44
Dichloromethane	0.73	0.42	0.49
Dimethylformamide	0.84	0.43	0.50
Toluene	0.98	0.43	0.55
Hexane	1.08	0.42	0.51

experimental measurement. It should be stressed that d_{kin} is widely used to investigate molecular rejection of solutes and gas diffusion, and is considered a key molecular property in the field of matter transport. Therefore, the *molecular topological analysis* algorithm developed here allows d_{kin} to be obtained in a homogeneous way for systems not listed in the literature and without resort to empirical or fitting parameters, but just by using the chemical formula of the target compounds.

12.2.2.2   *Pore molecular sieving algorithm*

The second algorithm, *pore molecular sieving*, was used in order to evaluate the rejection of functionalized nanopores. This algorithm was developed in a general way and it is applicable to any nanopore with a defined profile. In this case, CNTs were used to illustrate the computational method because they are interesting nanostructures and, in addition, can be regarded as ideal nanosized pores. The main elements of the *pore molecular sieving* algorithm are elucidated below.

A 3D, Cartesian coordinate system was fixed in such a way that the origin of the Cartesian axes matched the center of the CNT inlet, while the z-axis is along the CNT axis. A points grid is built on the CNT inlet: thus, the i-th point is located on the plane $z = 0$. The CNT radius and center angle, 2π, are divided into n and m integer numbers, which define the grid spacing. The van der Waals radius of carbon atoms was considered to account for the edge of the CNT. The atoms of the grafted functional groups (FGs) were treated as a set of rigid spheres always having as radius the van der Waals radius. These FGs, whose geometries were optimized at quantum level, cover part of the CNT inlet. In order to evaluate the covered portion, the projections of the FG atoms were considered on the $z = 0$ plane, modeled as disks with radius equal to their van der Waals radius. Finally, the CNT inlet area not covered by the FG atom projections defined the maximum free area, $A_{CNT,free}$, of the CNT. To calculate $A_{CNT,free}$, the following problem has to be solved: consider a circular grid of points and let D_{disk} be the diameter of a disk belonging to the set of disks whose intersections are not necessarily empty, then find the grid portion covered by the disks and evaluate its area. The solution of this problem gives the free grid points (those not covered) and those covered (labeled with '1' in Figure 12.3) by the projections of the FG atoms. Figure 12.3 shows an example of some disks covering a portion of the grid points. In order to evaluate the total area of the inlet's covered portion, the CNT inlet can be decomposed into circular (C_s) and annular (A_s) sectors, as shown in Figure 12.4. In particular, a C_s or A_s sector was considered to be 'covered' when all of its vertices were labeled with '1'. In this way, the total portion of the area covered by the grafted FGs can be calculated and the $A_{CNT,free}$ evaluated in terms of the difference from the well-defined area of the non-functionalized CNT.

It is important to recognize that the $A_{CNT,free}$ is not the area of the CNT inlet for the permeation of solutes. Only the steric hindrance of FGs in the molecule permeation path was considered in the $A_{CNT,free}$ calculation. In fact, a solute diffusing through a generic free grid point ($A_{CNT,free}$) could touch the FGs border and would be rejected. For this reason, the effective area available to solute permeation, $A_{mol,perm}$, is smaller than $A_{CNT,free}$. Thus, the rejection of solute by functionalized

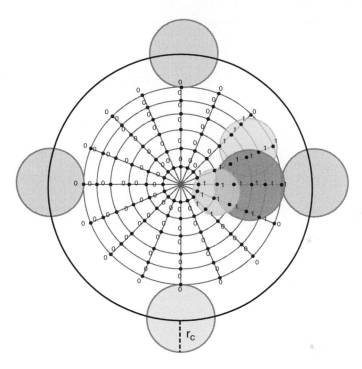

Figure 12.3.   Grid points labeled as free (0) or covered (1) by projections of functional group atoms.

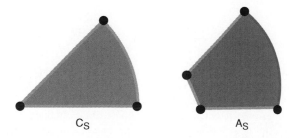

Figure 12.4.   Circular (C_s) and annular (A_s) sectors into which the CNT inlet is divided.

CNTs through molecular sieving was then investigated as a geometrical problem, taking into account the aspect ratios of penetrants. This molecular permeation area, $A_{mol,perm}$, is illustrated below for a target solute.

$A_{mol,perm}$ is the area through which a molecule can permeate without any steric hindrance by the CNT edge and/or anchored FGs. The $A_{mol,perm}$ depends on the $A_{CNT,free}$ and on the MCS of the permeant (as described above in Section 12.2.2.1). A simple example illustrates this dependence: two molecules with different MCSs will permeate in a different way through the same $A_{CNT,free}$ characterized by the same FGs, that is, the molecule with the smaller MCS will permeate more easily than the molecule with the larger MCS. In order to evaluate $A_{mol,perm}$, the area of a complex object lying in a known area, as shown in Figure 12.5, has to be calculated.

$A_{mol,perm}$ was evaluated by following a Monte Carlo approach (Metropolis and Ulam, 1949; Rosenbluth *et al.*, 1953), generally used in estimating the area of irregular objects. In Figure 12.5b, the top view of a functionalized CNT is shown in which the area covered by FGs is indicated in black and the $A_{CNT,free}$ and $A_{mol,perm}$ are shown in green and blue, respectively. Defining the ensemble of points belonging to the free portion and to the area available for solute permeation

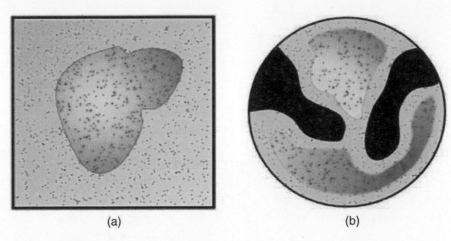

(a)　　　　　　　　　　　　　　　(b)

Figure 12.5.　Irregular 2D object areas (in blue): (a) general case; (b) region evaluated using a solute MCS. $A_{CNT,free}$ and FG-covered areas shown in green and black, respectively.

with $\sigma_{mol,perm}$ and $\sigma_{CNT,free}$, respectively, the ratio between $A_{mol,perm}$ and $A_{CNT,free}$ is equal to the ratio of $\sigma_{mol,perm}$ and $\sigma_{CNT,free}$ for assemblies composed of many points:

$$\frac{A_{mol,perm}}{A_{CNT,free}} = \frac{|\sigma_{mol,perm}|}{|\sigma_{CNT,free}|} \tag{12.15}$$

Thus, $A_{mol,perm}$ is evaluated as:

$$A_{mol,perm} = A_{CNT,free} \cdot \frac{|\sigma_{mol,perm}|}{|\sigma_{CNT,free}|} \tag{12.16}$$

The details by which the $\sigma_{mol,perm}$ points are calculated are reported in Bisignano (2014). The points belonging to $\sigma_{CNT,free}$ are the free (not covered) grid points that describe the $A_{CNT,free}$. As reported in several steric-hindrance pore models (Nakao and Kimura, 1982; Van der Bruggen et al., 2000), the sieving coefficient is always evaluated as a function of the solute and pore membrane radius: if the solute size is larger than the pore size, the solute cannot partition into the membrane, resulting in 100% rejection. Thus, inspired by this concept, the rejection of the target compounds is assumed in the *pore molecular sieving* algorithm to be proportional to $1 - \sum$, that is, $R \propto 1 - \sum$, with:

$$\sum = \frac{A_{mol,perm}}{A_{CNT,free}} \tag{12.17}$$

Without FGs on the border of the nanotube, $A_{CNT,free}$ is simply the inlet area excluding the thickness of the carbon, hydrogen and hydroxyl atoms. When FGs are present, however, then the $A_{CNT,free}$ is evaluated as the sum of the areas of the disjointed circular and annular grid sectors that are free of the projected FG atoms. Thus, the term $A_{CNT,free}$ is associated with the actual functionalized pore shape (Fig. 12.5b), obtained without resorting to approximate relationships through the fitting of experimental data (Van der Bruggen et al., 2000).

The $\sigma_{mol,perm}$ points, characterizing $A_{mol,perm}$, are a subset of $\sigma_{CNT,free}$ and depend on the shape and size of the rejected solute, which are evaluated at quantum mechanics level. Specifically, the $\sigma_{mol,perm}$ points are calculated by taking into consideration the free points in the inlet through which a target molecule can permeate with no steric hindrance due to either the CNT edge or the anchored FGs. It is to be expected that the effective area available to solute permeation, $A_{mol,perm}$, becomes smaller as the compound examined gets larger. Hence, the main role in $A_{mol,perm}$ is played by the dimensions of the penetrant, so that, as already noted, an accurate morphological analysis of solute geometries was carried out and, in particular, the molecular MCS of each target molecule was calculated to obtain $A_{mol,perm}$ and, in turn, Σ.

Thus, the proposed *pore molecular sieving* algorithm is based on the size of the solute, defined by its MCS (as derived using the *molecular topological analysis* algorithm), and on the steric hindrance of the FGs. The algorithm depends upon three assumptions:

- The molecule can be represented by its MCS and the CNT inlet by a circle of fixed diameter. The algorithm works well even if the pore border is not circular, although greater computational difficulties arise in this case. The MCS was chosen rather than the steric hindrance of a molecule because the latter may be minimal if the molecule is placed with its MCS parallel to the CNT inlet (Xiao and Wei, 1992). The validity of this assumption will be explained in the next section.
- The inlet area of the CNT, if covered by FGs, is not accessible to the molecule.
- The molecule is rejected if, when placed on a grid point of the CNT free area with its MCS parallel to the CNT inlet, it touches any covered grid point or the edge of the CNT. In this way, the points of the $\sigma_{mol,perm}$ set are assessed and, in turn, $A_{mol,perm}$ by means of Equation (12.16).

When the nanopore inlet is decorated with chemical groups, it can be challenging to obtain an unambiguous pore size value without resorting to simplifying assumptions and fitting procedures that risk losing the physical sense. To address this limitation, the proposed algorithm allows evaluation of CNT rejection without the introduction of any adjustable parameters. The algorithm is based on an accurate evaluation of the CNT inlet area not covered by FGs through which solutes can permeate without any steric hindrance. Thus, the selectivity of the CNTs is assessed using a procedure based only on a morphological analysis of the permeant and the functionalized pore entrance.

12.3   FUNCTIONALIZED CARBON NANOTUBES FOR WASTE WATER TREATMENT

The two new algorithms described in Section 12.2.2 have been applied to the study of the selectivity of non-functionalized and functionalized CNTs by exploiting the solute geometries and the solute-CNT interaction evaluated in the framework of DFT. Thus, this section is divided into two parts. First, solute rejection by non-functionalized CNTs is investigated, after which rejection of a sample range of molecules by functionalized NTs is illustrated. Some case studies of the procedure incorporating the new algorithms are highlighted, with the results and details of some findings shown from a recently published work (Bisignano *et al.*, 2015).

12.3.1   *Rejection in carbon nanotubes*

Rejection by non-functionalized CNTs of some uncharged organic solutes with low molecular weight, originating from industrial, pharmaceutical and olive mill wastewater, was investigated. In order to predict the optimal internal diameter of NTs able to reject the target compounds, a general criterion, completely free from empirical parameters, has been proposed. This criterion uses the molecular MCS, introduced in Section 12.2.2.1, and the optimized solute geometries evaluated at quantum mechanics level. In the design of innovative membranes for water treatment, the rejection of organic molecules with very low molecular weight is considered an aspect worthy of much consideration. CNTs can be considered as innovative materials because they combine remarkable hydrodynamic properties with the ability to reject very low molecular weight solutes thanks to their tunable internal nanometric diameters. A general criterion capable of predicting the *optimum internal diameter* of single- or multi-walled NTs (SW-NTs, MW-NTs) required to deliver a complete rejection of target solutes by a size exclusion mechanism is, therefore, desirable. In fact, such a criterion allows optimization of the preparation of novel composite CNT membranes and also provides information about the molecular separation mechanism.

Table 12.2.   Main molecular sizes of some target solutes, computed *ab initio*.

Compounds	H [Å]	d_{min} [Å]	d_{max} [Å]
Diclofenac	12.4	7.0	9.7
Triclosan	13.9	6.5	7.5
Vanillic acid	10.5	4.3	8.6
Tyrosol	10.4	5.5	6.7
p-Coumaric acid	12.1	3.5	7.3

Table 12.3.   MCD and correlated maximum internal diameters of armchair (columns 3 and 4) and zigzag (columns 5 and 6) SW-NTs.

Compounds	MCD [Å]	SW-NT (n,n)	d_{CNT} [Å]	SW-NT (n,0)	d_{CNT} [Å]
Diclofenac	14.0	(10,10)	13.6	(17,0)	13.3
Triclosan	12.5	(9,9)	12.2	(16,0)	12.5
Vanillic acid	12.4	(9.9)	12.2	(15,0)	11.7
Tyrosol	10.8	(8,8)	10.8	(13,0)	10.2
p-Coumaric acid	10.9	(8.8)	10.8	(13,0)	10.2

Near the pore inlet, steric hindrance is minimal if the molecule is oriented in such a way that its MCS is parallel to the CNT inlet. Based on this assumption, a geometric criterion, named maximum circumference diameter (MCD), was outlined for prediction of the optimal diameter of nanotubes required to obtain solute rejection. To put it another way, by using an internal diameter *greater* than the size predicted by the MCD criterion the target molecule will *not* be rejected. A circle that circumscribes the MCS was defined and it was shrunk until the first of the atoms projected onto the plane defining the MCS emerged from the circle. The diameter of this latter circumference defines the MCD. It is worth noting that by using this proposed criterion for a set of compounds having considerably different MCSs, it became easy to choose the diameters of the nanotubes needed to achieve rejection of the different solutes by composite CNT membranes within cascaded cut-off.

Diclofenac, triclosan, tyrosol, vanillic acid and *p*-coumaric acid were selected as model solutes. The main sizes of the target solutes, that is, H, d_{min} and d_{max}, shown in Table 12.2, were calculated along with the corresponding MCS and MCD. Molecular mechanics calculations were also carried out with the aim of analyzing the conformer space of the target solutes.

Universal force field and conjugated gradient algorithms were used for the geometry optimizations, with a convergence criterion of 10^{-8}. The conformational analysis was performed using the weighted rotor search method with the number of conformers equal to 100 (Hanwell *et al.*, 2012). This method was applied five times in a row in order to create a sampling of conformers for each solute. Based on this sampling, the energy of each conformer was calculated at quantum mechanics level. A Boltzmann distribution of the quantum energies was then analyzed and, for each solute, the conformer with the highest Boltzmann factor was favored over the others. Finally, a full geometry optimization of the most stable conformers was carried out in the context of a DFT-based method. The hybrid functional X3LYP with the 6-311G* atomic orbital basis set were used in these calculations (Valiev *et al.*, 2010). The use of this level of theory (i.e. functional and basis set) was justified by the fact that it gave good non-covalent interaction energies, at least at a qualitative level, as reported in Bisignano *et al.* (2015). Each geometry was optimized by using analytical energy gradients and a quasi-Newton optimization with approximate energy Hessian updates. Finally, the MCD for each solute was calculated, as shown in Table 12.3.

From the MCD criterion, the maximum diameters of armchair and zigzag SW-NTs able to reject the target solutes were then predicted. The diameter obtained from the MCD cannot be directly used to predict the internal size of the CNT as the NT symmetry must be taken into account. In fact, as shown in Table 12.3, the diameters for rejection of a given target molecule are different for armchair and zigzag SW-NTs. By analyzing the values reported in Table 12.3, it appears that a complete rejection of all the solutes in the sample occurs by using an (8,8) armchair SW-NT and a (13,0) zigzag SW-NT. The same computational analysis has also been carried out for charged solutes, such as dyes commonly used in textiles. It is important to emphasize that in this case, the arrangement of the water molecules in the first shell around the solutes, as well as the hydration energies, have been taken into account.

A validation of *ab initio* predictions has been carried out and the results are being published. Results from molecular dynamics (MD) simulations on tyrosol molecules and other similar solutes have been used here. The most interesting result of this validation concerns the distortions of the phenyl ring of tyrosol when it enters an (8,8) SW-NT. MD simulations have shown that tyrosol can diffuse in CNTs with diameters of 10.8 Å, in contrast to the prediction of the MCD criterion that instead suggests a rejection of this molecule, as indicated by the relevant value in the first column of Table 12.3. Nevertheless, in comparing the geometries of trapped tyrosol molecules with their structure when optimized in water, a strong distortion of the phenyl rings was found. In addition, a significant energy per molecule is necessary to relax the trapped geometries into the optimized structure. Hence, the diffusion of tyrosol in the (8,8) SW-NTs would be possible, according to MD simulations, but only on the basis of a distortion of the phenyl rings that is not very likely, and thus the MCD criterion yields the more correct prediction.

12.3.2 *Selectivity of functionalized carbon nanotubes with PIM-1 monomers*

Mixed-matrix membranes (MMMs), wherein nanoparticles are added to polymer matrices, are receiving huge focus because the nanostructure features can confer an increased efficiency to the MMMs in terms of selectivity or permeability. Among numerous nanoparticles, CNTs have attracted the most interest due to a combination of low transport resistance and chemical inertness.

The first published MD simulations, concerning the water flow inside a SW-NT with a diameter of 1 nm or less, showed flow rates that were orders of magnitude higher than the flows predicted by the standard no-slip Hagen-Poiseuille equation for cylindrical channels with equal diameters (Hummer *et al.*, 2001). This *flow enhancement* was initially described in terms of ballistic transport (Striolo, 2006) and associated with the hydrophobicity of the NT's wall (Whitby and Quirke, 2007). Further modeling and experimental results have used the flow enhancement, defined as the ratio of the measured or modeled flow rate to the no-slip Hagen-Poiseuille one, to evaluate the effective permeability of the CNT. Results have varied greatly, with flow enhancement values of up to 10,000 in tubes with diameters ranging from less than one to tens of nanometers (Thomas and McGaughey, 2009). This variation can be attributed to the fact that CNTs are a family of materials, whose properties can vary significantly as a function of the synthesis process, geometrical characteristics (i.e. diameter and chirality), surface structure (from turbostratic to highly organized graphitic) and chemistry (from hydrophobic to hydrophilic) (Whitby and Quirke, 2007). Moreover, a recent analysis by the authors showed that the orders of magnitude higher water flow rates observed in single tubes (i.e. the flow enhancement) do not automatically translate into orders of magnitude higher membrane permeability but rather smaller, though still significant, increases (Mattia *et al.*, 2015). This difference can be attributed primarily to the geometrical and structural characteristics of the CNT membranes. For example, in MW-NTs, the flow enhancement is associated with the flow in the innermost NTs, whereas the other concentric tubes do not contribute to the increased flow even though they occupy membrane surface positions. Hence, when MW-NT membranes are being considered, the balance between the flow rates through the pristine and modified membranes should be understood.

Bisignano *et al.* (2015) have recently developed a smart procedure to predict the enhancement of water permeability as a function of the membrane surface fraction occupied by MW-NTs, taking into account the effects previously highlighted. This computational procedure predicts the internal and external diameters of MW-NTs in order to define the increase in permeability of the pristine membrane when modified with the point out tubes. In the proposed procedure, the CNT flow enhancement is considered in a general mode although literature data were used. The procedure is applied to MW-NT-polymer membranes in which the NTs are vertically aligned with no voids between the external tubes and the polymeric matrix; in addition, the non-occupied polymer still contributes to the water permeability. The experimentally measured permeabilities of two CNT thin-film composite (TFC) membranes were used as a reference: an asymmetric polyester membrane (Wu *et al.*, 2013) and a porous chitosan membrane (Tang *et al.*, 2009). The synthesized membranes contain randomly aligned MW-NTs embedded in the two polymers with different morphologies. The smart procedure showed that aligning the tubes in this kind of membrane yields up to three orders of magnitude increase in flux compared to the virgin polyester TFC membranes, whereas randomly aligned tubes only double the flux with respect to the pristine systems. The method also revealed that the tube diameters of both the external and innermost tubes of the MW-NTs have to be thoroughly considered since the water cannot flow through the tubes' interlayers. This aspect was even more evident in the high-flux chitosan membrane, where the use of thick MW-NTs induced a decrease in permeability with respect to the virgin membrane, and the use of thin double-walled NTs generated only a modest permeability increase. By using MW-NTs with external and internal diameters equal to 35 nm and 4.44 nm, respectively, the water permeability increased 150-fold compared to the permeability of the unmodified polyester TFC membrane, with 50% of the surface occupied by nanotubes. However, by using external diameters equal to 15 nm and an equal internal size, the permeability increased 700-fold when compared to the pristine membrane. Thus, these simulations reveal the dramatic effect of the MW-NT's external diameter on the water permeability in CNT-TFC membranes.

In membrane-based wastewater treatment, permeability and selectivity are important performance parameters and, in general, an increase in the former comes at the expense of the latter. The simultaneous optimization of these features is not easy and is time-consuming. The *pore molecular sieving* algorithm described above can be exploited to achieve such an objective. This algorithm, in conjunction with the smart procedure just outlined, provides a computational framework for optimizing the permeability and selectivity of MW-NT membranes. According to the results soon as underlined, the functionalization of a nanotube with diameter 4.44 nm was carried out to maximize its selectivity while retaining higher permeability. Even though a large diameter was used, it was nonetheless possible to obtain a high rejection of small solutes from tip functionalization.

Monomers of polymers of intrinsic microporosity (PIMs) (McKeown and Budd, 2006) have been considered as FGs because of their controlled rigid structures. A detailed description of the FGs based on PIM-1 is provided in Bisignano *et al.* (2015). Equation (12.15) was used to evaluate the steric hindrance of PIM-1 monomers anchored on the selected CNT inlet. The geometries of FGs and solutes were optimized in a DFT framework using the exchange-correlation X3LYP (Xu *et al.*, 2005) and large basis sets (6-311++). Due to the potentially large size of the FG models, the optimization of their structures was carried out with relatively small basis sets (3-21G). For the CNT model, whose Cartesian coordinates were frozen during the optimization, a minimal basis set was also used. It is important to emphasize that the inclination of PIM-1 monomers on the CNT inlet is a key aspect, as it determines the steric hindrance of the functional chemical group. This is well illustrated by considering the black area in Figure 12.5b, which represents the FG projection on the CNT inlet. These projections affect both the $A_{CNT,free}$ and $A_{mol,perm}$ and hence the solute rejection. Based on the FG arrangements and the corresponding $1 - \sum$ values, a large number of chemical groups, such as oligo ethylene glycol and silanes (Chung *et al.*, 2007; Jirage *et al.*, 1997; Lee *et al.*, 2002), were screened and discarded as unsuitable. FGs based on PIM-1 monomers were chosen instead on the basis of providing the best combination of chain rigidity and coverage of the tube's inlet. The quantum optimization of their geometries provides equilibrium structures. As a result, the FG stiffness can be estimated by calculating the energy required to distort the FG from its equilibrium geometry. This analysis has shown that the anchored PIM-1 monomers provide good chain rigidity. Conversely, oligo ethylene glycol and silane groups have aliphatic chains with several rotation angles and thus, due to the existence of several rotamers, a conformational analysis was required in these cases. The multiple degrees of freedom in oligo ethylene glycol and silanes imply that the winding of these groups does not provide a sufficient steric hindrance in nanotubes of internal diameter 4.44 nm. It is important to emphasize that the assessment of the steric hindrance of the groups under test was carried out by comparing the $1 - \sum$ values of the non-functionalized and functionalized nanotubes.

Two small solutes, *rac*-fluoxetine and glucose, plus ethanol, whose separation from water also has industrial relevance, were chosen for testing the selectivity of the investigated functionalization models. Nanotubes with diameter 4.44 nm, selected to maximize permeability, have a somewhat detrimental effect on the selectivity for small molecules. Nonetheless, significant increases in rejection were obtained for functionalized tubes compared to non-functionalized ones where no rejection was observed. In particular, an increase in rejection by factors of 2.5 and 2 for *rac*-fluoxetine and glucose, respectively, compared to water was found when two or three FGs, formed by one PIM-1 monomer plus a molecular hook, were used. A smaller increase for the same FGs steric hindrance was found for ethanol rejection compared to water, due to their more comparable MCSs. Tetrafluoro-phthalonitrile was used to anchor these monomers on the CNT boundary. The choice of a particular structure for attaching the monomers is important because it affects both the inclination and the stiffness of the functional group. The $1 - \sum$ values provide accurate information concerning the structure-property relationship, in this case the functionalization model rejection. The proposed computational approach allows control of solute rejection by defining the number of PIM-1 monomers per FG and, in turn, the number of FGs anchored on the CNT inlet.

In order to optimize the permeability and selectivity of CNT membranes, both solute rejection and solvent (water) permeation should be maximized. This aspect (i.e. the water entrance effect) can be undertaken (modeled) at molecular scale by exploiting Equation (12.15). For example, while three FGs, each formed by two PIM-1 monomers and anchored using a 1,2-bis (2-bromoethyl)-4,5-difluorobenzene hook, yield the highest rejection for *rac*-fluoxetine, the best *rac*-fluoxetine-to-water rejection ratio is given instead by a configuration of three FGs formed by just one PIM-1 monomer. Similar conclusions can be drawn for glucose and ethanol.

Although commercial membranes with better solute-to-water rejection ratios exist, the principal aim of this work lies in the presentation of a novel methodology for optimizing selectivity and permeability of CNT-polymer membranes without recourse to any fitting. The novel method described here can be used to estimate, through a comparative but quantitative and rigorous analysis, the steric effect of different functional groups used for decorating nanopores with rigid or well-defined shapes. The combination of a large tube diameter and an optimal tip functionalization could yield higher water flow rates and solute rejection for a MW-NT-polymer membrane. Moreover, in recent years, many other PIM monomers have been synthesized with components that are more rigid than PIM-1 and, thus, the proposed procedure, based on Equation (12.15), allows prediction of the rejection capability of these novel PIM functional groups too.

12.4   CONCLUSION

In wastewater treatment, the rejection of lightweight solutes is an ongoing challenge. Membrane processes are considered to be a viable option for addressing this issue. Solute separation is intrinsically coupled with water permeability, because high rejection is often a cause of membrane fouling (e.g. pore blocking, etc.), which in turn dramatically reduces permeability. Thus, unfortunately, membrane fouling is inherently connected to solute rejection. As a result, the choice of innovative materials, such as nanostructures, should attend closely to these aspects, as well as to processability and cost of materials. Computational modeling can greatly accelerate this choice. In particular, modeling from first principles is very useful for the design of new nanostructures as it allows accurate assessment of key features without the use of empirical or fitting parameters.

In this chapter, *ab initio* modeling of the rejection of low molecular weight organic solutes, originating from cosmetic, textile and olive oil wastewaters, by MW-NTs was presented. A bespoke topological algorithm for optimizing the selectivity of functionalized CNTs was proposed, with the aim of achieving high molecular rejection without a marked fall in water permeability. This algorithm, free from adjustable parameters, allows prediction of the steric hindrance due to functional groups anchored on the CNT inlet. Although it has been used for studying the selectivity of functionalized CNTs, the algorithm can also be applied to predict the selectivity of any rigidly functionalized inlet pore.

First, rejection by non-functionalized NTs and the water permeability of the CNT composite membranes were analyzed in order to predict the permeability enhancement as a function of the inlet diameters and density of vertically aligned nanotubes. The permeability optimization showed that aligning the tubes could yield up to three orders of magnitude increase in flux compared to a virgin TFC membrane. Moreover, it was also shown that both the external and innermost tubes' diameters have to be considered to maximize permeability. Finally, effective functionalization models were investigated in order to maximize the organic solute rejection while retaining high permeability. MW-NTs with specific internal and external diameters matching commercially available tubes were functionalized with groups based on a PIM-1 monomer, and high rejection of light solutes was obtained by functionalizing the tip of MW-NTs of 4.4 nm diameter using these monomers.

ACKNOWLEDGEMENTS

The research described here has received funding from the European Community's Seventh Framework Programme as part of the BioNexGen project (Grant Agreement No. CP-FP 246039-2). The authors are grateful to the Cineca Consortium for the use of high-performance computers.

REFERENCES

Agenson, K.O., Oh, J.-I. & Urase, T. (2003) Retention of a wide variety of organic pollutants by different nanofiltration/reverse osmosis membranes: controlling parameters of process. *Journal of Membrane Science*, 225, 91–103.

Beck, A.D. (1993) Density-functional thermochemistry. III. The role of exact exchange. *Journal of Chemical Physics*, 98, 5648–5653.

Bisignano, F. (2014) Modeling of nanostructured membranes for wastewater purification. In: Dottorato in Scienza e Tecnica Bernardino Telesio Curriculum M3, XXVI ciclo, University of Calabria, Italy.

Bisignano, F., Mattia, D. & De Luca, G. (2015) Selectivity-permeability optimization of functionalized CNT-polymer membranes for water treatment: a modeling study. *Separation and Purification Technology*, 146, 235–242.

Bowen, T.C., Kalipcilar, H., Falconer, J.L. & Noble, R.D. (2003) Pervaporation of organic/water mixtures through B-ZSM-5 zeolite membrane on monolith supports. *Journal of Membrane Science*, 215, 235–247.

Braeken, L., Ramaekers, R., Zhang, Y., Maes, G., Van der Bruggen, B. & Vandecasteele, C. (2005) Influence of hydrophobicity on retention in nanofiltration of aqueous solutions containing organic compounds. *Journal of Membrane Science*, 252, 195–203.

Buekenhoudt, A., Bisignano, F., De Luca, G., Vandezande, P., Wouters, M. & Verhulst, M.K. (2013) Unravelling the solvent flux behavior of ceramic nanofiltration and ultrafiltration membranes. *Journal of Membrane Science*, 439, 36–47.

Cao, G. & Wang, Y. (2011) Introduction. In: Cao, G. & Wang, Y. (eds) *Nanostructures and Nanomaterials: Synthesis, Properties, and Applications*. 2nd edition, World Scientific, Singapore. pp. 1–17.

Chrystal, G. (1885) On the problem to construct the minimum circle enclosing n given points in a plane. *Proceedings of the Edinburgh Mathematical Society*, Third Meeting. p. 30.

Chung, T.-S., Jiang, L.Y., Li, Y. & Kulprathipanja, S. (2007) Mixed matrix membranes (MMMs) comprising organic polymers with dispersed inorganic fillers for gas separation. *Progress in Polymer Science*, 32, 483–507.

De Luca, G. (2013) Assessment of the key properties of materials used in membrane reactors by quantum computational approaches. In: Basile, A. (ed) *Handbook of Membrane Reactors*. Woodhead Publishing, Cambridge. pp. 598–626.

Eliosoff, J. & Unger, R. (1998) Minimal spanning circle of a set of points. Computational Geometry project, School of Computer Science, McGill University, Montreal. Available from: http://www.cs.mcgill.ca/~cs507/projects/1998/jacob/ [accessed November 2016].

Gulseren, O., Yildirim, T. & Ciraci, S. (2002) Systematic ab initio study of curvature effects in carbon nanotubes. *Physical Review* B, 65, 153405.

Hanwell, M.D., Curtis, D.E., Lonie, D.C., Vandermeersch, T., Zurek, E. & Hutchison, G.R. (2012) Avogadro: an advanced semantic chemical editor, visualization, and analysis platform. *Journal of Cheminformatics*, 4, 17.

Hearn, D.W. & Vijan, J. (1982) Efficient algorithms for the minimum circle problem. *Operations Research*, 30, 777–795.

Hummer, G., Rasaiah, J.C. & Noworyta, J.P. (2001) Water conduction through the hydrophobic channel of a carbon nanotube. *Nature*, 414, 188–190.

Iliyas, A., Eic, M., Zahedi-Niaki, M.H. & Vasenkov, S. (2007) Toward observation of single file diffusion using TZLC. *Journal of Physical Chemistry* B, 112(12), 3821–3825.

Jirage, K.B., Hulteen, J.C. & Martin, C.R. (1997) Nanotubule-based molecular-filtration membranes. *Science*, 278, 655–658.

Kiso, Y., Kitao, T., Kiyokatsu, J. & Miyagi, M. (1992) The effects of molecular width on permeation of organic solute through cellulose acetate reverse osmosis membrane. *Journal of Membrane Science*, 74(12), 95–103.

Lee, S.B., Mitchell, D.T., Trofin, L., Nevanen, T.K., Söderlund, H. & Martin, C.R. (2002) Antibody-based bio-nanotube membranes for enantiomeric drug separations. *Science*, 296, 2198–2200.

Mattia, D., Leese, H. & Lee, K.P. (2015) Carbon nanotube membranes: from flow enhancement to permeability. *Journal of Membrane Science*, 475, 266–272.

McKeown, N.B. & Budd, P.M. (2006) Polymers of intrinsic microporosity (PIMs): organic materials for membrane separations, heterogeneous catalysis and hydrogen storage. *Chemical Society Reviews*, 35, 675–683.

Metropolis, N. & Ulam, S. (1949) The Monte Carlo method. *Journal of the American Statistical Association*, 44, 335–341.

Nakao, S.-I. & Kimura, S. (1982) Models of membrane transport phenomena and their applications for ultrafiltration data. *Journal of Chemical Engineering of Japan*, 15, 200–205.

Ozaki, H. & Li, H. (2002) Rejection of organic compounds by ultralow pressure reverse osmosis membrane. *Water Research*, 36(1), 123–130.

Pacios, F. & Christiansen, P.A. (1985) Ab initio relativistic effective potentials with spin-orbit operators. I. Li through Ar. *Journal of Chemical Physics*, 82, 2664–2671.

Parr, R.G. & Yang, W. (1989) *Density-Functional Theory of Atoms and Molecules*. Oxford University Press, New York.

Perdew, J. P., Burke, K. & Ernzerhof, M. (1996) Generalized gradient approximation made simple. *Physical Review Letters*, 77, 3865–3868.

Rosenbluth, A., Rosenbluth, M., Teller, A. & Teller, E. (1953) Equation of state calculations by fast computing machines. *Journal of Chemical Physics*, 21, 1087–1092.

Saito, R., Dresselhaus, G. & Dresselhaus, M.S. (1998) *Physical Properties of Carbon Nanotubes*. Imperial College Press, London.

Santos, J.L.C., de Beukelaar, P., Vankelecom, I.F.J., Velizarov, S. & Crespo, J.G. (2006) Effect of solute geometry and orientation on the rejection of uncharged compounds by nanofiltration. *Separation and Purification Technology*, 50, 122–131.

Shao, P. & Huang, R.Y.M. (2007) Polymeric membrane pervaporation. *Journal of Membrane Science*, 287, 162–179.

Striolo, A. (2006) The mechanism of water diffusion in narrow carbon nanotubes. *Nano Letters*, 6, 633–639.

Szabo, A. & Ostlund, N. (1994) *Modern Quantum Chemistry: Introduction to Advanced Electronic Structure Theory*. Macmillan, London.

Tang, C., Zhang, Q., Wang, K., Fu, Q. & Zhang, C. (2009) Water transport behavior of chitosan porous membranes containing multi-walled carbon nanotubes (MW-NTs). *Journal of Membrane Science*, 337, 240–247.

Thomas, J.A. & McGaughey, A.J.H. (2009) Water flow in carbon nanotubes: transition to subcontinuum transport. *Physical Review Letters*, 102, 184502.

Valiev, M., Bylaska, E.J., Govind, N., Kowalski, K., Straatsma, T.P., van Dam, H.J.J., Wang, D., Nieplocha, J., Apra, E., Windus, T.L. & de Jong, W.A. (2010) NWChem: a comprehensive and scalable open-source solution for large scale molecular simulations. *Computer Physics Communications*, 181, 1477–1489.

Van der Bruggen, B., Schaep, J., Maes, W., Wilms, D. & Vandecasteele, C. (1998) Nanofiltration as a treatment method for the removal of pesticides from ground waters. *Desalination*, 117, 139–147.

Van der Bruggen, B., Schaep, J., Wilms, D. & Vandecasteele, C. (1999) Influence of molecular size, polarity and charge on the retention of organic molecules by nanofiltration. *Journal of Membrane Science*, 156, 29–41.

Van der Bruggen, B., Schaep, J., Wilms, D. & Vandecasteele, C. (2000) A comparison of models to describe the maximal retention of organic molecules in nanofiltration. *Separation Science and Technology*, 35, 169–182.

Whitby, M. & Quirke, N. (2007) Fluid flow in carbon nanotubes and nanopipes. *Nature Nanotechnology*, 2, 87–94.

White House. (2000) National nanotechnology initiative – leading to the next industrial revolution. Committee on Technology, National Science and Technology Council, Washington, DC. Available from: https://www.whitehouse.gov/files/documents/ostp/NSTC%20Reports/NNI2000.pdf [accessed November 2016].

Wu, H., Tang, B. & Wu, P. (2013) Optimization, characterization and nanofiltration properties test of MW-NTs/polyester thin film nanocomposite membrane. *Journal of Membrane Science*, 428, 425–433.

Xiao, J. & Wei, J. (1992) Diffusion mechanism of hydrocarbons in zeolites. I. Theory. *Chemical Engineering Science*, 47(5), 1123–1141.

Xu, S., Freund, R.M. & Sun, J. (2001) Solution methodologies for the smallest enclosing circle problem. Technical report, Singapore-MIT Alliance, National University of Singapore, Singapore.

Xu, X., Zhang, Q., Muller, R.P. & Goddard, W.A. (2005) An extended hybrid density functional (X3LYP) with improved descriptions of nonbond interactions and thermodynamic properties of molecular systems. *Journal of Chemical Physics*, 122, 14105.

Yangali-Quintanilla, V., Sadmani, A., McConville, M., Kennedy, M. & Amy, G. (2010) A QSAR model for predicting rejection of emerging contaminants (pharmaceuticals, endocrine disruptors) by nanofiltration membranes. *Water Research*, 44, 373–384.

CHAPTER 13

Recovery of bioactive compounds in citrus wastewater by membrane operations

Alfredo Cassano, Carmela Conidi & René Ruby-Figueroa

13.1 INTRODUCTION

Citrus fruits are very much appreciated for their distinctly pleasant flavors and aromas as well as their nutritional value and antioxidant activity (Talon and Gmitter, 2008). According to data published by FAO (2014), citrus represents the world's third most important fruit crop after apple and banana, with an annual (2013) production of about 136 million metric tons. Brazil and the USA are the world's leading producers, supplying over 40% of the world's current requirements (FAO, 2014). Other important citrus producers include Spain, Italy, Egypt, Mexico, China and India.

At large scale, juice and juice products represent a very important segment of the total processed-fruit industry. However, citrus juice processing is characterized by the production of large amounts of waste material such as peel and seeds. These wastes are a concern from an environmental point of view because plant material is usually prone to microbial spoilage. These byproducts are commonly reused for animal feeding or as fertilizers (Famyima and Ough, 1982; Nikolic et al., 1986). Intensive research in the field of citrus waste management suggests that these byproducts should be regarded as useful resources for the recovery of high added-value compounds such as soluble sugars, fiber, organic acids, amino acids, proteins, minerals, oils, lipids, flavonoids and vitamins, which can be recovered and reused in other market sectors (Braddock, 1995; Marín et al., 2002). In particular, the recovery of these compounds offers new opportunities for the formulation of products of interest in the food (dietary supplements and functional foods production), pharmaceutical (products with antibacterial, antiviral, anti-inflammatory, antiallergic and vasodilatory actions) and cosmetic industries (Benavente-García et al., 1997; Marín et al., 2002; Puupponen-Pimiä et al., 2002).

Over the past 15 years, different methodologies have been proposed for the extraction and recovery of valuable compounds from citrus byproducts. They are based on the use of resins (Di Mauro et al., 2000), organic solvents (Li et al., 2006a), enzymes (Li et al., 2006b), γ-irradiation (Oufedjikh et al., 2000) and heat treatment (Xu et al., 2007). These extraction methods are characterized by some drawbacks, such as the degradation of the compounds of interest due to high temperatures and long extraction times (as in solvent extractions), and health-related risks. The medical interest in drugs obtained from plants has led to an increased need for more ideal extraction methods, which could obtain the maximum of the bioactive constituents in a shorter processing time at a low cost.

The recovery of antioxidant compounds from agricultural byproducts is currently conducted in five principal stages: (i) macroscopic pretreatment; (ii) macro- and micro-molecular separation; (iii) extraction; (iv) isolation and purification; (v) product formation.

Pressure-driven membrane operations include microfiltration (MF), ultrafiltration (UF), nanofiltration (NF) and reverse osmosis (RO). Between them, they cover most parts of the "five-stage universal recovery processing" strategy because they can be used in macroscopic pretreatment (MF), in macro- and micro-molecular separation (UF), in purification (NF) and in

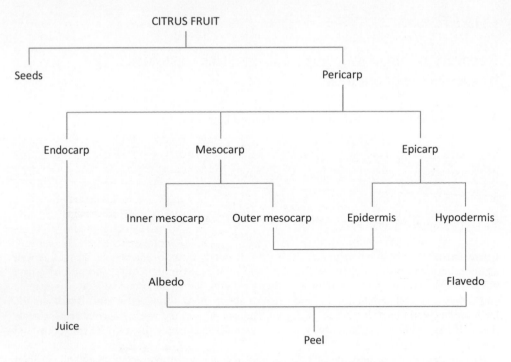

Figure 13.1.   Structure of citrus fruits (adapted from Izquierdo and Sendra, 1993).

concentration (RO) (Galanakis, 2012). All of these processes offer several advantages over con-ventional methods in terms of higher separation efficiency, lower energy requirements, milder operating conditions, an absence of additives and phase transitions, simpler equipment and easier scale-up.

This chapter focuses on the application of membrane processes to citrus production, mainly concentrating on the recovery of bioactive compounds from citrus byproducts. The first section of the chapter will provide an overview of traditional processes in the citrus industry. Selected applications related to the use of pressure-driven membrane operations and the immobilization of β-cyclodextrins in polymeric membranes will be analyzed and discussed in the second section.

13.2   CITRUS PRODUCTION AND COMPOSITION

Fruit juices and products play a major role in human nutrition and in the food industry around the world. For example, citrus juices comprise almost 70% of all fruit or vegetable juices consumed in the USA. Every continent except Antarctica is endowed with citrus-growing regions. Within the citrus juice industry, orange and tangerine/mandarin products dominate the market. Based on data published by the US Department of Agriculture, global orange production for 2015/16 was expected to be 45.8 million metric tons (a slight fall compared to that of the previous year) (USDA, 2016).

13.2.1   *Citrus structure*

Citrus fruits consist of several parts, depicted schematically in Figure 13.1. Juice vesicles, located in the endocarp, contain the juice, which constitutes about 50% of the total weight of a typical orange fruit. The peel is formed by the flavedo (epicarp and outer mesocarp) and the albedo (inner mesocarp). Flavedo and albedo account, respectively, for about 10% and 25% of the whole fruit. The flavedo contains peel oil in oil sacs and the albedo contains most of the pectins. After juice extraction, rag, pulp and seeds account for the rest of the total fruit weight. Most of the fruit, almost 90%, is water. The rest mainly takes the form of sugars and acids. Minerals, amino acids, aroma compounds and pectins are present in smaller proportions. These percentages vary greatly in other citrus fruits, but they also vary within orange fruits, depending on variety, ripeness, area and agricultural practice (Izquierdo and Sendra, 1993).

13.2.2 *Citrus bioactive compounds*

Citrus fruits are considered a nutrient-dense food with a relatively low caloric content and an impressive list of bioactive compounds. These range from vitamins and minerals to health-promoting substances including fibers, carotenoids and phenolic compounds (mainly flavonoids). Epidemiological studies on diets rich in citrus fruits have associated them with a reduced risk of cardiovascular disease, stroke, cancer, diabetes and other inflammatory diseases. Reported effects include reduction in oxidative stress, inflammation and/or tumor suppression, glucose control, inhibition of LDL oxidation, promotion of plaque stability, improved vascular endothelial function and decreased tendency for thrombosis (Gonzalez-Molina *et al.*, 2010). Table 13.1 summarizes the general classes of bioactive compounds in citrus fruits and their byproducts, distributions and reported health effects.

13.2.2.1 *Phenolic compounds*
Most of the pharmacological properties of the citrus fruits and their byproducts are attributed to the presence of phenolic compounds (Escarpa and Gonzalez, 2001).

The two main types of polyphenols identified in citrus are flavonoids and non-flavonoids. The latter include structurally simple molecules such as phenolic acids (synaptic, ferulic, *p*-hydroxybenzoic, vanillic, *p*-coumaric, caffeic and gallic) and stilbenes, and highly complex molecules such as stilbene oligomers and tannins (Rapisarda *et al.*, 1998). Flavonoids include more than 9000 identified compounds. They are characterized by a common benzo-γ-pyrone structure, which has been reported to act as an antioxidant in various biological systems. Multiple combinations of hydroxyl groups, sugars, oxygens and methyl groups attached to this structure generate the various classes of flavonoids: flavanols, flavanones, flavones, flavan-3-ols (catechins), anthocyanins and isoflavones (Khan *et al.*, 2014).

Citrus flavanones are present in the glycoside or aglycone forms. Among the aglycone forms, naringenin and hesperetin are the most important flavanones. Among the glycoside forms, two types are identified: neohesperidosides (naringin, neohesperidin and neoeriocitrin) and rutinosides (hesperidin, narirutin and didymin) (Tripoli *et al.*, 2007).

The flavonoid composition of the different *Citrus* species is not always the same. For example, in the sweet orange (*Citrus sinensis*), the most abundant flavonoid component is hesperidin, followed by narirutin and dydimin. The 'Blood' variety of sweet orange is characterized by the presence of anthocyanins, typical coloring compounds of flowers and fruits, the level of which is strongly dependent on the degree of maturation Gattuso *et al.* (2007).

13.2.2.2 *Carotenoids*
Citrus fruits and byproducts are one of most complex sources of carotenoids, with a large diversity among the different species and cultivars in terms of types and amounts. In general, β-carotene, α-carotene, lutein, β, β-xanthophylls (particularly violaxanthin, β-cryptoxanthin and their esters) are common to most citrus fruits. It is interesting to note that citrus fruits contain a group of genus-specific C-30 apocarotenoids, mainly located in the peel (Kato *et al.*, 2004). The most abundant C-30 apocarotenoid is β-citraurin, which provides the intense orange-reddish coloration to the peel of some oranges, mandarins and hybrids (Matsumoto *et al.*, 2007).

All carotenoids contain a system of conjugated double bonds which allow them to interact efficiently with reactive oxygen species. β-carotene, α-carotene and lutein, as well as several other citrus carotenoids, are efficient quenchers of singlet molecular oxygen and scavengers of peroxyl radicals. In human studies, numerous associations between a low carotenoid intake or status and

Table 13.1. Bioactive compounds of citrus fruits, their byproducts and their potential health benefits.

Bioactive compounds		Examples	Species	Potential health benefits	References
Phenolic compounds	Flavonoids	Hesperidin, neohesperidin, naringin	*Citrus limon Citrus bergamia*	AI, AOX	Gattuso *et al.* (2007)
		Naringenin, narirutin	*Citrus paradisi*	AI, AOX	Gattuso *et al.* (2007)
		Anthocyanins, dydimin	*Citrus paradisi*	AC, AI, AOX, CHB, CR	Benavente-Garcia and Castillo (2008)
			Citrus sinensis		
		Neoeriocitrin, bruteridin, melitidin	*Citrus bergamia*	AD, AI, AOX, CHB, CR	Di Donna *et al.* (2009)
	Phenolic acids	Eriocitrin, diosmin	*Citrus limon*	AC, AI, AOX, CHB, CR	Nakajima *et al.* (2014)
		Synaptic, ferulic, *p*-hydroxybenzoic, vanillic, *p*-coumaric, caffeic, gallic	*Citrus limon Citrus bergamia*	AC, AI, AOX, CHB, CR	Gonzalez-Molina *et al.* (2010)
			Citrus paradisi Citrus sinensis, *Citrus reticulate*		
Carotenoids		β-carotene, α-carotene, lutein, β,β-xanthophylls, violaxanthin, β-cryptoxanthin, β-citraurin	*Citrus limon Citrus bergamia*, *Citrus paradisi Citrus sinensis* *Citrus reticulate*	AC, AI, AOX, CHB	Kato *et al.* (2004); Matsumoto *et al.* (2007)
Fibers		Pectin, gums, cellulose, hemicellulose	*Citrus limon Citrus bergamia*, *Citrus paradisi Citrus sinensis*, *Citrus reticulate*	AD, CR	O'Shea *et al.* (2012); Abirami *et al.* (2014)
Vitamins		Vitamin C, vitamin B (B9)	*Citrus limon Citrus bergamia*, *Citrus paradisi Citrus sinensis*, *Citrus reticulate*	AI, AOX, NFP	Escobedo-Avellaneda *et al.* (2014)

Legend: AC: anticancer; AD: antidiabetic; AI: anti-inflammatory activity; AOX: antioxidant; CHB: cardiovascular health benefits; CR: cholesterol reduction; NFP: neural function protection.

an increased risk of cancer, age-related macular degeneration, cataract and cardiovascular disease have been observed.

13.2.2.3 *Fibers*

Citrus fruits and byproducts are considered an excellent source of dietary fibers, including both water-soluble (pectin and gums) and water-insoluble (cellulose and hemicelluloses) compounds. Pectin is the predominant type of water-soluble fiber in citrus (as much as 65–70% of the total fiber) (O'Shea *et al.*, 2012). The remaining fibers consist of cellulose, hemicellulose and trace amounts of gums. Citrus also contains lignin, a fiber-like component. Dietary incorporation of pectin appears to affect several metabolic and digestive processes: pectin has been associated with numerous physiological effects, including decreases in glucose absorption and improvements in insulin response, the lowering of plasma LDL cholesterol concentrations, and binding to minerals to decrease their bioavailability (Elleuch *et al.*, 2011). In addition, recent studies have indicated that citrus pectin demonstrates an anticancer potential and an immune-modulatory effect (Abirami *et al.*, 2014).

13.2.2.4 *Vitamins*

Citrus fruits and their byproducts are a rich source of vitamins for human diet. Vitamin C is considered the most prevalent vitamin in citrus; other vitamins present in minor quantities are A and B-group (B2, B3, B9) vitamins (Escobedo-Avellaneda *et al.*, 2014).

Vitamin C is an essential water-soluble compound with antioxidant activity and excellent reducing properties. Vitamin C acts against oxidation of lipids, proteins and DNA, protecting their structure and biological function. The oxidation of these biomolecules generates measurable reaction products, such as 8-oxo-deoxyguanosine from DNA, F2-isoprostanes from lipids, and carbonyl derivatives from proteins. In addition, vitamin C acts as an electron donor for different enzymes; it is involved in collagen hydroxylation, synthesis of carnitine, biosynthesis of norepinephrine from dopamine and modulation of tyrosine metabolism.

Among the vitamins of group B, the most represented in citrus is folic acid, a water-soluble vitamin known as B9, which plays an important role in human nutrition (Öhrvik and Witthöft, 2008) Folate compounds are important for DNA synthesis, supporting the preservation of genetic information, and for the production of amino acids. Folic acid also plays an important role in lowering the levels of homocysteine in the blood. Folate-rich diets have been associated with a decreased risk of cardiovascular disease.

There is strong scientific evidence to support a link between folic acid intake and the prevention of neural tube defects in infants. The US Center for Disease Control and Prevention recommends a consumption of about 0.4 mg day^{-1} by all women of childbearing age, and especially those who are planning a pregnancy.

Nowadays, folic acid-fortified beverages are used as a method of increasing the intake of this vitamin in the pregnant population and are widely recommended (Pérez Prieto *et al.*, 2006).

13.3 CITRUS INDUSTRY

13.3.1 *Traditional processing*

Traditional citrus juice processing is depicted in Figure 13.2. In the *cleaning/washing* step, dirt and foreign materials are removed from the fruits. A sanitation step can be implemented to decrease the microbial load, as well as to eliminate undesirable flavors or mycotoxin contamination.

Fruits that have been rinsed and dried pass through graders in which bad fruits are removed while the remaining fruits are automatically segregated by size prior to extraction (*sorting/culling*), which depends upon the fruit being of an appropriate size.

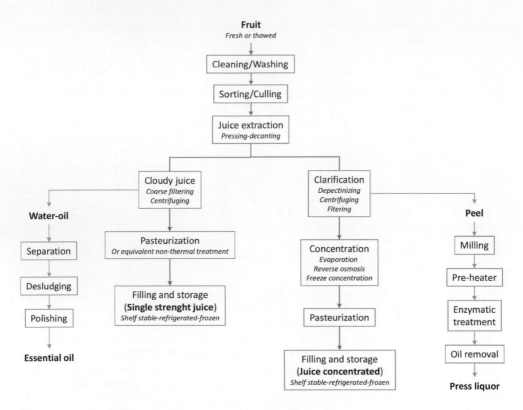

Figure 13.2.   Flowchart of industrial orange juice processing.

The principal unit operation in citrus production is the juice extraction: its efficiency allows the production of a finished drink with the desired characteristics. Thus peel, containing bitter resins, must be carefully removed to avoid contamination of the sweeter juice.

The extracted juice is then treated according to the characteristics required of the final product. For cloudy juices, further clarification might be unnecessary or may involve coarse filtration or controlled centrifugation to remove only larger, insoluble particles. For clear juices, complete depectinization, by addition of enzymes, fine filtration, or high-speed centrifugation, will be required to achieve visual clarity (Downing, 1996).

The production of clear juices involves a *clarification* step, which can be accomplished in one of two general ways: enzymatically and non-enzymatically. The enzyme-based depectinization has two effects: it degrades the viscous soluble pectin and, at the same time, causes the aggregation of cloud particles and their accumulation on the bottom of the tank. In acidic conditions, as created by citrus juices, pectin molecules carry a negative charge, allowing them to repel each other. Pectinases, the enzymes typically used for this process, degrade the pectin and expose part of the positively charged protein, reducing the electrostatic repulsion between cloud particles which clump together. These larger particles will eventually settle out but, to improve the process, flocculating or fining agents, such as sparkolloid, gelatin, kieselsol, bentonite and isinglass, must be added (Lozano, 2003). Non-enzymatic clarification involves disruption of the emulsion by other means, such as heat (Smock and Neubert, 1950), mechanical separation by the use of decanters and finishers, centrifugation, diatomaceous earth filtration, as well as membrane filtration (Kilara and Van Buren, 1989).

The *concentration* step is commonly carried out by thermal evaporation with equipment known as a thermally accelerated short-time evaporator (TASTE). Concentration processes offer significant advantages in terms of reduction of bulk, thereby reducing storage volumes and transportation costs too. In addition, storage of the cold concentrate is less likely to support yeast growth due to its high sugar concentration and consequently lower water activity. Many different evaporator designs can be employed for juice concentration, including evaporators with single-pass recirculating concentrators, single or multiple stages, single or multiple effects, and with or without a surface-mixing form of condenser (Hartel, 1992). Notwithstanding several advantages, concentration, when performed by evaporation, is an energy-intensive process which gives rise to high operating costs and environmental challenges (Bichsel and Sandre, 1982).

In the case of clear juices, concentration can be achieved by using a combination of RO and evaporation. In this regard, crossflow membrane systems are ideally suited for this application due to the self-cleaning turbulence effect. RO is effective in concentrating a low-solids juice (7 to 8 °Bx) by two to three times, after which evaporation technology has to be applied.

Citrus juice has some natural protection against the growth of bacteria, yeast and mold due to its low pH (about 4). Nevertheless, a thermal process named *pasteurization* is still required to avoid further spoilage. This process not only helps to reduce microbial levels but also inactivates certain enzymes that cause the pulp to separate from the juice, which would result in an aesthetically undesirable beverage. Pasteurization of the finished juice is accomplished by hot-filling or bottled juice pasteurization. Hot-filling is conducted by passing the final juice through a heat exchanger and raising the temperature to 88–95°C. A holding time of at least three minutes is normally applied to the juice, prior to cooling. A shelf life in high-quality storage can range from nine to 12 months in glass bottles or in high-barrier containers (McLellan and Padilla-Zakour, 2004).

Alternatives to thermal pasteurization have become a reality in the food industry. The production of cold-treated juices offers the advantage of fresh-like characteristics with extended shelf life. Technologies that are being utilized or developed include ultraviolet light (UV), high-pressure processing, pulsed electric fields, electron-beam irradiation, high carbon dioxide processing and chemical sterilants (Knorr *et al.*, 2002; Odebo, 2001; Worobo *et al.*, 1998). Iu *et al.* (2001) demonstrated that a combination of pulsed electric fields and heat treatments was very effective against *Escherichia coli* O157:H7 in apple cider. The use of UV at 14 mJ cm^{-2}, a method approved by the FDA to deliver safe refrigerated apple juice, has been described by Tandon *et al.* (2003). The UV-treated juice retained the fresh flavor and was favorably compared to juice pasteurized at 71°C for 6 s.

To ensure sterility, the pasteurized juice should be filled while still hot (*filling and storage*). Where possible, metal or glass bottles and cans can be preheated. The primary functions of food packaging are to retard or prevent the loss of quality, to contain the food adequately, and to give protection against environmental contamination (Crosby, 1981; Paine and Paine, 1983).

Manufacture of citrus juices produces large quantities of liquid and solid waste. The disposal of these materials is regulated by law and can represent a significant cost in the juice plant operation. In the following section, the typical wastes produced in citrus juice processing and the potential use of these byproducts will be described in detail.

13.3.2 *Wastes from citrus production*

Residuals are a necessary consequence of processing agricultural raw materials. These residuals, dumped at local landfills and discharged to natural streams, have now become a major environmental concern. Environmental protection standards are becoming more stringent, causing a remarkable growth in disposal costs. Meeting the combined challenge of environmental protection and economic competitiveness requires a fresh look at the management of fruit-processing residuals.

Nowadays, the challenge has been focused not only on the treatment of these residues in order to reduce waste-disposal problems but also to generate value-added products or additives that can be used in other industrial sectors, including cosmeceutical, pharmaceutical and food companies. Possible value-added products from solid waste are fuels, proteins and biochemicals. In other cases, the effluents will need to be substantially processed before they can be safely discharged into public waste management systems (Hang, 1999).

The effluents from the processing lines fall into three broad categories: wash water (lightly polluted), process water (highly polluted) and solid waste (extremely highly polluted).

Lightly polluted liquid effluents, with a biological oxygen demand (BOD) of less than $200 \, \text{mg L}^{-1}$, should be treated in aerobic systems such as aerated lagoons. Anaerobic biological treatment systems are better suited to highly polluted process water, with a BOD of up to $20,000 \, \text{mg L}^{-1}$. Efficiencies of 70% to 95% in the reduction of BOD are expected with these systems. For higher reduction rates, the treated effluent might need to be further processed in an aerobic system (Koevoets, 2002). The activated sludge process is another widely used system for wastewater treatment. The treatment unit consists of a bioreactor that provides an environment for converting soluble waste solids into insoluble microbial cells under aerobic conditions, and a clarifier where microbial cells are allowed to settle. The settled cells or sludge may be returned to the bioreactor or handled as waste. The reduction in BOD by this method ranges from 80% to 99%, depending on the waste characteristics, loading rate and other operating conditions (Hang, 1999). Solid wastes pose a more difficult dilemma due to reduced opportunities for the disposal of solids. In particular, solid citrus wastes composed largely of citrus peel have been extensively used as animal feed or fertilizer. In fact, these byproducts can be found as a feed supplement for sheep, pigs, horses and deer. Alternatively, they can be disposed of in local landfills (Koevoets, 2002). Although proper composting offers an excellent opportunity to dispose of a waste product and create a valuable end-product, the current challenge is focused on the recovery of valuable components for the development of products of interest for food, cosmetic and pharmacological applications.

13.3.3   *Potential applications of citrus byproducts*

The correct management of citrus byproducts is a real concern for industrial companies and government institutions. Paradoxically, citrus residues possess a range of natural components, such as sugars, pectin, carotenoids, flavonoids, vitamin C and folic acid, that can be very useful in different industrial sectors (Patil *et al.*, 2009). Nowadays, any improvement in the treatment of final byproducts to prevent serious contamination problems and to formulate new added-value products is of great interest.

Citrus peel is the main waste or byproduct obtained in juice production. Usually, residues such as peel, rag, pulp and seed are destined for animal feeds. Unfortunately, this waste is susceptible to spoilage owing to its high moisture content (78–90%); therefore, it must be consumed quickly, creating additional costs for transportation and storage (Lanza, 2003).

Pectin is located mainly in the albedo and juice sacs. It has a high molecular weight (100–200 kDa) and consists of long chains of polygalacturonic acid units linked together by α (1–4) glycosidic bonds with side chains of rhamnose, arabinans, galactans and xylose. Pectin is appreciated in the food industry as a gelling, thickening and stabilizing agent as well. Applications in the pharmaceutical industry have also been reported, basically as an ingredient for preparation of anti-diarrheal and detoxifying drugs, as well as for preparation of suspensions with potential applications in the delivery of controlled-release drugs (Liu *et al.*, 2003; Piriyaprasarth and Sriamornsak, 2011). In addition, some fragments of pectins have also been shown to exert a positive effect on reducing metastatic stages both *in vitro* and *in vivo* (Maxwell *et al.*, 2012). In the food industry, pectin is mainly used for the preparation of specific products such as jam, marmalade and jelly.

Peel oil is a valuable byproduct recovered from citrus peel, recognized for its flavor and fragrance but also for its antimicrobial and antioxidant properties; when it is added to food subjected to thermal treatment for preservation, no alteration of the food's organoleptic properties is produced (Tyagi *et al.*, 2014). For this reason, citrus essential oils could be considered as a suitable alternative to the use of chemical additives in the food industry, attending to the need for food safety and satisfying the demand of consumers for natural food components

(Viuda-Martos *et al.*, 2008). Among numerous compounds, the major component is D-limonene, which is used in the electronics industry as well as in the manufacture of synthetic resins and adhesives (Lanza, 2003).

Among the citrus peels, the most important essential oil extracted is from the bergamot orange (*Citrus bergamia*). Bergamot essence is widely used in the pharmaceutical industry for its antibacterial and antiseptic properties, in the cosmetic industry for the production of perfumes, body lotions, soaps, etc., and in food industries as an aroma in the preparation of sweets, liquors and tea.

Phenolic compounds and flavonoids are particular examples of the attractive components of citrus byproducts. Epidemiological studies have correlated their consumption with a lower risk of different types of cancer and cardiovascular disease (Kris-Etherton *et al.*, 2002), thus demonstrating their antioxidant, anti-inflammatory and radical-scavenging properties (Barreca *et al.*, 2011). In this regard, various efforts have been made to recover phenolic compounds from citrus byproducts (Ruby-Figueroa *et al.*, 2012). The recovery of these compounds will be a profitable investment if they can eventually be used to develop high added-value products in the food, pharmaceutical and cosmetic industries.

13.4   MEMBRANE OPERATIONS IN CITRUS PROCESSING

Nowadays, membrane processes are established separation techniques in several industries due to their large number of advantages, including high efficiency, simple equipment and low energy consumption (Jing and Howard, 2010). Over the last two decades, the worldwide market for membrane technology in the food industry has increased to a market value of about €800–850 million and it is now the second biggest industrial market for membranes after water and wastewater treatment, which includes desalination (Lipnizki, 2010). Unit operations, such as clarification, stabilization, depectinization and concentration, are typical steps in which membrane processes, such as MF, UF, NF, RO, osmotic distillation (OD) and membrane distillation (MD), have been successfully utilized as alternative technologies to the conventional operations of the fruit juice processing industry (Jiao *et al.*, 2004). In this regard, several alternatives have been reported for the application of membrane operations, not only as a valid alternative to thermal evaporation processes but also for the classic filtration of citrus juices based on the use of fining agents (gelatin, diatomaceous earth, bentonite and silica sol). In addition, integrated membrane systems represent a useful approach for the treatment of citrus byproducts in order to recover attractive compounds for the development of high added-value products destined for the pharmaceutical, cosmetic and food industries (Cassano *et al.*, 2014a).

A large variety of synthetic membranes have been reported in scientific and patent literature. The differences may be generated by material partitioning during membrane formation or by particular post-formation surface treatments. Membrane chemistry determines important properties such as hydrophilicity or hydrophobicity, the presence or absence of ionic charges, chemical and thermal resistances, binding affinities for solutes or particles, and biocompatibility (Cheryan, 1998). Membrane materials must be chemically resistant to both feed and cleaning solutions, mechanically and thermally stable, and characterized by high selectivity and permeability. The materials used for the preparation of membranes can be synthetic polymers, cellulose derivatives, ceramics, inorganics and metals; they may be neutral or carry electrical charges.

Although over 130 materials have been used to manufacture membranes, only a few have achieved commercial status, and fewer still have obtained regulatory approval for use in food. Polyethylene, polypropylene, polycarbonate and cellulose acetate are all widely used in the manufacture of MF and UF membranes.

In pressure-driven membrane processes, the selection of the most effective membrane for a specific application plays an important role in determining the level of separation that can be obtained. However, for an efficient utilization of membranes, design of the process is equally important. The process design is defined by different aspects, including the membrane configuration, filtration methods (dead-end and crossflow configurations) and the process configuration. Process design is also important for the control of concentration polarization and fouling phenomena, determining to a large extent the most useful membrane for a specific separation. In this regard, the membrane module concept denotes the device in which the membrane must be installed to perform the separation process. At a large industrial scale, membrane modules are available in six basic designs: cartridge, hollow fiber, spiral-wound, tubular, plate-and-frame and capillary. These are quite different in their design, mode of operation, production costs and the energy requirements for pumping the feed solution through the module.

Membranes can be operated in either dead-end or crossflow configurations. In the dead-end mode the feed flow is perpendicular to the membrane surface. It is forced through the membrane, which causes the retained particles to accumulate and form a type of cake layer at the membrane surface. The thickness of the cake layer increases with filtration time, and the permeation rate decreases as the thickness of the cake layer increases. Dead-end filtration is often used as a method to estimate the specific cake resistance for crossflow filtration and usually gives reasonable data for spherical- and ellipsoidal-shaped cells (Tanaka *et al.*, 1994). On the other hand, in crossflow filtration the fluid flows in a direction parallel to the membrane surface and permeates through the membrane due to an imposed transmembrane pressure difference. Unlike dead-end filtration, rejected particles form a cake layer on the membrane surface which does not build up indefinitely, so the cake formed is of a limited thickness (Tiller and Cooper, 1962). The cake structure will be affected by different phenomena, such as the collapse of the pore structure, pore compression and pore distortion. This set of phenomena will affect to differing extents the porosity, the pore size and the pore tortuosity of the filtration cake (Mota *et al.*, 2002).

In pressure-driven membrane operations, process configurations include a total recycling configuration (where permeate and retentate streams are recycled back to the feed reservoir so that a steady state is attained in the fixed concentration of the feed), a batch concentration configuration (where the retentate is returned to the feed reservoir and the permeate is collected separately), a feed-and-bleed configuration (where the permeate is removed from the system, together with a small part of the retentate), and diafiltration (which involves the addition of water to the retentate to overcome low permeate fluxes at high concentrations or to improve the separation of permeable compounds) (Cassano *et al.*, 2014b). The selection of an appropriate process configuration will depend on the specific application and its goals.

Membrane performance is evaluated mainly in terms of filtration rate and membrane separation properties, generally measured in terms of membrane rejection. A brief description of the main pressure-driven membrane processes applied in the citrus industry follows, together with selected applications in the recovery of phenolic compounds from citrus byproducts. Some additional applications, investigated at laboratory scale, for juice debittering and the removal of pesticides from citrus essential oils are also reported and discussed.

13.4.1 *Microfiltration and ultrafiltration*

MF and UF are pressure-driven processes, considered as the most important in food technology and biotechnology. They are simple membrane-based filtration systems in which the separation principle is a form of sieving (Jonsson, 1998).

The difference between membrane filtration and sieving is the cutoff point, which in the case of membrane filtration is determined by the pore size of the membrane. The driving force for the mass transport across the membrane is a pressure gradient in the range 0.05–0.4 MPa. Because only large particles, in the range of 200 nm to 20 μm, can be separated by the membrane, the diffusion of particles and the osmotic pressure difference between the feed and filtrate solution are negligibly low. Mass transport in MF membranes takes place by viscous flow through the membrane pores. The solid membrane matrix can be considered as completely impermeable. Therefore, the hydrodynamic permeability can be expressed in terms of the membrane pore size, the porosity and the solution viscosity according to Hagen-Poiseuille's law, which is given by:

$$J_{\mathrm{v}} = \frac{\varepsilon r^2}{8\eta\tau} \frac{\Delta p}{\Delta z}$$

(13.1)

where J_v is the volumetric flux, ε the membrane porosity, r the pore radius, η the viscosity, τ the tortuosity factor, Δp the pressure difference across the membrane, and Δz the membrane thickness (Strathmann *et al.*, 2006).

Another parameter of interest in the practical application of MF is the separation characteristic of the membrane. This is generally expressed by the retention or rejection of a particle of a certain size, which is given by:

$$R_i = 1 - \frac{C_i^p}{C_i^f} \tag{13.2}$$

where R is the rejection coefficient, C is the concentration, the subscript i refers to a given component in the feed (f) and the permeate (p) (Strathmann *et al.*, 2006).

In UF, as in MF, the driving force is a pressure gradient in the range 0.2–1 MPa. UF membranes are typically asymmetric, with the smallest pores on the surface facing the feed solution; UF membrane pores are significantly smaller than those of MF membranes. Their diameters at the feed side of the membrane are between 2 and 10 nm and retained compounds have a molecular weight between 5000 and several million Daltons. Because UF membranes also retain some relatively low molecular weight solutes, osmotic pressure differences between the feed and the filtrate can be significant and diffusive fluxes of the solutes across the membrane are no longer negligibly low (Digens, 2007; Strathmann *et al.*, 2006). The flux through a UF membrane can be described as a function of the pressure difference between the feed and permeate solutions, the hydrodynamic permeability for the viscous flow, the osmotic pressure difference between feed and permeate solutions, and the phenomenological coefficient determining the diffusive flow of water through the membrane pores:

$$J_w \cong J_v = \overline{V}_w^2 L_w \frac{\Delta p - \Delta \pi}{\Delta z} + L_v \frac{\Delta p}{\Delta z} \tag{13.3}$$

where J_v and J_w are the total volume and solvent fluxes, respectively, Δp and $\Delta \pi$ are, respectively, the hydrostatic and the osmotic pressure gradients across the membrane, L_v is the hydrodynamic permeability and L_w the diffusive permeability of the solvent, and Δz is the thickness of the selective barrier of the membrane. Generally, the osmotic pressure can be neglected in UF processes. However, in the case of solutions that contain a high concentration of retained components with relatively low molecular weight, the osmotic pressure must be considered.

Permeate fluxes and permeate quality are the most important aspects in the selection of a proper MF or UF membrane for citrus juice clarification. Pretreatment methods such as treatment with pectinase or fining agents and centrifugation can reduce the particulate matter in the juice, leading to a remarkable improvement of permeate fluxes and the attainment of higher concentration factors (Rai *et al.*, 2007).

Permeate fluxes are strongly affected by operating parameters, the nature of the membrane and the juice composition. They generally increase with transmembrane pressure (TMP), up to a limiting value depending on the physical properties of the juice and axial velocity. An increase in the feed flow rate produces a linear increase of permeate flux due to the effect of shear stress at the membrane surface, which enhances the rate of removal of deposited particles and reduces the thickness of the polarized layer. Increasing the operating temperature produces a reduction in juice viscosity, together with an increase of the diffusion coefficient of macromolecules and a consequent enhancement of the permeate flux (Cassano *et al.*, 2007).

Polysulfone, poly(ether sulfone) (PES), poly(vinylidene fluoride) (PVDF), polyamide and polypropylene membranes are typical polymeric membranes used in citrus juice clarification. However, the major drawback of polymeric membranes is their low stability in extreme pH conditions and, consequently, limited shelf life for juice processing applications. By contrast, ceramic membranes offer many advantages over polymeric membranes in terms of high flux, easy cleaning, temperature stability, high pressure resistance, resistance to concentrated caustic acids and very high abrasion resistance.

13.4.1.1   *Membrane fouling*

Clarification of citrus juice and citrus byproducts is a prerequisite for concentration and fractionation by membrane processes. The use of MF and UF membranes offers many advantages over conventional clarification procedures based on the use of fining agents and filtration by diatomaceous earth. These include the possibility of operation at room temperature, the elimination of filter aids, increased juice yields, easy cleaning and maintenance of the equipment, and a reduction in waste products (Köseoglu *et al.*, 1990). The overall extraction efficiency is around 95–97%, which compares to 90–93% for traditional processing (Swientek, 1986). However, the major limiting factor in membrane-based clarification processes is flux decline over time due to membrane fouling. This phenomenon can be attributed to the adsorption of non-permeating solutes (such as pectic materials, proteins and fibers) on the membrane surface and to the precipitation of smaller solutes that are normally permeable in the membrane pores. The high concentration of macromolecules at the surface of the membrane tends to form a gel-type layer, which acts as an active membrane layer of fine porosity.

Membrane fouling is a key factor affecting the commercial viability of the process because it reduces productivity, leading to more frequent membrane cleaning and replacement and thereby increasing operating costs. In addition, the deposition of a gel layer and its subsequent growth on the membrane surface leads to a modification of membrane selectivity, which brings about changes in product quality.

Several methods are available for the evaluation of membrane fouling, including the measurement of flux decline, fouling time, modified fouling index, fouling-layer resistance and deposit thickness (Rai and De, 2009).

An analysis of membrane fouling in the clarification of orange juice using PVDF MF membranes revealed that the separation process is controlled by cake filtration mechanisms at relatively low velocity (i.e. Reynolds number = 5,000) and low TMPs. At a higher Reynolds number (15,000), an increase in applied TMP allows the limiting permeate flux to increase by a factor of about four. In these conditions the filtration process is controlled by a complete pore-blocking fouling mechanism and flux decay is negligible (Todisco *et al.*, 1996).

Different mathematical models have been proposed in order to describe the flux decline, based on equations applied to dead-end filtration mechanisms at constant pressure (Hérmia, 1982) and opportunely modified in order to describe the fouling mechanism involved in crossflow filtration. In the Field model (Field *et al.*, 1995), the permeate flux decline over time is described according to the following differential equation:

$$-\frac{dJ}{dt} = k(J - J_{\lim})J^{2-n} \tag{13.4}$$

where $J_{\lim}$ represents the limit value of the permeate flux obtained in steady-state conditions, and k and n are phenomenological coefficient and general index, respectively, depending on fouling mechanism. The proposed model allows, on the basis of experimental data, indication of the fouling mechanism involved in the filtration process, according to the estimated value for n as follows:

- *Complete pore blocking (n = 2)*
  This situation corresponds to a condition of complete pore obstruction by means of sealing (blocking), which occurs when the particles are larger than the pore size.
- *Partial pore blocking (n = 1)*
  In this condition a dynamic situation of blocking/unblocking occurs in which particles may bridge a pore by obstructing the entrance but not completely blocking it.
- *Cake filtration (n = 0)*
  When solid particles or macromolecules do not enter the pores they form a cake on the membrane surface. In this case the overall resistance to the mass transport is composed of a cake resistance and a membrane resistance which is assumed to remain unchanged.

- *Internal pore blocking (n = 1.5; $J_{lim} = 0$)*

  Particles enter the pores, reducing pore volume. In this case membrane resistance increases as a consequence of pore size reduction. In these conditions the fouling phenomenon becomes independent of crossflow velocity and a limit value for the permeate flux is not reached, i.e. $J_{lim} = 0$.

The above model was used to analyze the flux decline in the UF of blood orange juice with a tubular PVDF membrane having a molecular weight cutoff (MWCO) of 15 kDa (Cassano *et al.*, 2007). The analysis revealed that in fixed conditions of TMP and temperature the fouling mechanism evolves from partial pore blocking ($n = 1$) to complete pore blocking ($n = 2$) in dependence of the axial velocity. Theoretical predictions in terms of permeate flux as a function of time were in agreement with the experimental data.

Recently, a suitable lab-scale filtration strategy was investigated in order to predict membrane fouling during crossflow MF of orange juice through a relevant evaluation of its fouling behavior. A broad range of operating conditions of dead-end filtration was explored through a D-optimal experimental design. The results showed that prediction of orange juice fouling propensities in pilot-scale MF was possible. Therefore, dead-end filtration was proposed as a process-specific tool to predict membrane fouling in the clarification of orange juice by crossflow MF (Dahdouh *et al.*, 2015).

13.4.2 *Reverse osmosis (RO)*

The industrial concentration of fruit juices is usually performed by multistage vacuum evaporation processes, in which water is removed at high temperatures followed by recovery and concentration of volatile flavors and their reintroduction to the concentrated product (Jiao *et al.*, 2004).

Fruit juice concentration by RO has been of interest to the fruit-processing industry for about 30 years. In this process a hydraulic pressure greater than the osmotic pressure is applied through a semi-permeable membrane in order to enable a flow of water from the side with high solute concentration to the low concentration one. Sugars and organic acids are the osmo-active substances that contribute to the osmotic pressures of citrus juices.

The main advantages of RO, when compared with traditional thermal evaporation, are the generation of higher-quality products due to the use of low operating temperatures, resulting in an increased retention of nutritional and aroma compounds, reduced energy consumption and lower equipment costs (Girard and Fukumoto, 2000). On the other hand, the main disadvantage of RO is its inability to match the concentrations of standard products produced by evaporation because of osmotic pressure limitations. Indeed, when fruit juices are concentrated by RO with polymeric membranes, the maximum achievable concentration without affecting permeate flux and solute recovery is about 25–30 °Bx. High operating pressures (ranging from 1 to 20 MPa) are needed in order to overcome the osmotic pressure of the concentrated juice. This suggests the use of RO as a pre-concentration step in combination with other concentration techniques, such as freeze concentration or evaporation, in order to reduce energy consumption and increase production capacity.

Oil-soluble aroma compounds of orange juice are more easily retained by cellulose acetate membranes than water-soluble ones. High recoveries of sugars (greater than 98%), acids (up to 85%) and volatile flavor compounds can be obtained by using spiral-wound polyamide membranes (Medina and Garcia, 1988). The differential rejection of RO membranes with regards to sugars and organic acids produces an increase in the Brix/acid ratio with a consequent reduction in juice bitterness.

13.4.3 *Nanofiltration*

NF is very similar to UF and RO, because in all three processes a hydrostatic pressure is applied to transport a molecular mixture to the surface of a membrane. The main difference between NF and UF is the pore size of the membrane and thus the molecular weight of the components that are retained. NF membranes have a pore size in the range of 0.5–2 nm, corresponding to a MWCO of about 400–500 Da. In this case, given that NF membranes have smaller pore sizes in comparison to UF, osmotic pressure may be significant. The pressure applied is higher than UF and lower than RO, usually in the range 0.3–3 MPa.

The main difference between NF and RO is the transport mechanism through the membrane. In NF, it is assumed that all components permeate the membrane exclusively through geometrically well-defined pores, whereas in RO the components permeate the membrane by diffusion through a more or less homogeneous polymer matrix and the separation is a result of the solubility and diffusivity of solutes in the polymer matrix (Strathmann *et al.*, 2006). In addition, NF membranes often carry positive or negative electrical charge, whereas UF and RO membranes are neutral. As a consequence, the separation mechanism is not only determined by size exclusion but also by the so-called Donnan exclusion, which postulates that ions carrying the same charge as the membrane (so-called co-ions) will be excluded from the membrane. Details relating to mass transport in NF membranes have been widely reported in the literature (Bowen and Mukhtar, 1996; Schäfer, 2005; Tsuru *et al.*, 1991; Wang *et al.*, 1995).

Although NF has also been investigated in the concentration of other fruit juices, such as apple, pear and blackcurrant (Bánvölgyi *et al.*, 2006; Warczok *et al.*, 2004), the fractionation of citrus juices and citrus byproducts by NF appears a very promising molecular separation technology for the recovery of bioactive compounds as an alternative to conventional systems.

13.4.4   *Selected applications*

13.4.4.1   *Orange press liquor*

Membrane technologies appear an attractive approach for the recovery and concentration of bioactive compounds (BACs) from agro-food wastes, thus transforming the effluents or byproducts into source material for high-value compounds (Crespo and Brazinha, 2010; Mudimu *et al.*, 2012; Tsibranska and Tylkowski, 2014).

Traditional treatments of citrus peel involve the use of lime followed by milling and pressing. The resulting press liquor is enriched in BACs, such as flavonoids and phenolic acids, recognized for their beneficial implications for human health due to their antioxidant activity and free radical-scavenging ability (Bocco *et al.*, 1998). The orange press liquor has, on average, a total soluble solids (TSS) content of 10 °Bx; it can be concentrated up to 65–70 °Bx by multiple-effect evaporation to obtain citrus molasses, which can be used in the production of beverage alcohol and as a cattle feed. However, thermal evaporation is characterized by high energy consumption and the thermal degradation of BACs (Cassano and Jiao, 2014). Membrane processes, such as RO, can only be used for pre-concentration of citrus press liquor, because the maximum concentration obtained with RO is still far from the value required to obtain citrus molasses (72 °Bx).

García *et al.* (2002) investigated the use of a spiral-wound RO membrane (Filmtec SW30-2540, Dow Chemical, USA) to pre-concentrate model solutions of sucrose (10, 20 and 30 °Bx) at different pressures, feed flows and temperatures. This work was aimed at evaluating the potential of the RO process in the production of a permeate stream (which can be reused in the juice production) and a retentate which can be thermally concentrated to produce citrus molasses from orange press liquor. Garcia-Castello *et al.* (2006) also established that RO, used as a pre-concentration step, is able to reduce energy consumption by a factor of about 7.7 in comparison to a traditional multiple-effect evaporator.

RO has also been used to pre-concentrate a synthetic orange press liquor with and without the addition of pectins, as part of an evaluation of the necessity for pretreatment (e.g. depectinization) (Garcia-Castello *et al.*, 2011). The synthetic solution without pectins was well concentrated in all of the conditions investigated. On the other hand, the high viscosity of the solution containing pectins, as well as strong membrane fouling, allowed concentration of the solution only up to a volume reduction factor (VRF) of 1.2 at the maximum TMP studied (5 MPa), suggesting the necessity of a pretreatment step before RO.

Membrane processes are also attractive technologies for the production of pectins from citrus peel when compared with traditional extraction processes based on the use of large amounts of ethanol. In this sense, the use of MF/UF can be a useful technique to separate pectins from BACs. Xie *et al.* (2008) studied the performance of a tubular ceramic UF membrane (ZrO_2, 30 kDa) in the treatment of a pectin-containing solution extracted from citrus peel. They observed a pectin rejection greater than 90%, whereas the rejection of phenolic compounds was less than 20%. These promising results prove that the decolorization, separation and purification of pectin can be achieved simultaneously through the use of ceramic UF membranes.

Ruby-Figueroa *et al.* (2011) evaluated the performance of a UF membrane (polysulfone, 100 kDa, China Blue Star Membrane Technology Co. Ltd., China) in the clarification of orange press liquor using the response surface methodology (RSM) approach. The effect of operating conditions, such as TMP, temperature and feed flowrate, on the permeate flux and fouling index was investigated. Optimization of multiple responses permitted the establishment of operating conditions that simultaneously gave maximum permeate flux and minimum fouling index. According to the results obtained, the maximum permeate flux (of about $0.007 \, kg \, m^{-2} \, s^{-1}$) and the minimum fouling index (48%) were obtained in optimized operating conditions of 0.14 MPa TMP, 15°C and flow of $0.046 \, L \, s^{-1}$. In other work, the operating conditions were optimized in order to minimize the rejection of polyphenols in the clarification process by using the same UF membrane. A minimum rejection of 28.4% was obtained at a TMP of 0.02 MPa, a temperature of 19.85°C and a feed flowrate of $0.068 \, L \, s^{-1}$ (Ruby-Figueroa *et al.*, 2012).

Recently, the clarification of orange press liquors was investigated by using PVDF hollow fiber membranes prepared through the dry/wet spinning technique (Simone *et al.*, 2016). Some of the produced fibers exhibited high permeability (pure water permeability $\sim$$5300 \, L \, m^{-2} \, h^{-1} \, MPa^{-1}$), coupled to good mechanical resistance and pore size in the range of MF membranes. In optimized operating conditions, the selected fibers produced steady-state fluxes of about $41 \, L \, m^{-2} \, h^{-1}$, with rejection towards polyphenols and total antioxidant activity at 4.1% and 1.4%, respectively. The produced fibers exhibited a fouling index of 55.6%. A first cleaning with distilled water permitted recovery of about 56% of the initial water permeability ($2960 \, L \, m^{-2} \, h^{-1} \, MPa^{-1}$), due to the removal of the reversible polarized layer; a chemical cleaning with a $50 \, mg \, L^{-1}$ sodium hypochlorite solution (at 40°C for 60 min) permitted recovery of about 87% of the water permeability. This incomplete recovery of water permeability was attributed to an irreversible component of fouling.

NF is a useful approach for separating and concentrating BACs of orange press liquors previously clarified by MF or UF. Conidi *et al.* (2012) evaluated the potential of four spiral-wound NF membranes of different materials (polyamide, polypiperazine amide and PES) and MWCOs (180, 300, 400 and 1000 Da) in the separation of phenolic compounds of orange press liquor from sugar compounds. A correlation between sugar rejection and MWCO was identified. Indeed, a strong reduction in the average rejection of sugars was observed when MWCO was increased. On the other hand, the rejection of anthocyanins was higher than 89%, independent of the MWCO of the selected membrane. The PES10 NF membrane, with a MWCO of 1000 Da, showed the lowest average rejection of sugar compounds and high rejections of anthocyanins (89.2%) and flavonoids (70%). Permeate flux values at lower TMPs were also notably higher than for the other NF membranes investigated.

Recently, Cassano *et al.* (2014c) investigated the potential of an integrated membrane process for the recovery and concentration of flavonoids from orange press liquor. The press liquor was previously clarified by using a UF membrane module in hollow fiber configuration (polysulfone, 100 kDa, China Blue Star Membrane Technology Co. Ltd., China) in particular operating conditions (TMP: 0.054 MPa; temperature: 25±1°C; feed flowrate: $0.14 \, L \, s^{-1}$). This process allowed the removal of all suspended solids from the raw press liquor, while flavonoids and anthocyanins were recovered in the clarified liquor (rejections of flavanones and anthocyanins were lower than 1%). The clarified press liquor was then pre-concentrated by NF using a spiral-wound PES membrane (NF-PES 10, 2440C, Microdyn-Nadir, Germany). The NF process was operated at 0.8 MPa and 20°C in a batch concentration mode to reach a VRF of 5. In these conditions, steady-state permeate fluxes of $6 \, L \, m^{-2} \, h^{-1}$ were obtained. The rejections by the NF membrane of flavanones and anthocyanins were 97.4% and 98.9%, respectively.

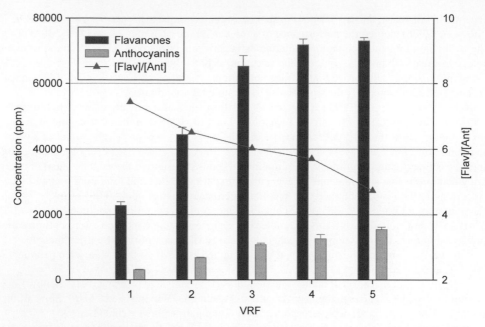

Figure 13.3.   Concentration of flavanones and anthocyanins in the retentate as a function of VRF in the nanofiltration of clarified bergamot juice with a PES10 membrane.

Furthermore, the ratio between flavanones and anthocyanins was decreased by increasing the VRF (Fig. 13.3). These results allow balancing of the flavonoid content in relation to that of anthocyanins and, consequently, the modification of the characteristics of the final product in terms of bittering capacity (from the presence of flavonoids) and coloring power (from the presence of anthocyanins). A final concentration of the liquor was reached by using OD. According to the proposed process scheme, depicted in Figure 13.4, flavanones and anthocyanins are pre-concentrated in the NF step with the production of a permeate stream containing sugars and minerals. The final treatment of the NF retentate by OD produces a concentrated solution of great interest for food and pharmaceutical applications. Indeed, flavanones are widely acknowledged for their pharmacological properties (a favorable effect on capillary fragility and treatment of inflammatory states), which arise from their antioxidant activity. In addition, the extract can be used in food coloring, avoiding the use of artificial colorants.

13.4.4.2   *Bergamot juice*

Bergamot is the common name of the fruit *Citrus bergamia* Risso. This is a natural hybrid fruit, derived from bitter orange and lemon, produced almost exclusively in the Ionic area of Calabria (Italy), where soil characteristics and pH (<6.5–7.5) are particularly suitable for its cultivation. It is mainly used in the production of the essential oil from its peel, which is widely employed in the pharmaceutical (antibacterial and antiseptic activity), cosmetic (perfumes, body lotions, soaps, etc.) and food (aroma, sweets, liquors and tea) industries (Verzera *et al.*, 2003).

The juice, however, has not so far found a real use in the food industry due to its bitter taste and is, therefore, considered as a waste product of the essential oil production. However, current studies on bergamot juice have revealed a remarkable content of flavonoids, known for their beneficial effects on human health (Di Donna *et al.*, 2011, 2014), and the extraction of bioactive compounds from bergamot juice is, today, considered a research topic of great interest in terms of its potential application in various market areas (Pernice *et al.*, 2009).

The recovery and concentration of polyphenols from bergamot juice has been also investigated by using membrane operations in a sequential design. In particular, an integrated membrane

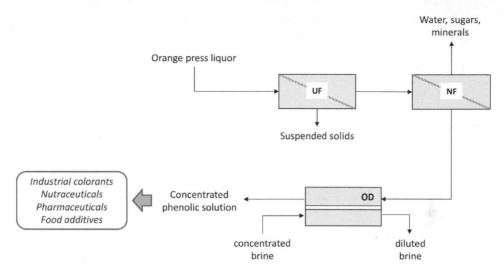

Figure 13.4. Proposed process scheme for the separation and concentration of phenolic compounds from orange press liquor.

process based on the use of UF and NF membranes was investigated by Conidi *et al.* (2011). In the first step, the depectinized juice was clarified by using UF hollow fiber membranes (polysulfone, 100 kDa, China Blue Star Membrane Technology Co. Ltd., China) under specific operating conditions (TMP: 0.07 MPa; temperature: 24°C; feed flow rate: 0.032 L s^{-1}). This step allowed the removal of suspended solids and macromolecules, such as pectins, giving a clear juice with a sugar content unchanged in relation to the fresh juice. A majority of the organic acids, flavonoids and polyphenols were recovered in the UF permeate, as demonstrated by the total antioxidant activity of the clarified juice which was only 9% lower than that of the depectinized juice. The clarified juice was then treated with a UF membrane (Etna 01PP, flat-sheet fluoropolymer, 1000 Da, Alfa Laval, Sweden) and two different ceramic NF membranes (monotubular TiO$_2$ membranes, 750 and 450 Da, Inopor, Germany) in order to evaluate the effect of the MWCO on the rejection by the membranes of sugars, organic acids and polyphenols. The results relating to the effect of the MWCO on the rejection of sugars and flavonoids are depicted in Figure 13.5.

The results obtained indicated that the best separation of polyphenols and sugars occurred with the membrane of 450 Da (higher rejection of flavonoids and moderate rejection of sugars). As a consequence, the NF retentate can be considered as a source of bioactive compounds of potential interest in the pharmaceutical industry, particularly given the recent statin-like active principles that have been identified in bergamot juice (Di Donna *et al.*, 2009).

Recently, different NF membranes in a spiral-wound configuration (NF PES10, N30F and NF270), with different MWCOs and polymeric materials, were evaluated for the treatment of clarified bergamot juice in terms of productivity and selectivity towards flavonoids, total phenolic compounds, sugars, and total antioxidant activity (Conidi *et al.*, 2014). In general, the NF PES10 membrane, a PES membrane with a MWCO of 1,000 Da, showed the largest gap between the rejection coefficients affecting TSS (mainly sugars) and phenolic compounds (Table 13.2). The retentate fractions produced, characterized by high antioxidant activity, can therefore be considered to be of interest in nutraceutical applications. For all the membranes investigated, an increase in TMP from 0.4 to 1.6 MPa produced an increased rejection of total polyphenols.

13.4.4.3 *Immobilization of β-cyclodextrins in polymeric membranes*
β-cyclodextrins (β-CDs) are cyclic oligosaccharides consisting of seven α-D-glucose units, which are 1,4-linked. These molecules assume a cone-like structure with a hydrophobic cavity able to

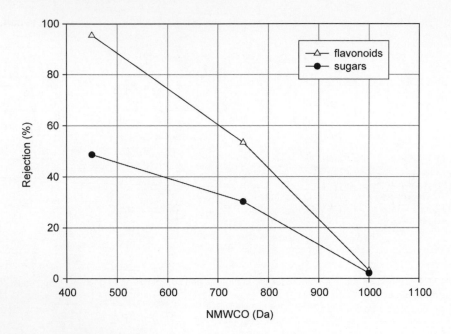

Figure 13.5.   Effect of nominal MWCO on the rejection of sugars and flavonoids by UF and NF membranes.

Table 13.2.   Total antioxidant activity (TAA), total soluble solids (TSS), total polyphenols and associated rejection rates (R) in bergamot juice treated with different NF membranes.

Membrane type	Sample	TAA [mM Trolox]	R_{TAA} [%]	TSS [°Bx]	R_{TSS} [%]	Total polyphenols [mg L^{-1}]	$R_{polyphenols}$ [%]
NF PES10	F	8.6 ± 0.3		5.8 ± 0.2		660.2 ± 9.6	
	P	3.3 ± 0.1	80.2	3.85 ± 0.4	35.8	302.7 ± 5.4	71.6
	R	16.7 ± 2.0		6.0 ± 0.8		1067 ± 10.1	
N30F	F	7.2 ± 0.4		6.0 ± 1.2		462.2 ± 9.35	
	P	3.7 ± 0.2	76.8	2.84 ± 0.5	64.9	294.4 ± 8.34	69.7
	R	16 ± 1.1		8.1 ± 0.45		972.2 ± 17.34	
NF 270	F	9.4 ± 0.1		4.06 ± 0.2		602.7 ± 15.45	
	P	2.1 ± 0.3	88.2	1.0 ± 0.05	87.1	207.2 ± 12.25	81.9
	R	17.9 ± 1.3		7.8 ± 0.8		1148.1 ± 16.34	

Legend: F, feed; P, permeate; R, retentate.

totally or partially encapsulate guest molecules through noncovalent bonds such as hydrogen bonds, electrostatic interactions and van der Waals forces (Del Valle, 2004; Szejtli, 2004). The hydroxyl groups are oriented on the external part of the structure, making it hydrophilic. The immobilization of β-CD in polymeric materials can be exploited to trap organic pollutants or to remove unwanted compounds from different sources (Kozlowski and Sliwa, 2008). Some applications of membranes based on this material that have been investigated at laboratory scale and are of potential interest in the citrus processing industry are discussed below.

13.4.4.3.1   *Removal of pesticides from citrus essential oils*
Citrus peel oils are rich in aroma and flavor components and, therefore, are generally used at industrial level as flavoring additives in both food (bakery goods, soft drinks and citrus juices) and

Table 13.3.   Analysis of pesticides (imazalil, thiabendazole, o-phenylphenol and chlorpyrifos) in fractions of citrus essential oils treated with hollow fiber membranes loaded with triacetyl-β-cyclodextrins.

Essential oil type	Sample	Imz [$\mu g L^{-1}$]	Tbz [$\mu g L^{-1}$]	Opp [$\mu g L^{-1}$]	Clor [$\mu g L^{-1}$]
Lemon	F	798	470	2535	1050
	P	401	240	2513	1044
	R	2308	1545	3044	1437
Orange	F	470	–	8500	–
	P	220	–	8140	–
	R	1830	–	8687	–

Legend: F, feed; P, permeate; R, retentate.

non-food (perfumes, soaps, cosmetics, etc.) products. Volatiles form 85% to 99% of the entire oil; these consist of monoterpene and sesquiterpene hydrocarbons and their oxygenated derivatives, as well as aliphatic aldehydes, alcohols and esters. A monocyclic terpene, D-limonene, constitutes up to 68% of citrus essential oils. The non-volatile fraction, constituting between 1% and 15% of the oil, is composed of hydrocarbons, sterols, fatty acids, waxes, carotenoids, coumarins, psoralens and flavonoids (Tranchida *et al.*, 2012).

Unfortunately, low but significant levels of pesticides can often be found in citrus peel oil too. Indeed, these compounds build up preferentially on the peel and are concentrated in the essences extracted from the flavedo.

Several legislations, including EU directives (Council directives 76/895/EEC, 86/362/EEC, 86/363/EEC and 90/642/EEC) and Swiss regulation (RS 817.021.23), have established maximum residue limits (MRL) for pesticides in foodstuffs in order to ensure the protection of human health from exposure to pesticide residues. Although the development of biodegradable pesticides appears promising, the use of non-biodegradable pesticides is expected to continue in the near future. Methods to reduce the level of pesticides in raw citrus peel oils, without modifying their aroma and flavor characteristics, would therefore be desirable.

Essentially pesticide-free citrus oil can be obtained by mild distillation of raw citrus oil in one or more short-path distillation columns, from which the essentially pesticide-free citrus oil is collected as the distillant; all the pesticides from the feed material are collected in the residue stream from the last column in the series (Muraldihara, 1996).

Recently, polyacrylonitrile (PAN) hollow fiber membranes, loaded with triacetyl-β-cyclodextrins, were prepared and tested for the rejection of pesticides from mandarin, lemon and orange essential oils (Cassano *et al.*, 2013). The prepared membranes exhibited good stability towards the treated essential oils and average permeate fluxes between 15 and 30 kg m^{-2} h^{-1} in the selected operating conditions. They enabled removal of about 49% of imazalil and thiabendazole from lemon essential oil, and about 50% of imazalil from orange and mandarin essential oils. No significant removal of o-phenylphenol or chlorpyrifos was observed in the permeate stream of the process (Table 13.3). The regeneration of hollow fiber membranes with ethanol did not modify the selectivity of the membranes towards the analyzed pesticides. Moreover, the removal of the pesticides did not affect the organoleptic and physico-chemical characteristics of the processed oils.

13.4.4.3.2   *Debittering of citrus juices*
The consumer acceptability of citrus juice is negatively affected by the presence of bitter compounds such as limonin (a limonoid) and naringin (a flavonoid), which typically increase after squeezing or heating (Stinco *et al.*, 2013). Therefore, several techniques have been used to decrease the concentration of bitter compounds in citrus juices, including biodegradation by enzymes (Cavia-Saiz *et al.*, 2011), UF (Todisco *et al.*, 1998), molecularly imprinted polymeric

membranes (Trotta *et al.*, 2002), use of adsorbent resins (Kola *et al.*, 2010; Liu and Gao, 2015) and β-cyclodextrin polymers (Binello *et al.*, 2008).

Mixed polymers with covalently linked cyclodextrin units can be obtained by condensation with bifunctional or polyfunctional cross-linking agents such as aldehydes, ketones, isocyanates and epoxides. These polymers are effective in removing limonin and naringin from orange and grapefruit juices (Shaw *et al.*, 1984). Flat-sheet membranes made of an amorphous polyether ether ketone known as PEEK-WC and a β-CD derivative (O-octyloxycarbonyl-β-CD) were prepared through the diffusion-induced phase separation method (Fontananova *et al.*, 2004). The β-CD derivative entrapped in the polymeric membrane was able to form inclusion complexes with naringin. The amount of naringin retained was increased by increasing the amount of the β-CD derivative entrapped in the membrane: membranes prepared with 7.5 wt% of β-CD derivative retained 3.74 μgmol of naringin per gram of membrane.

The entrapment of the β-CD derivative in the polymeric membrane optimized its interaction with naringin and increased the stability of the host/guest complex. Indeed, a low capacity of the β-CD derivative to form a host/guest complex with naringin was observed when the β-CD derivative was simply dispersed in an aqueous solution of naringin.

13.5   CONCLUDING REMARKS AND FUTURE TRENDS

Membrane operations represent useful approaches for the recovery of high added-value compounds from citrus byproducts when compared with traditional separation systems. Promising results obtained at laboratory scale indicate that the conventional industrial cycle of the citrus juice industry can be redesigned through the introduction of these technologies, leading to several advantages in terms of energy saving, product quality, simplification of depolluting processes, and reductions in water consumption and disposal costs. In addition, the selection of proper membrane-based recovery procedures offers interesting perspectives for the exploitation of citrus byproducts in the provision of value-added natural antioxidants, antimicrobial agents, vitamins, etc., of great interest in food, pharmaceutical and cosmetic applications.

The production of more selective and permeable membranes and improvements in process engineering, including module and process design, as well as the development of molecularly imprinted membranes with molecular recognition properties for target compounds, represent research topics for future investigation that might provide a significant contribution to the improvement of protocols for making the most of citrus fruits.

REFERENCES

Abirami, A., Nagarani, G. & Siddhuraju, P. (2014) Measurement of functional properties and health promoting aspects-glucose retardation index of peel, pulp and peel fiber from Citrus hystrix and Citrus maxima. *Bioactive Carbohydrates and Dietary Fibre*, 4, 16–26.

Bánvölgyi, S., Horváth, S., Békássy-Molnár, E. & Vatai, G. (2006) Concentration of blackcurrant (*Ribes nigrum* L.) juice with nanofiltration. *Desalination*, 200, 535–536.

Barreca, D., Bellocco, E., Caristi, C., Leuzzi, U. & Gattuso, G. (2011) Flavonoid profile and radical-scavenging activity of Mediterranean sweet lemon (*Citrus limetta* Risso) juice. *Food Chemistry*, 129, 417–422.

Benavente-García, O. & Castillo, J. (2008) Update on uses and properties of citrus flavonoids: new findings in anticancer, cardiovascular, and anti-inflammatory activity. *Journal of Agricultural and Food Chemistry*, 56, 6185–6205.

Bichsel, S.E. & Sandre, A.M. (1982) Application of membrane technology to juice concentration. *International Sugar Journal*, 84, 266–268.

Binello, A., Robaldo, B., Barge, A., Cavalli, R. & Cravotto, G. (2008) Synthesis of cyclodextrin-based polymers and their use as debittering agents. *Journal of Applied Polymer Science*, 107, 2549–2557.

Bocco, A., Cuvelier, M.E., Richard, H. & Berset, C. (1998) Antioxidant activity and phenolic composition of citrus peel and seed extract. *Journal of Agricultural and Food Chemistry*, 46, 2123–2129.

Bowen, W.R. & Mukhtar, H. (1996) Characterization and prediction of separation performance of nanofiltration membranes. *Journal of Membrane Science*, 112, 263–274.

Braddock, R.J. (1995) By-products of citrus fruits. *Food Technology* (Chicago), 49, 74–77.

Cassano, A. & Jiao, B. (2014) Integrated membrane operation in citrus processing. In: Cassano, A. & Drioli, E. (eds) *Integrated Membrane Operation in the Food Industry*. Walter de Gruyter & Co, Berlin. pp. 87–112.

Cassano, A., Marchio, M. & Drioli, E. (2007) Clarification of blood orange juice by ultrafiltration: analyses of operating parameters, membrane fouling and juice quality. *Desalination*, 212, 15–27.

Cassano, A., Tasselli, F., Conidi, C., Drioli, E., Timpone, R., D'Avella, M., Badalamenti, F. & Corleone, V. (2013) PAN hollow fiber membranes with triacetyl-b-cyclodextrin for the removal of pesticides from citrus essential oils. *Separation and Purification Technology*, 116, 124–130.

Cassano, A., Conidi, C. & Drioli, E. (2014a) Integrated membrane operation in fruit juice processing. In: Cassano, A. & Drioli, E. (eds) *Integrated Membrane Operation in the Food Industry*. Walter de Gruyter & Co, Berlin. pp. 59–86.

Cassano, A., Ruby-Figueroa, R. & Drioli, E. (2014b) Membrane separation. In: Varzakas, T. & Tzia, C. (eds) *Food Engineering Handbook – Food Process Engineering*. CRC Press, Boca Raton, FL. pp. 1–29.

Cassano, A., Conidi, C. & Ruby-Figueroa, R. (2014c) Recovery of flavonoids from orange press liquor by an integrated membrane process. *Membranes*, 4, 509–524.

Cavia-Saiz, M., Muniz, P., Ortega, N. & Busto, M.D. (2011) Effect of enzymatic debittering on antioxidant capacity and protective role against oxidative stress of grapefruit juice in comparison with adsorption on exchange resin. *Food Chemistry*, 125, 158–163.

Cheryan, M. (1998) *Ultrafiltration and Microfiltration Handbook*. Technomic Publishing, Lancaster, PA.

Conidi, C. & Cassano, A. (2015) Recovery of phenolic compounds from bergamot juice by nanofiltration membranes. *Desalination and Water Treatment*, 56, 3510–3518.

Conidi, C., Cassano, A. & Drioli, E. (2011) A membrane-based study for the recovery of polyphenols from bergamot juice. *Journal of Membrane Science*, 375, 182–190.

Conidi, C., Cassano, A. & Drioli, E. (2012) Recovery of phenolic compounds from orange press liquor by nanofiltration. *Food and Bioproducts Processing*, 90, 867–874.

Crespo, J.G. & Brazinha, C. (2010) Membrane processing: natural antioxidants from winemaking by-products. *Filtration & Separation*, 47, 32–35.

Crosby, N.T. (1981) *Food Packaging Materials: Aspects of Analysis and Migration of Contaminants*. Applied Science Publishers, London.

Dahdouh, L., Wisniewski, C., Ricci, J., Kapitan-Gnimdu, A., Dornier, M. & Delalonde, M. (2015) Development of an original lab-scale filtration strategy for the prediction of microfiltration performance: application to orange juice clarification. *Separation and Purification Technology*, 156, 42–50.

Del Valle, E.M.M. (2004) Cyclodextrins and their uses: a review. *Process Biochemistry*, 39, 1033–1046.

Di Donna, L., De Luca, G., Mazzotti, F., Napoli, A., Salerno, R., Taverna, D. & Sindona, G. (2009) Statin-like principles of bergamot fruit (*Citrus bergamia*): isolation of 3-hydroxymethylglutaryl flavonoid glycosides. *Journal of Natural Products*, 72, 1352–1354.

Di Donna, L., Gallucci, G., Malaj, N., Romano, E., Tagarelli, E. & Sindona, G. (2011) Recycling of industrial essential oil waste: Brutieridin and Melitidin, two anticholesterolaemic active principles from bergamot albedo. *Food Chemistry*, 125, 438–441.

Di Donna, L., Iacopetta, D., Cappello, A.R., Gallucci, G., Martello, E., Fiorillo, M., Dolce, E. & Sindona, G. (2014) Hypocholesterolaemic activity of 3-hydroxy-3-methyl-glutaryl flavanones enriched fraction from bergamot fruit (*Citrus bergamia*): "In vivo" studies. *Journal of Functional Foods*, 7, 558–568.

Digens, B. (2007) Membrane and filtration technologies and the separation and recovery of food processing waste. In: Waldron, K. (ed) *Handbook of Waste Management and Co-product Recovery in Food Processing*. Woodhead Publishing, Cambridge. pp. 258–281.

Di Mauro, A., Fallico, B., Passerini, A. & Maccarone, E. (2000) Waste water from citrus processing as a source of hesperidin by concentration of styrene-divinylbenzene resin. *Journal of Agricultural and Food Chemistry*, 48, 2291–2295.

Downing, D.L. (1996) Processing procedures for canned food products. A complete course in canning and related processes. CTI Publications (Cooling Technology Institute), Baltimore, MD.

Elleuch, M., Bedigian, D., Roiseux, O., Besbes, S., Blecker, C. & Attia, H. (2011) Dietary fiber and fiber-rich by-products of food processing: characterization, technological functionality and commercial applications: a review *Food Chemistry*, 124, 411–421.

Escarpa, A. & Gonzalez, M.C. (2001) An overview of analytical chemistry of phenolic compounds in foods. *Critical Reviews in Analytical Chemistry* 31, 57–139.

Escobedo-Avellaneda, Z., Gutierrez-Uribe, J., Valdez-Fragoso, A., Torres, J.A. & Welti-Chanes, J. (2014) Phytochemicals and antioxidant activity of juice, flavedo, albedo and comminuted orange. *Journal of Functional Foods*, 6, 470–481.

Famyima, O. & Ough, C. (1982) Grape pomace: possibilities as animal feed. *American Journal of Enology and Viticulture*, 33, 44–46.

FAO (2014) FAOSTAT. Food and Agricultural Organization of the United Nations, Rome. Available from: http://www.fao.org/faostat/en/#data/QC/visualize [accessed November 2016].

Fontananova, E., Basile, A., Cassano, A. & Drioli, E. (2003) Preparation of polymeric membranes entrapping β-cyclodextrins and their molecular recognition of naringin. *Journal of Inclusion Phenomena and Macrocyclic Chemistry*, 47, 33–37.

Galanakis, C.M. (2012) Recovery of high added-value components from food wastes: conventional, emerging technologies and commercialized applications. *Trends in Food Science & Technology*, 26, 68–87.

García, E., Gonzálves, J.M. & Lora, J. (2002) Use of reverse osmosis as a preconcentration system of waste leaching liquid from the citric juice production industry. *Desalination*, 148, 137–142.

Garcia-Castello, E.M., Lora-Garcia, J., Garcia-Garrido, J. & Rodriguez-Lopez, A.D. (2006) Energetic comparison for leaching waste liquid from the citric juice production using both reverse osmosis and multi-effect evaporation. *Desalination*, 191, 178–185.

Garcia-Castello, E.M., Mayor, L., Chorques, S., Arguelles, A., Vidal-Brotons, D. & Gras, M.L. (2011) Reverse osmosis concentration of press liquor from orange juice solid wastes: flux decline mechanisms. *Journal of Food Engineering*, 106, 199–205.

Gattuso, G., Barreca, D., Gargiulli, C., Leuzzi, U. & Caristi, C. (2007) Flavonoids composition of citrus juice. *Molecules*, 12, 1641–1673.

Girard, B. & Fukumoto, L.R. (2000) Membrane processing of fruit juices and beverages: a review. *Critical Reviews in Biotechnology*, 20, 109–175.

Gonzalez-Molina, E., Dominguez-Perles, R., Moreno, D.A. & Garcia-Viguera, C. (2010) Natural bioactive compounds of *Citrus limon* for food and health. *Journal of Pharmaceutical and Biomedical Analysis*, 51, 327–354.

Hang, Y.D. (1999) Waste management: fruits and vegetables. In: Francis, F.J. (ed) *Encyclopedia of Food Science and Technology*. John Wiley & Sons, New York.

Hartel, R.W. (1992) Evaporation and freeze concentration. In: Heldman, D.R. & Lund, D.B. (eds) *Handbook of Food Engineering*. Marcel Dekker, New York. pp. 343–393.

Iu, J., Mittal, G.S. & Griffiths, M.W. (2001) Reduction in levels of *Escherichia coli* O157:H7 in apple cider by pulsed electric fields. *Journal of Food Protection*, 64 (2001), 964–969.

Izquierdo, L. & Sendra, J.M. (1993) Composition and characterization. In: Macrae, R., Robinson, R.K. & Sadler, M.J. (eds) *Encyclopedia of Food Science, Food Technology and Nutrition*. Academic Press, San Diego, CA. pp. 1335–1341.

Jiao, B., Cassano, A. & Drioli, E. (2004) Recent advances on membrane processes for the concentration of fruit juices: a review. *Journal of Food Engineering*, 63 (2004), 303–324.

Jing, L. & Howard, A.C. (2010) Application of membrane techniques for purification of natural products. *Biotechnology Letters*, 32, 601–608.

Jonsson, G. (1980) Overview of theories for water and solute transport in UF/RO membranes. *Desalination*, 35, 21–38.

Kato, M., Ikoma, Y., Matsumoto, H., Sugiura, M., Hyodo, H. & Yano, Y. (2004) Accumulation of carotenoids and expression of carotenoid biosynthesis genes during maturation in citrus fruit. *Plant Physiology*, 134, 824–837.

Khan, M.K., Huma, Z. & Dangles, O. (2014) A comprehensive review on flavanones, the major citrus polyphenols. *Journal of Food Composition and Analysis*, 33, 85–104.

Kilara, A. & Van Buren, J. (1989) Clarification of apple juice. In: Downing, D.L. (ed) *Processed Apple Products*. Van Nostrand Reinhold, New York. pp. 83–96.

Knorr, D., Ade-Omowaye, B.I.O. & Heinz, V. (2002) Nutritional improvement of plant foods by nonthermal processing. *Proceedings of the Nutrition Society*, 61, 311–318.

Koevoets, W.A.A. (2002) Reliable and sustainable treatment solution for fruit processing effluents. *Fruit Processing*, 12, 102–106.

Kola, O., Kaya, C., Duran, H. & Altan, A. (2010) Removal of limonin bitterness by treatment of ion exchange and adsorbent resins. *Food Science and Biotechnology*, 19, 411–416.

Köseoglu, S.S., Lawhon, J.T. & Lusas, E.W. (1990) Use of membranes in citrus processing. *Food Technology*, 44, 90–97.

Kozlowski, C.A. & Sliwa, W. (2008) The use of membranes with cyclodextrin units in separation process: recent advances. *Carbohydrate Polymers*, 74, 1–9.

Kris-Etherton, P.M., Hecker, K.D., Bonanome, A., Coval, S.M., Binkoski, A.E., Hilpert, K.F., Griel, A.E. & Etherton, T.D. (2002) Bioactive compounds in foods: their role in the prevention of cardiovascular disease and cancer. *American Journal of Medicine*, 113, 71–88.

Lanza, C.M. (2003) Processed and derived products of oranges. In: Caballero, B., Trugo, L. & Finglas, P. (eds) *Encyclopedia of Food Science and Nutrition*. Elsevier, London. pp. 1346–1354.

Li, B.B., Smith, B. & Hossain, Md.M. (2006a) Extraction of phenolics from citrus peels. I. Solvent extraction method. *Separation and Purification Technology*, 48, 182–188.

Li, B.B., Smith, B. & Hossain, Md.M. (2006b) Extraction of phenolics from citrus peels. II. Enzyme-assisted extraction method. *Separation and Purification Technology*, 48, 189–196.

Lipnizki, F. (2010) Cross-flow membrane applications in the food industry. In: Peinemann, K.V., Pereira Nunes, S. & Giorno, L. (eds) *Membranes for Food Applications*. Wiley-VCH, Weinheim, Germany. pp. 1–24.

Liu, L., Fishman, M.L., Kost, J. & Hicks, K.B. (2003) Pectin-based systems for colon-specific drug delivery via oral route. *Biomaterials*, 24, 3333–3343.

Liu, Q. & Gao, Y. (2015) Binary adsorption isotherm and kinetics on debittering process of ponkan (*Citrus reticulata* Blanco) juice with macroporous resins. *LWT-Food Science and Technology*, 63, 1245–1253.

Lohrasbi, M., Pourbafrani, M., Niklasson, C. & Taherzadeh, M.J. (2010) Process design and economic analysis of a citrus waste biorefinery with biofuels and limonene as products. *Bioresource Technology*, 101, 7382–7388.

Lozano, J.E. (2003) Separation and clarification. In: Caballero, B., Trugo, L. & Finglas, P. (eds) *Encyclopedia of Food Science and Nutrition*. Elsevier, London. pp. 5187–5196.

McLellan, M.R. & Padilla-Zakour, O.I. (2004) Juice processing. In: Barrett, D., Somogyi, L. & Ramaswamy, H. (eds) *Processing Fruits. Science and Technology*. CRC Press, Boca Raton, FL. pp. 73–96.

Marín, F.R., Martínez, M., Uribesalgo, T., Castillo, S. & Frutos, M.J. (2002) Changes in nutraceutical composition of lemon juices according to different industrial extraction systems. *Food Chemistry*, 78, 319–324.

Matsumoto, H., Ikoma, Y., Kato, M., Kuniga, T., Nakajima, N. & Yoshida, T. (2007) Quantification of carotenoids in citrus fruit by LC-MS and comparison of patterns of seasonal changes for carotenoids among citrus varieties. *Journal of Agricultural and Food Chemistry*, 55, 2356–2368.

Maxwell, E.G., Belshaw, N.J., Waldron, K.W. & Morris, V.J. (2012) Pectin – An emerging new bioactive food polysaccharide. *Trends in Food Science & Technology*, 24, 64–73.

Medina, B.G. & Garcia, A. (1988) Concentration of orange juice by reverse osmosis. *Journal of Food Process Engineering*, 10, 217–230.

Mota, M., Teixeira, J.A., Yelshin, A. (2002) Influence of cell-shape on the cake resistance in dead-end and cross-flow filtrations. *Separation and Purification Technology*, 27, 137–144.

Mudimu, O.A., Peters, M., Brauner, F. & Braun, G. (2012) Overview of membrane processes for the recovery of polyphenols from olive mill wastewater. *American Journal of Environmental Sciences*, 8, 195–201.

Muraldihara, H.S. (1996) Removal of pesticides from citrus peel oil. US patent 5558893A.

Nakajiima, V.M., Macedo, G.A. & Macedo, J.A. (2014) Citrus bioactive phenolics: roles in the obesity treatment. *LWT- Food Science and Technology*, 59, 1205–1212.

Nikolic, J.A., Cuperlovic, M., Milijic, C., Djordjevic, D. & Krsmanovic, J. (1986) The potential nutritive value for ruminants of some fibrous residues from the food processing industry. *Acta Veterinaria* (Belgrade), 36, 13–22.

Odebo, U. (2001) Fresher under pressure. A fully commercial cold pasteurization method for fruit products. *Fruit Processing*, 11(6), 220–221.

Öhrvik, V. & Witthöft, C. (2008) Orange juice is a good folate source in respect to folate content and stability during storage and simulated digestion. *European Journal of Nutrition*, 47, 92–98.

O'Shea, N., Arendt, E.K. & Gallagher, E. (2012) Dietary fiber and phytochemical characteristics of fruit and vegetable by-products and their recent applications as novel ingredients in food products. *Innovative Food Science and Emerging Technologies*, 16, 1–10.

Oufedjikh, H., Mahrouz, M., Amiot, M.J. & Lacroix, M. (2000) Effect of γ-irradiation on phenolic compounds and phenylalanine ammonia-lyase activity during storage in relation to peel injury from peel of *Citrus clementina* Hort. ex. Tanaka. *Journal of Agricultural and Food Chemistry*, 48, 559–565.

Paine, F.A. & Paine, H.Y. (1983) *A Handbook of Food Packaging*. Leonard Hill, Glasgow.

Patil, B.S., Jayaprakasha, G.K., Chidambara Murthy, K.N. & Vikram, A. (2009) Bioactive compounds: historical perspectives, opportunities, and challenges. *Journal of Agricultural and Food Chemistry*, 57(18), 8142–8160.

Pernice, R., Borriello, G., Ferracane, R., Borrelli, R., Cennamo, F., Ritieni, A. (2009) Bergamot: a source of natural antioxidants for functionalized fruit juices. *Food Chemistry*, 112, 545–550.

Pérez Prieto, S., Cancho Grande, B., García Falcón, S. & Simal Gándara, J. (2006) Screening for folic acid content in vitamin-fortified beverages. *Food Control*, 17, 900–904.

Piriyaprasarth, S. & Sriamornsak, P. (2011) Flocculating and suspending properties of commercial citrus pectin and pectin extracted from pomelo (*Citrus maxima*) peel. *Carbohydrate Polymers*, 83, 561–568.

Puupponen-Pimiä, R., Aura, A.M., Oskman-Caldentey, K.M., Myllärinen, P., Saarela, M., Mattila-Sandholm, T. & Poutanen, K. (2002) Development of functional ingredients for gut health. *Trends in Food Science and Technology*, 13, 3–11.

Rai, P. & De, S. (2009) Clarification of pectin-containing juice using ultrafiltration. *Current Science*, 96, 1361–1371.

Rai, P., Majumdar, G.C., Das Gupta, S. & De, S. (2007) Effect of various pretreatment methods on permeate flux and quality during ultrafiltration of mosambi juice. *Journal of Food Engineering*, 78, 561–568.

Rapisarda, P., Carollo, G., Fallico, B., Tomaselli, F. & Maccarone, E. (1998) Hydroxycinnamic acids as markers of Italian blood orange juices. *Journal of Agricultural and Food Chemistry*, 46, 464–470.

Ruby-Figueroa, R., Cassano, A. & Drioli, E. (2011) Ultrafiltration of orange press liquor: optimization for permeate flux and fouling index by response surface methodology. *Separation and Purification Technology*, 80, 1–10.

Ruby-Figueroa, R., Cassano, A. & Drioli, E. (2012) Ultrafiltration of orange press liquor: optimization of operating conditions for the recovery of antioxidant compounds by response surface methodology. *Separation and Purification Technology*, 98, 255–261.

Schäfer, A.I., Fane, A.G. & Waite, T.D. (2005) *Nanofiltration: Principles and Applications*. Elsevier, London.

Shaw, P.E., Tatum, J.H. & Wilson, C.W. (1984) Improved flavor of navel orange and grapefruit juices by removal of bitter components with beta-cyclodextrin polymer. *Journal of Agricultural and Food Chemistry*, 32, 832–836.

Simone, S., Conidi, C., Ursino, C., Cassano, A. & Figoli, A. (2016) Clarification of orange press liquors by PVDF hollow fiber membranes. *Membranes*, 6, 9.

Smock, R.M. & Neubert, A.M. (1950) *Apples and Apple Products*. Interscience Publishers, New York.

Stinco, C.M., Fernández-Vázquez, R., Hernanz, D., Heredia, F.J., Meléndez-Martínez, A.J. & Vicario, I.M. (2013) Industrial orange juice debittering: impact on bioactive compounds and nutritional value. *Journal of Food Engineering*, 116, 155–161.

Strathmann, H., Giorno, L. & Drioli, E. (2006) An introduction to membrane science and technology. National Research Council, Rome, Italy.

Swientek, R.J. (1986) Ultrafiltration expanding role in food and beverage processing. *Food Processing*, 47, 71–83.

Szejtli, J. (2004) Past, present, and future of cyclodextrin research. *Pure and Applied Chemistry*, 76, 1825–1845.

Talon, M. & Gmitter, F.G.Jr. (2008) Citrus genomics. *International Journal of Plant Genomics*, 2008, 52836.

Tanaka, T., Abe, K.I., Asakawa, H., Yoshida, H. & Nakanishi, K. (1994) Filtration characteristics and structure of cake in cross-flow filtration of bacterial suspension. *Journal of Fermentation and Bioengineering*, 78, 455–461.

Tandon, K., Worobo, R.W., Churey, J.J. & Padilla-Zakour, O.I. (2003) Storage quality of pasteurized and UV treated apple juice. *Journal of Food Processing and Preservation*, 27, 21–34.

Tiller, F.M. & Cooper, H.R. (1962) The role of porosity in filtration. Part V. Porosity variation in filter cakes. *AIChE Journal*, 8, 445–449.

Todisco, S., Peña, L., Drioli, E. & Tallarico, P. (1996) Analysis of the fouling mechanism in microfiltration of orange juice. *Journal of Food Processing and Preservation*, 20, 453–466.

Todisco, S., Tallarico, P. & Drioli, E. (1998) Modelling and analysis of ultrafiltration effects on the quality of freshly squeezed orange juice. *Italian Food & Beverage Technology*, XII(May), 3–8.

Tranchida, P.Q., Bonaccorsi, I., Dugo, P., Mondello, L. & Dugo, G. (2012) Analysis of citrus essential oils: state of the art and future perspectives: a review. *Flavour and Fragrance Journal*, 27, 98–123.

Tripoli, E., La Guardia, M., Giammanco, S., Di Majo, D. & Giammanco, M. (2007) Citrus flavonoids: molecular structure, biological activity and nutritional properties: a review. *Food Chemistry*, 104, 466–479.

Trotta, F., Drioli, E., Baggiani, C. & Lacopo, D. (2002) Molecular imprinted polymeric membrane for naringin recognition. *Journal of Membrane Science*, 201, 77–84.

Tsibranska, I. & Tylkowski, B. (2014) Concentration of polyphenols by integrated membrane operation. In: Cassano, A. & Drioli, E. (eds) *Integrated Membrane Operation in the Food Industry*. Walter de Gruyter & Co, Berlin. pp. 269–293.

Tsuru, T., Nakao, S. & Kimura, S. (1991) Calculation of ion rejection by extended Nernst-Planck equation with charged reverse osmosis membranes for single and mixed electrolyte solutions. *Journal of Chemical Engineering of Japan*, 24, 511–517.

Tyagi, A.K., Gottardi, D., Malik, A. & Guerzoni, M.A. (2014) Chemical composition, in vitro anti-yeast activity and fruit juice preservation potential of lemon grass oil. *LWT- Food Science and Technology*, 57, 731–737.

USDA (2016) Citrus: World Markets and Trade, July 2016. Foreign Agricultural Service, United States Department of Agriculture, Washington, DC. Available from: http://www.fas.usda.gov/ [accessed November 2016].

Verzera, A., Trozzi, A., Gazea, F., Cicciarello, G., Cotroneo, A. (2003) Effects of rootstock on the composition of bergamot (*Citrus bergamia* Risso et Poiteau) essential oil. *Journal of Agricultural and Food Chemistry*, 51, 206–10.

Viuda-Martos, M., Ruiz-Navajas, Y., Fernández-López, J. & Pérez-Álvarez, J. (2008) Antifungal activity of lemon (*Citrus lemon* L.), mandarin (*Citrus reticulata* L.), grapefruit (*Citrus paradisi* L.) and orange (*Citrus sinensis* L.) essential oils. *Food Control*, 19, 1130–1138.

Wang, X.L., Tsuru, T., Nakao, S. & Kimura, S. (1995) Electrolyte transport through nanofiltration membranes by the space charge model and comparison with the Theorell Meyer Sievers model. *Journal of Membrane Science*, 103, 117–133.

Warczok, J., Ferrando, M., Lopez, F. & Guell, C. (2004) Concentration of apple and pear juices by nanofiltration at low pressure. *Journal of Food Engineering*, 63, 63–70.

Worobo, R.W., Churey, J.J. & Padilla-Zakour, O. (1998) Apple cider: treatment options to comply with new regulations. *Journal of the Association of Food and Drug Officials*, 62, 19–26.

Xie, L., Li, X. & Guo, Y. (2008) Ultrafiltration behaviours of pectin-containing solution extracted from citrus peel on a ZrO_2 ceramic membrane pilot unit. *Korean Journal of Chemical Engineering*, 25, 149–153.

Xu, G.H., Ye, X., Chen, J. & Liu, D. (2007) Effect of heat treatment on the phenolic compounds and antioxidant capacity of citrus peel extract. *Journal of Agricultural and Food Chemistry*, 55, 330–335.

CHAPTER 14

Integration of membrane bioreactors with nanofiltration and reverse osmosis for wastewater reclamation and reuse

Nalan Kabay, Samuel Bunani, Taylan Ö. Pek, Gökhan Sert, Arash Arianfar, Elmuntaser Eltayeb, Müşerref Arda, Özdemir Egemen & Mithat Yüksel

14.1 INTRODUCTION

According to the comprehensive assessment of water management in agriculture by the International Water Management Institute, one third of the world's population faces some form of water scarcity. The World Water Development Report by the UN predicted that population growth, climate change, widespread mismanagement and increasing demand for energy could lead to a major global water crisis. The UN's Food and Agriculture Organization indicated that 1800 million people will be living in countries or regions with absolute water scarcity, and two thirds of the world's population could be living under water stress conditions by 2025. Thus, a range of practical solutions are urgently needed for the sustainable protection of water resources and the production of alternatives.

Due to the concerns regarding global warming and increasing fuel costs, alternative desalination processes have been proposed for the production of safe water. Membrane processes, which are said to be 'green' technologies, are preferred over thermal alternatives. Reverse osmosis (RO) is the best-known desalination process among a range of membrane-based processes such as electrodialysis and electrodeionization. In the case of RO, when pressure greater than osmotic pressure is applied, the membrane allows the passage of water but rejects almost all the ions and salts present, resulting in a concentrated retentate on the feed side of the membrane, and the product (or permeate) on the other side of the membrane. Thus, the most permeable component gets enriched in the permeate stream while the least permeable component gets enriched in the retentate stream (Koltuniewicz and Drioli, 2008).

Advanced technologies in wastewater treatment have also become important due to the fact that discharge standards became more strict and there is a need for water recovery and reuse. Recently, membrane bioreactor (MBR) systems have been used for the treatment of municipal and industrial wastewater. MBR technology is a treatment method in which a low-pressure membrane filtration is integrated as the last stage of the system instead of a precipitation basin. Compared to the activated sludge process, the MBR process is superior in a number of ways. First, the hydraulic retention time in MBR systems is short due to the high concentration of suspended solids. Thus, the cost of installation decreases due to the decrease in reactor volume. Second, the separation of biomass is independent of the precipitation of activated sludge because it occurs by microfiltration (MF) or ultrafiltration (UF). Thus, there is no need for a precipitation basin at the last stage and some of the associated problems cannot occur. Third, because it is possible to work with much older sludges in MBR systems, the amount of waste sludge in MBR systems is also much lower than that in the conventional activated sludge process. In addition, MBR systems are also able to provide disinfection as a result of membrane filtration with pore sizes as small as 0.01–0.1 μm. Overall, MBR systems seem to be good alternatives for reusing treated wastewater as they produce a good quality of water. However, the salinity of the water produced must be decreased in some cases if the water is to be recovered and reused for irrigation and other processes. For this, advanced membrane treatment processes such as nanofiltration (NF) and RO will be needed (Li et al., 2008).

14.2   IMPORTANCE OF WASTEWATER RECLAMATION AND REUSE

Reuse of wastewater conserves fresh water resources and also contributes to environmental protection. Reclamation is the treatment and recovery of water to make it available for reuse (Asano *et al.*, 2007). According to the recent literature, wastewater recovery and reuse in industry is no longer just an option but an absolute necessity, although solid waste generation and energy expenditure are also taken into consideration at the same time. On the other hand, growing worldwide water scarcity and stringent water development regulations to protect the environment are two major challenges for water professionals when implementing the integrated management of available and alternative water resources.

The necessary quality and level of treatment required for recycled water depends on the following factors (Lazarova and Bahri, 2005):

- Water quality requirements and regulations.
- Degree of worker and public exposure.
- Type of distribution and irrigation systems.
- Soil characteristics.
- Any crops being irrigated.

Water pollution control efforts in many countries have made treated municipal and industrial wastewaters suitable for economical augmentation of the existing water supply, when compared to increasingly expensive and environmentally destructive new water resource developments that include dams and reservoirs. The use of treated municipal wastewater is now considered to be a beneficial, competitive and viable water source option (Lazarova and Bahri, 2005).

Reclaimed wastewater effluents constitute an alternative source of water for a wide variety of applications, including landscape and agricultural irrigation, toilet flushing, industrial processing, power plant cooling, the creation, restoration and maintenance of wetland habitat, and groundwater recharge (Bunani *et al.*, 2015a). Obviously, water quality refers to the characteristics of a water supply that will influence its suitability for a specific use. For agricultural irrigation, for instance, the chemical constituents of concern in reclaimed water are salts (i.e. salinity), sodium, trace elements, excessive chlorine residuals and biological nutrients, together with any pathogens that may be present (USEPA, 2004).

Salinity, which is the amount of salt dissolved in water, directly affects plant growth, generally has an adverse effect on agricultural crop performance, and can also affect soil properties. Consequently, without knowledge of both soil and water salinity and correspondingly appropriate management, long-term irrigated crop productivity can decrease. To evaluate the usability of a water supply as agricultural irrigation water, salinity, reduced water infiltration, toxicity and the effects related to a group of miscellaneous water constituents have to be assessed (Bunani, *et al.*, 2015a). Salinity is here considered in terms of electrical conductivity (EC_w) and water infiltration in terms of sodium adsorption ratio (SAR). The sodium hazard is assessed according to EC_w, SAR, residual sodium carbonate (RSC), soluble sodium percentage (SSP) and the exchangeable sodium percentage (ESP). Other toxicity effects and miscellaneous water constituents are evaluated according to the content of ions such as sodium, chloride, nitrate and bicarbonate (Ayers, 1985).

Membrane-based technologies offer many advantages over conventional techniques for water and wastewater treatment. These advantages are summarized as follows (Judd and Jefferson, 2003):

- They are energy-efficient processes since there is no phase change during separation.
- Processes can be operated continuously because no regeneration is needed during the process.
- Little or no addition of chemicals is required during operation.

The use of membrane technologies, especially for industrial wastewater reclamation and reuse, has historically been limited due to the costs involved. However, significant improvements in the efficiencies and cost-effectiveness of membrane techniques, as well as increasing stresses on freshwater supplies, have allowed the competitiveness of recycling treated wastewater to increase over that of discharge (Judd and Jefferson, 2003). Today, membrane separation technologies form a set of efficient tools for solving the problem of water shortage. Many of the advantages of membrane separation processes are connected to strategies for intensification of water treatment, decrease of chemicals consumption, reduction of waste production and reasonable use of energy. Generally, these features of membrane technologies render them environmentally friendly methods, and their modularity and compactness allow them to be readily combined with existing installations (Koltuniewicz and Drioli, 2008).

14.3 MEMBRANE BIOREACTOR (MBR) SYSTEMS

MBR technology was first introduced, to the Japanese market, in the 1970s. In the late 1970s, commercial aerobic MBR processes first appeared in North America, and then in the 1980s in Japan, before being introduced in Europe in the mid-1990s. MBR systems have been used in sanitary and industrial applications since the 1990s. Globally, over 500 commercial MBR processes are now in operation. More than 60% of the world's MBR processes were installed in Japan, where the number of commercial MBRs has increased. Most of these systems combine the membrane separation process with an aerobic biological process. Approximately half of the commercial MBRs are of the *submerged* type and the remainder are *side-stream* (Stephenson *et al*., 2006).

Nowadays, the use of MBR systems is popular all over the world, especially in Japan, for industrial and domestic wastewater treatment and reuse. The Japanese government established an investment program for the development of a water recycling process with a small footprint and a high-quality water product. Japan's Kubota Corporation was one of the companies that participated in the program and it developed a submerged MBR system with flat-sheet membranes. In 1982, US company Dorr-Oliver introduced the membrane anaerobic reactor system (MARS) for the treatment of food industry wastewaters. The process used an external UF membrane with a high loading of chemical oxygen demand (COD), achieving up to 99% removal. In the early 1980s, two systems were developed in the UK with either UF or MF membranes: membrane aeration bioreactors (MABRs) and extractive membrane bioreactors (EMBRs) were operated at pilot scale (Stephenson *et al*., 2006).

Each commercial MBR achieves transmembrane pressure (TMP), the pressure needed for permeation, differently. Kubota's MBR implements a hydrostatic head above the membrane unit, which is submerged at depth, during normal operation. A combination of the hydrostatic head and a vacuum applied to the permeate side of the membrane is used in the ZenoGem (Zenon Environmental Inc., Ontario, Canada) MBR and the Kubota process during peak hydraulic loading. Orelis (Salindres, France) and Wehrle-Werk (Emmendingen, Germany) MBR processes throttle the pressure in the recirculation line to which the side-stream membrane unit is attached (Stephenson *et al*., 2006).

In the biochemical stage of wastewater treatment, organic carbon and nutrients are removed from wastewater by microbes. These microbes live and grow enmeshed in extracellular polymeric substances (EPS) that bind them into discrete microcolonies forming three-dimensional, aggregated microbial structures called flocs.

The ability of microorganisms to form flocs is vital for the activated sludge treatment of wastewater. The floc structure enables not only the adsorption of soluble substrates but also the adsorption of the colloidal matter and macromolecules additionally found in wastewaters (Liwarska-Bizukojc and Bizukojc, 2005; Shuler and Kargi, 2002). The diversity of the microbial community in activated sludge is very large, containing prokaryotes (bacteria), eukaryotes (protozoa, nematodes, rotifers) and viruses. In this complex microsystem, bacteria dominate the microbial population and play a key role in the degradation process (Shuler and Kargi, 2002).

MBR technology with biochemical and sludge-separation stages integrated into one step implies a continuous generation of new sludge through the consumption of an organic material

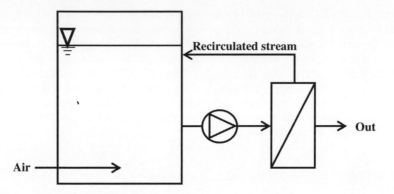

Figure 14.1.   Side-stream MBR configuration.

feed, while some sludge mass is decayed by endogenous respiration. Endogenous respiration involves consumption of cell-internal substrates, which leads to a loss of activity and slightly reduced biomass. It can be both aerobic and anoxic, although under anoxic conditions it is a lot slower and protozoa, in particular, are considerably less active under denitrifying conditions (slower predation) (Gujer *et al.*, 1999).

14.3.1   *Types of MBRs*

MBRs combine membrane technology with biological reactors for the treatment of wastewaters. MBRs are used for separation and retention of solids, bubble-less aeration within the bioreactor and for extraction of the principal organic pollutants from industrial wastewaters. MBR processes are often used as a substitute for sedimentation. They can also be used for the mass transfer of gases (such systems are not to be confused with so-called 'membrane aerators', which is a term used for some fine-bubble diffusers), usually oxygen for aerobic processes. Other uses of MBR processes are the controlled transfer of nutrients into a bioreactor and the extraction of pollutants from wastewaters that are untreatable by conventional biological processes (Stephenson *et al.*, 2006).

Side-stream MBRs
The first MBRs were developed commercially by Dorr-Oliver in the late 1960s, combining UF with a conventional activated sludge process (CASP), for application to shipboard sewage treatment.

These systems were based on what have come to be known as 'side-stream' (sMBR) configurations (Fig. 14.1) (Judd *et al.*, 2006).

Immersed MBRs
The most widely implemented membranes in biomass rejection applications are immersed MBRs, in which the membrane is immersed (or submerged) in the bioreactor (Fig. 14.2). The key advantage of immersed MBRs is their low specific energy consumption. The energy demand of an immersed MBR is lower by a factor of two in comparison to that of a side-stream MBR. Aeration is a significant parameter in immersed MBR application and it has a large influence on hydraulic and biological process performance. Aeration provides oxygen to biomass, increasing biodegradability, and it maintains solids in suspension. The following are the main factors that affect both the design and operation of immersed MBRs (Judd *et al.*, 2006):

- The membrane design and the sustaining of permeability.
- Feed water characteristics and pretreatment.
- Aeration of both the membrane and the bulk biomass.
- Sludge withdrawal and retention time.
- Bioactivity and nature of the biomass.

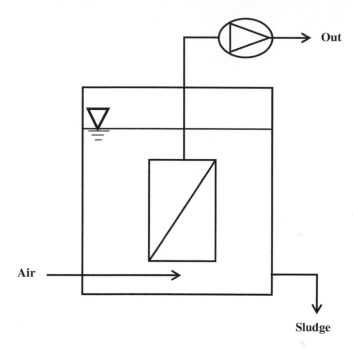

Figure 14.2.   Immersed MBR configuration.

These elements are clearly interrelated. The sludge withdrawal rate determines the retention time, which obviously affects the biomass concentration. All of the biological and physical properties are affected by the biomass concentration. In all water treatment processes, feed water properties have the biggest impact upon the design and operation of the process. Furthermore, for immersed MBRs, membrane fouling is likely to have a significant effect on feed water chemistry (Judd *et al.*, 2006).

14.3.2   *Advantages and disadvantages of MBR systems*

It is clear that the combination of membrane processes with biological wastewater treatment processes results in advantages that were once exclusive to the former; in particular, small footprint, process intensification, modularisation and retrofit potential (Stephenson *et al.*, 2006). The advantages of MBR systems also include their complete removal of solids from effluent, effluent disinfection, combined COD, solids and nutrient removal in a single unit, high loading rate capability and low sludge production. However, MBRs also present some disadvantages, such as aeration limitations, the need for membrane washing due to fouling, and high membrane costs.

14.3.3   *Integration of MBR with RO/NF*

Membrane technology is considered to offer the best technological prospect in the 21st century for the development of wastewater treatment and reclamation systems. One of the most promising wastewater reclamation systems is a combination of MBR followed by either RO or NF, termed MBR-RO or MBR-NF integrated membrane systems, because they have many advantages compared with conventional systems, such as better treated water quality, smaller footprint, lower sludge production and more convenient operation and maintenance (Halder and Kullmann, 2011).

Table 14.1.   Characteristics of NF and RO membranes used by Bunani *et al.* (2014) (Dow, 2013; GE, 2013).

Designation	Manufacturer	Polymer	MWCO [Da]	T_{max} [°C]	p_{max} [MPa]	pH range
NF-270	Dow Filmtec	PTFC	200–300	45	4.1	3–10
NF-90	Dow Filmtec	PTFC	200	45	4.1	3–10
AK-BWRO	GE Osmonics	PTFC	Dense	50	2.8	4–11
BW30-RO	Dow Filmtec	PTFC	Dense	50	4.1	3–10

Legend: MWCO, molecular weight cut-off; PTFC, polyamide thin-film composite.

The RO process will remove most dissolved solids, so that water can be recycled into many industrial processes or secondary uses, or even form an indirect part of a potable reuse scheme. However, there are some major challenges for RO/NF systems operating on biologically treated wastewater. Many RO/NF systems today use membrane filtration as a pretreatment to remove suspended solids. The membrane system does an excellent job of providing water with low levels of suspended solids to feed an RO/NF system. However, the membrane system requires additional space and also does not sufficiently reduce the amount of dissolved contaminants such as organics that are fed to the RO/NF system. To remove dissolved organics, a biological wastewater treatment is required. However, instead of separating the biological treatment process and the UF step, it has become state of the art to combine both processes in a MBR (Yang *et al.*, 2009).

NF is a pressure-driven membrane process that lies between UF and RO in terms of its ability to reject molecular or ionic species. NF membranes usually provide good retention of small organic molecules and inorganic salts, especially if multivalent ions are involved (Drioli and Giorno, 2009).

NF membrane separation characteristics are principally based on the sieve effect. But most commercial NF membranes are also charged, so ion rejection by NF membranes results from the combination of electrostatic and steric interactions associated with charge shielding, Donnan exclusion and the degree of ion hydration (Peeters *et al.*, 1998; Pontalier *et al.*, 1997). NF membranes can be applied to the advanced treatment of the effluents from conventional biological treatments or MBRs.

MBR systems can be broadly defined as systems integrating the biological degradation of wastewaters with membrane filtration (Cicek *et al.*, 1998). These systems have proven quite effective in removing both organic and inorganic contaminants as well as biological entities from wastewater. Since high or changing salinity presents a challenge for bio-treatment processes, a MBR treatment system may not be adequate if the treated wastewater is expected to be reused again. However, both NF and RO can remove salinity and various inorganic and organic species in water. Bunani *et al.* (2014) studied the treatment of MBR effluent with two different NF (NF-270 and NF-90) membranes and an RO membrane (AK-BWRO). Another study reported the treatment of MBR effluent with BW30RO membrane (Sert, 2015). In both works, the reusability for agricultural irrigation of the water produced was assessed.

The bio-treated wastewater was obtained from the ITOB-OSB (İzmir, Turkey) wastewater treatment plant in which MBR technology with Kubota membranes was employed. The characteristics of the NF/RO membranes used are summarized in Table 14.1.

The results of volumetric permeate flux tests revealed that in the tested NF membranes, the NF-270 membrane exhibited the maximum flux of $128 \, \text{L m}^{-2} \, \text{h}^{-1}$ and the NF-90 membrane gave a maximum flux of $42 \, \text{L m}^{-2} \, \text{h}^{-1}$. The AK-BWRO membrane showed an average flux of $31.4 \, \text{L m}^{-2} \, \text{h}^{-1}$ (Bunani *et al.*, 2014, 2015b). The fluxes *vs.* time plots are shown in Figure 14.3. The volumetric permeate flux results obtained were in good agreement with the membranes' properties and MWCOs shown in Table 14.1.

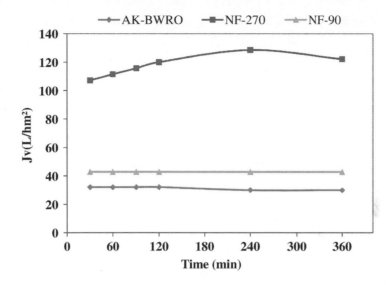

Figure 14.3. Permeate flux *vs.* time for used membranes (adapted from Bunani *et al.*, 2014, 2015b).

Table 14.2. MBR-NF and MBR-RO membranes' product water quality (adapted from Bunani *et al.*, 2015b).

Parameters	Feed values (min–max)	NF-90 Permeate average	NF-270 Permeate average	AK-BWRO Permeate average	Irrigation water standards (Ayers, 1985)
pH	8.21–8.60	6.87	8.28	7.17	6.0–8.50
Na^+ [mg L^{-1}]	1025–1110	76.1	559	43.7	0–920
Ca^{2+} [mg L^{-1}]	149–165	0.06	32.9	1.07	0–400
Mg^{2+} [mg L^{-1}]	34.0–39.6	<1.00	7.69	0.5	0–60.0
EC [μS cm^{-1}]	6130–7960	500	3560	286	0–3000
Salinity [psu]	3.44–4.60	0.26	1.91	0.14	0–1.94[a]
SAR	–	16.0	22.8	8.71	–

[a](Solak, 2010)

The quality of the water product obtained from different NF membranes is summarized in Table 14.2. As can be seen in that table, the NF-90 membrane produced better water quality than the NF-270 membrane. Consequently, NF-90 delivered better rejection performance.

The difference in performance of the membranes observed here is attributable to the properties of the membranes. Those properties are pore size, represented by MWCO, membrane surface charge, hydrophobic/hydrophilic characteristics, and the structure or composition of the active layer of the NF membranes. As the best-performing membrane, the product water quality obtained with NF-90 was compared to an irrigation water standard to establish its reusability as agricultural irrigation water. Table 14.2 reveals how well NF-90 product water quality fits with agricultural irrigation standards.

Although MBR-RO effluent has a high quality, its use in agricultural irrigation presents infiltration hazards due to its lower salinity values (as measured by EC$_w$) and SAR (see Tables 14.3, 14.4 and 14.5). To manage this problem, the combination of MBR-RO effluent with MBR effluent is strongly indicated. Thus, the SAR and salinity values that result from different mixing ratios are also shown in Tables 14.3 and 14.4, while the guidelines for interpreting irrigation water quality

Table 14.3. Water quality obtained from NF-90 and AK-BWRO membranes using various combinations of MBR and RO/NF effluents (adapted from Bunani *et al.*, 2015b).

Membranes	MBR effluent proportion	0	0.1	0.2	0.3	0.4	0.5	0.6	0.7	0.8	0.9	1
	MBR-RO effluent proportion	1	0.9	0.8	0.7	0.6	0.5	0.4	0.3	0.2	0.1	0
NF-90	SAR	16.0	16.4	16.7	17.1	17.4	17.8	18.1	18.5	18.8	19.2	19.5
(1.0 MPa)	EC_w [dS m^{-1}]	0.52	1.08	1.64	2.20	2.76	3.32	3.88	4.45	5.01	5.57	6.13
AK-BWRO	SAR	8.71	9.85	11.0	12.1	13.2	14.4	15.6	16.7	17.8	19.0	20.1
(1.0 MPa)	EC_w [dS m^{-1}]	0.29	0.94	1.59	2.24	2.89	3.54	4.19	4.84	5.49	6.14	6.80

Table 14.4. Water quality obtained from NF-90 and BW30-RO membranes using various combinations of MBR and RO/NF effluents (reproduced with permission from Sert, 2015).

Membranes	MBR effluent proportion	0	0.1	0.2	0.3	0.4	0.5	0.6	0.7	0.8	0.9	1
	MBR-RO effluent proportion	1	0.9	0.8	0.7	0.6	0.5	0.4	0.3	0.2	0.1	0
NF-90	SAR	13.8	14.9	16.1	17.3	18.4	19.6	20.7	21.9	23.0	24.2	25.3
(1.0 MPa)	EC_w [dS m^{-1}]	0.38	0.96	1.54	2.12	2.70	3.29	3.87	4.45	5.03	5.61	6.19
BW30-RO	SAR	14.0	15.2	16.3	17.5	18.7	19.9	21.0	22.2	234	24.6	25.7
(1.0 MPa)	EC_w [dS m^{-1}]	0.17	0.71	1.24	1.78	2.32	2.86	3.40	3.94	4.47	5.01	5.55

Table 14.5. Guideline for interpreting water quality for irrigation with regards to infiltration hazards indicated by salinity and SAR values (Ministry of Environment and Forestry, 2010).

			Degree of restriction on use		
Potential irrigation problem		Units	None	Slight to moderate	Severe
Salinity *(affects crop water availability)*					
	EC_w	[dS m^{-1}]	<0.7	0.7–3.0	>3.0
Infiltration *(affects infiltration rate of water into the soil; evaluate using EC_w and SAR together)*					
SAR	0–3	EC_w	[dS m^{-1}]	>0.7	0.7–0.2	<0.2
	3–6			>1.2	1.2–0.3	<0.3
	6–12			>1.9	1.9–0.5	<0.5
	12–20			>2.9	2.9–1.3	<1.3
	20–40			>5.0	5.0–2.9	<2.9

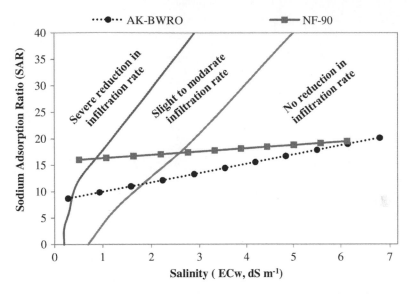

Figure 14.4.   Diagram of infiltration rate of mixed water quality at different proportions of MBR and RO/NF effluents as affected by SAR and salinity (laboratory-scale results) (adapted from Bunani *et al.*, 2015b).

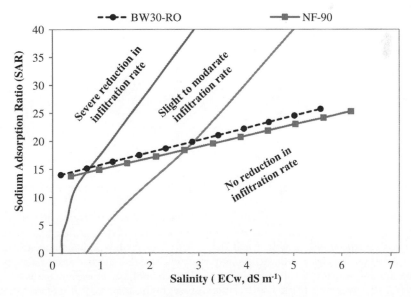

Figure 14.5.   Diagram of infiltration rate of mixed water quality at different proportions of MBR and RO/NF effluents as affected by SAR and salinity (field study results) (reproduced with permission from Sert, 2015).

with respect to the infiltration hazard are summarized in Table 14.5. The subdivision of infiltration rate zone due to this combination is shown on Figures 14.4 and 14.5.

In the field work, the practicality of membrane separation methods such as NF and RO in the reuse of wastewater treated with the MBR process was also investigated. To that end, tests were carried out with the spiral-wound NF/RO systems that are installed at the wastewater treatment facility where MBR technology is already used in an Organized Industrial Zone. On the basis of the results of previous studies on NF-90 (Bunani *et al.*, 2014) and BW30 membranes (Sert, 2015),

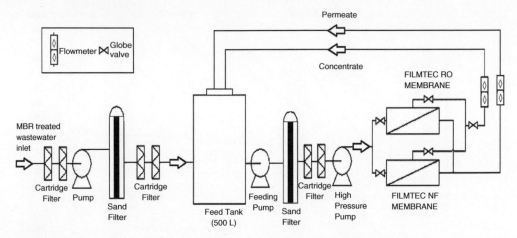

Figure 14.6.   Pilot-scale spiral-wound RO/NF membrane system (reproduced with permission from Sert, 2015).

Table 14.6.   Comparison of permeate characteristics for NF-90 and BW30-RO membranes (reproduced with permission from Sert, 2015).

Parameters	Feed average	NF-90 Permeate average	BW30-RO Permeate average
pH	7.57	6.52	6.15
Na^+ [mg L^{-1}]	1231	86.0	39.0
Ca^{2+} [mg L^{-1}]	119	1.43	0.30
Mg^{2+} [mg L^{-1}]	24.0	0.26	0.00
EC [μS cm^{-1}]	6880	520	210
Salinity [psu]	3.72	0.00	0.00
SAR	–	13.8	14.0

Dow Filmtec NF90-2540 NF and BW30-2540 RO membranes were used in the field with a pilot-scale system.

In a batch system (Fig. 14.6), the results revealed that the average permeate flux of the NF-90 membrane was greater than that of the BW30 membrane. The permeate flux was measured as 44 L m^{-2} h^{-1} for the NF-90 membrane and 28 L m^{-2} h^{-1} for the BW30 membrane (Sert, 2015). Permeate recovery increases with increasing permeate flux, and the permeate recovery of the NF-90 membrane was greater than that of the BW30 membrane. However, the water quality obtained from the integrated MBR-BWRO system is very high compared to that of the MBR-NF90 system (Table 14.6).

14.3.4   *Fouling in integrated MBR-RO treatment system*

In the combination of MBR and RO for water reuse, the protection of the RO system from components that can cause fouling is critical. There are four categories of fouling that need to be considered (Halder and Kullmann, 2011):

- Particles that might block the brine spacers of the RO.
- Fouling of the RO membranes by adsorption of soluble organics in the feed.
- Biofouling caused by bacteria.
- Inorganic fouling by the precipitation of salts.

The rejection of particles is the biggest advantage of using a combination of biological treatment with membrane filtration. Using UF on biologically treated water generates a permeate in which suspended solids are less than $1 \, mg \, L^{-1}$ and particle size is below $0.1 \, \mu m$.

Reducing the amount of soluble organics in the wastewater before RO is a second advantage of coupling it with an MBR system. The combination of biological treatment and RO in the integrated process of an MBR significantly increases the overall efficiency of biodegradation.

Biofouling is the most critical issue for RO. In combination, factors such as residence time, higher temperatures and, sometimes, light have created ideal conditions to promote bio-growth. Several parameters of system design and plant operation have proven to be effective in reducing bio-growth (Halder and Kullmann, 2011):

- The pipework between MBR and buffer tank should be as short as possible without any dead zones. Safety filters on the RO inlet can protect the RO but must be frequently changed or cleaned.
- Chemical cleaning of the MBR membranes should be carried out frequently. A more frequent dosing with low concentrations of chlorine prevents bio-growth in the permeate of the MBR system.
- Inorganic fouling is a well-known problem for RO applications. Therefore, the use of antiscalants is necessary. When treating industrial water with a large amount of dissolved components, it is important to monitor pH changes in the MBR system. Increasing pH facilitates precipitation of less soluble salts such as calcium phosphate or calcium sulfate. The denitrification process and the module aeration also increase the pH of the water, the latter as a result of stripping out CO_2. In such cases, it is recommended that a membrane module that requires little aeration for air scouring is chosen (Halder and Kullmann, 2011).

14.4   APPLICATION OF NANOTECHNOLOGY FOR MEMBRANE TREATMENT OF WASTEWATER

Several technical barriers exist that reduce the overall robustness of the MBR system:

- Fouling is a significant limiting factor in a wide range of MBR applications because it requires additional technologies and involves extra costs for cleaning, maintenance and membrane replacement.
- Filter-cake formation on MBR membranes reduces the overall performance of the system over time.

Thus, although MBR systems have a high operating efficiency in terms of costs, maintenance and quality when it comes to the treatment of wastewater containing high levels of biodegradable organic compounds, they are inefficient with respect to waters with fouling problems. Therefore, MBR systems can only be applied to wastewaters with reduced fouling tendencies or with the use of intensive cleaning and/or membrane replacement regimes.

Developments in nanotechnology have also been very helpful for water and wastewater treatment (Das *et al.*, 2014). Many recent achievements in materials science have offered new-generation membranes by incorporating nanoparticles and nanocomposites into membrane structures (Goh *et al.*, 2016). A high water flux and salt rejection ability have been obtained by incorporating carbon nanotubes into desalination membranes. Ultra-fast water flux is achieved through the highly accessible nanopores of graphene monolayers incorporated into membrane structures, and antifouling properties are obtained with the use of graphene oxide. In addition, biomimetic and bioinspired materials with a great efficiency for desalination have been considered as potential membranes (Goh *et al.*, 2016).

It was reported that zeolite nanoparticles mixed with polymer matrix to form a thin-film RO membrane increased the water transport and salt retention ability of the resulting membranes. Silica nanoparticles doped with RO polymer matrices used for desalination improved polymeric networks, pore diameters and transport properties (Das *et al.*, 2014).

Kochkodan and Hilal (2015) published an extensive review paper on the preparation and properties of composite polymer membranes that have a great resistance to biofouling. They mentioned that such resistance was conferred on membranes using interfacial polymerization, surface grafting, coating with a polymeric protective layer, and the surface modification of polymer membranes with nanoparticles.

Most recently, Galiano *et al.* (2015) published an exciting research paper on a novel surface modification method based on polymerizable bicontinuous microemulsions (PBM). They implied that such hydrophilic and antifouling coating materials were developed especially for wastewater treatment applications. It was reported that the improved hydrophilicity, smoother surface, channel-like structure and antimicrobial activity of the resulting PBM membranes make them potential materials for use in MBR systems where antifouling and anti-biofouling properties become very important (Galiano *et al.*, 2015).

14.5   OTHER RECENT STUDIES OF NOVEL MBR SYSTEMS

Recently, some studies of novel high-retention (HR) MBR systems have also been published in the literature (Luo *et al.*, 2014). While traditional MBRs integrate low-pressure membrane filtration processes such as MF or UF with activated sludge processes, HR-MBR systems are based on NF, forward osmosis (FO), RO and membrane distillation. According to the literature, it is possible to efficiently remove trace organic chemicals, especially hydrophilic and persistent ones, by using HR-MBR systems. Hybrid UF-osmotic MBR was investigated by Holloway *et al.* (2015) in order to obtain a high-quality RO permeate stream and nutrient-rich UF permeate stream from a single, integrated stream.

14.6   CONCLUSIONS

In terms of quantity, quality and usage by a diverse range of sectors, water resources are, nowadays, confronted by many problems. New and varied practical solutions are needed for the sustainable protection of water resources as the ability to increase fresh water resources is currently limited in terms of both technical and economic aspects. Many of our water resources have been polluted and cannot be used as a result of wastewater discharges, especially from industrial zones, with insufficient treatment. The use of advanced technologies in wastewater treatment have become increasingly important due to the fact that discharge standards became more strict and there is a need for water recovery and reuse. MBR systems have been extensively used for the treatment of municipal and industrial wastewaters. However, membrane fouling is still a major operational problem in MBR applications in the field. Thus, there is an urgent need to study membrane fouling and find effective strategies for controlling it in MBRs (Wang *et al.*, 2014). In addition, new hybrid MBR systems should be developed for direct and indirect potable water reuse applications (Holloway *et al.*, 2015). The commercialization of high-retention, low-fouling MBR membranes will be an important issue in the future.

ACKNOWLEDGEMENTS

The authors would like to thank the Ministry of Science, Industry and Technology of the Turkish Republic for financial support to our research project (Project no. 0330.STZ.2013-2) and to TÜBİTAK for the grant (Project no. 114Y500). Also S. Bunani is grateful to TÜBİTAK for PhD scholarship (TÜBİTAK-2235).

We acknowledge Ege University for supporting our research projects (2012/MUH/035, 2013/SUF/004, EU-2013-ÇSUAM-003 and 2014/MUH/040) on wastewater reclamation and reuse issue. We acknowledge ITOB-OSB for their great support to our studies in the wastewater treatment field. We also thank M. Akçay, G. Serin, E. Altiok and İ. Parlar for their supports.

REFERENCES

Asano, T., Burton, F.L. & Leverenz, H.L., Tsuchihashi, R. & Tchobanoglous, G. (2007) *Water Reuse, Issues, Technologies, and Applications.* McGraw Hill, New York.

Ayers, R.S. & Westcot, D.W. (1985) Water Quality for Agriculture. FAO Irrigation and Drainage Paper 29. Food and Agriculture Organisation of the United Nations, Rome.

Bunani, S., Yörükoğlu, E., Yüksel, Ü., Kabay, N., Yüksel, M., Sert, G. & Pek, T.Ö. (2014) Application of nanofiltration for reuse of wastewater. *International Journal of Global Warming*, 6(2/3), 325–338.

Bunani, S., Yörükoğlu, E., Yüksel, Ü., Kabay, N., Yüksel, M. & Sert, G. (2015a) Application of reverse osmosis for reuse of secondary treated urban wastewater in agricultural irrigation. *Desalination*, 364, 68–74.

Bunani, S., Yörükoğlu, E., Sert, G., Kabay, N., Yüksel, Ü., Yüksel, M., Egemen, Ö. & Pek, T.Ö. (2015b) Utilization of reverse osmosis (RO) for reuse of MBR-treated wastewater in irrigation-preliminary tests and quality analysis of product water. *Environmental Science and Pollution Research*. doi:10.1007/s11356-015-4199-y.

Cicek, N., Winnen, H., Suidan, M.T., Wrenn, B.E., Urbain, V. & Manem, J. (1998) Effectiveness of the membrane bioreactor in the biodegradation of high molecular weight compounds. *Water Research*, 32(5), 1553–1563.

Das, R., Ali, M.E., Abd Hamid, S.B., Ramakrishna, S. & Chowdhury, Z.Z. (2014) Carbon nanotube membranes for water purification: a bright future in water desalination. *Desalination*, 336, 97–109.

Dow. (2013) Filmtec RO membrane product specifications. Available from: http://www.lenntech.com/products/membrane/filmtec-.htm [accessed November 2016].

Drioli, E. & Giorno, L. (2009) *Membrane Operations: Innovative Separations and Transformations.* Wiley-VCH Verlag, Weinheim, Germany.

Galiano, F., Figoli, A., Deowan, S.A., Johnson, D., Altinkaya, S.A., Veltri, L., De Luca, G., Mancuso, R., Hilal, N., Gabriele, B. & Hoinkis, J. (2015) A step forward to a more efficient wastewater treatment by membrane surface modification via polymerizable bicontinuous microemulsion. *Journal of Membrane Science*, 482, 103–114.

GE. (2013) GE Osmonics Desal RO membrane product specifications. Available from: http://www.lenntech.com/products/membrane/osmonics/osmonics.htm [accessed November 2016].

Goh, P.S., Ismail, A.F. & Hilal, N. (2016) Nano-enabled membranes technology: Sustainable and revolutionary solutions for membrane desalination? *Desalination*, 280, 100–104.

Gujer, W., Henze, M., Mino, T. & Van Loosdrecht, M. (1999) Activated sludge model. *Water Science and Technology*, 39, 183–193.

Halder, J. & Kullmann, C. (2011) Integrated membrane bioreactors (MBR) and reverse osmosis (RO) for water reuse. *F & S International Edition*, 12/2012, Koch Membrane Systems, Aachen, Germany. pp. 40–43.

Holloway, R.W., Wait, A.S., Da Silva, A.F., Herron, J., Schutter, M.D., Lampi, K. & Cath, T.Y. (2015) Long-term pilot scale investigation of novel hybrid ultrafiltration-osmotic membrane bioreactors. *Desalination*, 363, 64–74.

Judd, S. & Jefferson, B. (2003) *Membranes for Industrial Wastewater Recovery and Re-use.* Elsevier, Oxford.

Judd, S. & Judd, C. (2006) *The MBR Book: Principles and Applications of Membrane Bioreactors for Water and Wastewater Treatment.* Elsevier, Oxford.

Kochkodan, V. & Hilal, N. (2015) A comprehensive review on surface modified polymer membranes for biofouling mitigation. *Desalination*, 356, 187–207.

Koltuniewicz, A.B. & Drioli, E. (2008) *Membranes in Clean Technologies – Theory and Practice*, Vol. 1–2. Wiley-VCH, Weinheim, Germany.

Lazarova, V. & Bahri, A. (2005) *Water Reuse for Irrigation: Agriculture, Landscapes, and Turf Grass.* CRC Press, Boca Raton, FL.

Li, N.N., Fane, A.G., Hu, W.S.W. & Matsuura, T. (2008) *Advanced Membrane Technology and Applications.* John Wiley & Sons, Hoboken, NJ.

Liwarska-Bizukojc, E. & Bizukojc, M. (2005) Digital image analysis to estimate the influence of sodium dodecyl sulphate on activated sludge flocs. *Process Biochemistry*, 40, 2067–2072.

Luo, W., Hai, F.I., Price, E.W., Guo, W., Ngo, H.H., Yamamoto, K. & Nghiem, L.D. (2014) High retention membrane bioreactors: challenges and opportunities. *Bioresource Technology*, 167, 539–546.

Ministry of Environment and Forestry (2010) Report on technical methods for wastewater treatment plants. Turkish Republic Ministry of Environment and Forestry, Ankara, Turkey.

Peeters, J.M.M., Boom, J.P., Mulder, M.H.V. & Strathmann, H. (1998) Retention measurements of nanofiltration membranes with electrolyte solutions. *Journal of Membrane Science*, 145, 199–209.

Pontalier, P.Y., Ismail, A. & Ghoul, M. (1997) Mechanisms for the selective rejection of solutes in nanofiltration membranes. *Separation and Purification Technology*, 12, 175–181.

Sert, G. (2015) *Application of Membrane Technologies for Reuse of Industrial Wastewater treated by MBR Process.* PhD Thesis, Ege University, Izmir, Turkey.

Shuler, M.L. & Kargi, F. (2002) *Bioprocess Engineering: Basic Concepts.* 2nd edition, Prentice-Hall International, Upper Saddle River, NJ.

Solak, S. (2010) *Quality Analysis of Water Produced from Seawater by Reverse Osmosis Method-Studying the Effect of Membrane Type on Water Quality.* MS Thesis, Department of Chemistry, Ege University, Izmir, Turkey.

Stephenson, T., Judd, S., Jefferson, B. & Brindle, K. (2006) *Membrane Bioreactors for Wastewater Treatment.* IWA Publishing, London.

USEPA (2004) Guidelines for Water Reuse. EPA/625/R-04/108, US Environmental Protection Agency, Washington, DC.

Wang, F., Zhang, M., Peng, W., He, Y., Lin, H., Chen, J., Hong, H., Wang, A. & Yu, H. (2014) Effects of ionic strength on membrane fouling in a membrane bioreactor. *Bioresource Technology*, 156, 35–41.

Yang, Y.F., Huang, S.S., Takabatake, H. & Hanada, S. (2009) Development of MBR RO integrated membrane system for wastewater reclamation. *Journal EICA*, 14(2–3), 87–90.

Subject index

Sustainable Water Developments

Book Series Editor: Jochen Bundschuh

ISSN: 2373-7506

Publisher: CRC Press/Balkema, Taylor & Francis Group

1. Membrane Technologies for Water Treatment: Removal of Toxic Trace
   Elements with Emphasis on Arsenic, Fluoride and Uranium
   Editors: Alberto Figoli, Jan Hoinkis & Jochen Bundschuh
   2016
   ISBN: 978-1-138-02720-6 (Hbk)

2. Innovative Materials and Methods for Water Treatment:
   Solutions for Arsenic and Chromium Removal
   Editors: Marek Bryjak, Nalan Kabay, Bernabé L. Rivas & Jochen Bundschuh
   2016
   ISBN: 978-1-138-02749-7 (Hbk)

3. Membrane Technology for Water and Wastewater Treatment, Energy and Environment
   Editors: Ahmad Fauzi Ismail & Takeshi Matsuura
   2016
   ISBN: 978-1-138-02901-9 (Hbk)

4. Renewable Energy Technologies for Water Desalination
   Editors: Hacene Mahmoudi, Noreddine Ghaffour, Mattheus A. Goosen & Jochen Bundschuh
   2017
   ISBN: 978-1-138-02917-0 (Hbk)

5. Application of Nanotechnology in Membranes for Water Treatment
   Editors: Alberto Figoli, Jan Hoinkis, Sacide Alsoy Altinkaya & Jochen Bundschuh
   2017
   ISBN: 978-1-138-89658-1 (Hbk)

Printed and bound by CPI Group (UK) Ltd, Croydon, CR0 4YY

29/10/2024

01780489-0001